SAS/IML® Software: Usage and Reference

Version 6
First Edition

SAS Institute Inc.
SAS Campus Drive
Cary, NC 27513

The correct bibliographic citation for this manual is as follows: SAS Institute Inc., *SAS/IML® Software: Usage and Reference, Version 6, First Edition,* Cary, NC: SAS Institute Inc., 1989. 501 pp.

SAS/IML® Software: Usage and Reference, Version 6, First Edition

The SAS® System is an integrated system of software providing complete control over data access, management, analysis, and presentation. Base SAS software is the foundation of the SAS System. Products within the SAS System include SAS/ACCESS® SAS/AF® SAS/ASSIST® SAS/CALC® SAS/CPE® SAS/DMI® SAS/EIS® SAS/ENGLISH® SAS/ETS® SAS/FSP® SAS/GRAPH® SAS/IML® SAS/IMS-DL/I® SAS/INSIGHT® SAS/LAB® SAS/OR® SAS/PH-Clinical® SAS/QC® SAS/REPLAY-CICS® SAS/SHARE® SAS/STAT® SAS/TOOLKIT® SAS/CONNECT™ SAS/DB2™ SAS/LOOKUP™ SAS/NVISION™ SAS/SQL-DS™ and SAS/TUTOR™ software. Other SAS Institute products are SYSTEM 2000® Data Management software, with basic SYSTEM 2000, CREATE™ Multi-User™ QueX™ Screen Writer™ and CICS interface software; NeoVisuals® software; JMP® JMP IN® JMP Serve® and JMP *Design*™ software; SAS/RTERM® software; and the SAS/C® Compiler and the SAS/CX® Compiler. MultiVendor Architecture™ and MVA™ are trademarks of SAS Institute Inc. SAS Video Productions™ and the SVP logo are service marks of SAS Institute Inc. SAS Institute also offers SAS Consulting® Ambassador Select™ and On-Site Ambassador™ services. *Authorline® SAS Communications® SAS Training® SAS Views®* the SASware Ballot® and *Observations*™ are published by SAS Institute Inc. All trademarks above are registered trademarks or trademarks of SAS Institute Inc. in the USA and other countries. ® indicates USA registration.

The Institute is a private company devoted to the support and further development of its software and related services.

Other brand and product names are registered trademarks or trademarks of their respective companies.

DOC S19, VER 1.55, 121289

Contents

Reference Aids

Figures

Tables

Credits

Documentation

Composition	Cynthia M. Hopkins, Nancy Mitchell, Pamela A. Troutman, David S. Tyree, Anita Yefko
Graphics	Creative Services Department
Proofreading	Kevin A. Clark, Gwendolyn T. Colvin, Jennifer M. Ginn, Reid V. Harris, III, Beth A. Heiney, Hanna P. Hicks, Beryl C. Pittman, Josephine P. Pope, Toni P. Sherrill, John M. West, Anna B. Williams, Susan E. Willard
Technical Review	Linda W. Binkley, Rajen H. Doshi, Alan R. Eaton, Eddie Routten, Sandra D. Schlotzhauer, Craig Simpson, Donna E. Woodward
Writing and Editing	Jacqueline F. Allen, James J. Ashton, Gary R. Meek

Software

Development

John P. Sall is the original designer of SAS/IML software. Rajen H. Doshi is the principal developer responsible for support and continuing development of SAS/IML software. Other major contributors include Claire S. Cates, Marc-david Cohen, Alan R. Eaton, Wolfgang M. Hartmann, Leigh A. Ihnen, Warren F. Kuhfeld, Katherine Ng, Warren S. Sarle, and Terry Woodfield.

Technical Support

Eddie Routten, Donna E. Woodward

Testing

Linda W. Binkley, Craig Simpson

Acknowledgments

Many people make significant and continuing contributions to the development of SAS Institute's software products. First among them are SAS users who have served as chairpersons for the SAS Users Group International (SUGI). They are Sally Carson, Helene Cavior, Dr. Michael Farrell, Rudolf J. Freund, Bob Hamer, Dr. Ronald W. Helms, Donald Henderson, Gerry Hobbs, Julian Horwich, Kenneth L. Koonce, Rich La Valley, Dr. Ramon C. Littell, J. Philip Miller, Pat Hermes Smith, Dr. Rodney Strand, and Dr. William Wilson.

Others who have contributed to the development and implementation of Release 6.06 of SAS/IML software include Ronald Helms, University of North Carolina at Chapel Hill; Forrest W. Young, University of North Carolina at Chapel Hill; and Phil Spector, University of California, Berkeley.

The final responsibility for the SAS System lies with SAS Institute alone. We hope that you will always let us know your feelings about the system and its documentation. It is through such communication that the progress of SAS software continues to be accomplished.

Using This Book

Purpose

SAS/IML Software: Usage and Reference, Version 6, First Edition documents the features, functions, statements, and subroutines available with Release 6.06 of SAS/IML software. The usage chapters are task-oriented, and the reference chapter provides syntax and descriptions.

For mainframe and minicomputer sites using Release 6.06 software, *SAS/IML Software: Usage and Reference* supersedes the *SAS/IML User's Guide, Version 5 Edition*. For microcomputers and UNIX® workstations, this book replaces the *SAS/IML User's Guide, Release 6.03 Edition*.

If you are using the SAS System with PC DOS, you should continue to use the *SAS/IML User's Guide, Release 6.03 Edition*.

You should read the remainder of "Using This Book" to learn what assumptions are made in this book and what conventions are used to present information about SAS/IML software.

Audience

SAS/IML Software: Usage and Reference is intended to assist programmers and data analysts who use SAS/IML software. This book assumes you are familiar with basic SAS System concepts such as creating SAS data sets with the DATA step.

Prerequisites

The following table summarizes the SAS System concepts that you need to understand in order to use SAS/IML software.

If you need to know how to	Refer to
invoke the SAS System at your site	instructions provided by the SAS Software Consultant at your site
use base SAS software	*SAS Introductory Guide, Third Edition* for a brief introduction, or *SAS Language and Procedures: Usage, Version 6, First Edition* for a more thorough introduction
create SAS data sets, and use base SAS functions and formats	*SAS Language: Reference, Version 6, First Edition*

UNIX is a registered trademark of AT&T.

How to Use This Book

The following sections provide an overview of what information is contained in this book and how it is organized.

Organization

SAS/IML Software: Usage and Reference is divided into two parts.

Usage

Usage provides a task-oriented approach to learning IML. The first three chapters provide introductory material on the capabilities of SAS/IML software, a language overview, and a tutorial session:

Chapter 1, "Introduction to SAS/IML Software"

Chapter 2, "Understanding the SAS/IML Language"

Chapter 3, "Tutorial: A Module for Linear Regression"

Chapters 4 through 14 provide a task-oriented treatment of a comprehensive set of topics:

Chapter 4, "Working with Matrices"

Chapter 5, "Programming"

Chapter 6, "Working with SAS Data Sets"

Chapter 7, "File Access"

Chapter 8, "Applications: Statistical Examples"

Chapter 9, "Introduction to SAS/IML Graphics"

Chapter 10, "Graphics Applications"

Chapter 11, "Windows and Display Features"

Chapter 12, "Storage Features"

Chapter 13, "Using SAS/IML Software to Generate IML Statements"

Chapter 14, "Further Notes"

Reference

Reference provides descriptions and necessary syntax for all SAS/IML operators, functions, statements, and subroutines.

Chapter 15, "SAS/IML Language Reference"

Appendices

Four appendices follow Chapter 15:

Appendix 1, "SAS/IML Quick Reference"

Appendix 2, "SAS/IML Software Compared with the MATRIX Procedure"

Appendix 3, "The MATIML Procedure"

Appendix 4, "Alternative Characters"

What You Should Read

Refer to the following table to determine which sections of *SAS/IML Software: Usage and Reference* are appropriate for your level of experience.

Level of Experience	Suggested Reading
no previous experience with SAS/IML software	Begin by reading Chapters 1 through 3. Follow by reading the chapters of interest and referring to Chapter 15 for detailed information.
experience with SAS/IML software	Begin by reading "Changes and Enhancements" for a list of features that are new or changed from the previous release of SAS/IML software. Then refer to the appropriate chapters and sections of Chapter 15.

Reference Aids

SAS/IML Software: Usage and Reference includes a variety of features to help locate the information you need. The following sections are provided for your easy reference:

Changes and Enhancements	provides information about features of SAS/IML software that have been added or modified. This section follows "Using This Book."
References	provides references to articles and books in the literature.
Index	references pertinent terms and concepts.

Each operator, function, subroutine, and statement is referenced in Chapter 15. Each element has the following structure, making information easy to find.

Note: In the following list, *Main Heading* can be replaced with the name of an operator, function, call, or statement.

Main Heading	provides the name of the element, what it does, and what it returns (where applicable).
Syntax	provides the form of the statement and descriptions of arguments (where applicable).
Description	provides a detailed description, references when appropriate, and examples of valid statements.

The graphic on the inside front cover lists software products available in the SAS System and groups them according to categories. The graphic on the inside back cover gives an overview of the capabilities and features of SAS/IML software.

Conventions

This section covers the conventions used in this book, including typographical conventions, syntax conventions, and conventions used in presenting an interactive session.

Typographical Conventions

This book uses several type styles. The following list summarizes style conventions:

roman	is the basic type style used for most text.
UPPERCASE ROMAN	is used for references in text to SAS/IML language elements.
italic	is used in margin notes to provide cautionary information, in text to define terms, and in formulas to indicate variables and subscripts.
bold	is used in headings, in text to indicate very important points, and in formulas to indicate matrices and vectors.
`monospace`	is used to show example code and output. In most cases, this book uses lowercase type for SAS code. You can enter your own SAS code in lowercase, uppercase, or a mixture of the two. The SAS System changes your variable names to uppercase.

Syntax Conventions

Syntax conventions are used to show the general form of a SAS/IML statement. This book uses the following conventions for syntax:

UPPERCASE BOLD	indicates the names of operators, functions, subroutines, and statements. These must be spelled as shown.
UPPERCASE ROMAN	indicates elements that must be spelled as shown.
italic	indicates elements that you supply (for example, VAR=*names*).

< > (angle brackets)	indicate optional elements.
| (vertical bar)	indicates a choice of one item from the group formed by the vertical bar or bars. Items separated by bars are either mutually exclusive or aliases.
. . . (ellipsis)	indicates items that can be repeated indefinitely.

Punctuation such as parentheses, commas, semicolons, and square brackets must be typed as shown.

The following example illustrates some of these syntax conventions:

EDIT *SAS-data-set* <VAR *operand*> <WHERE(*expression*)>;

□ **EDIT** is in uppercase bold because it is the name of a SAS/IML statement.

□ *SAS-data-set* is in italic because you supply the name of an existing SAS data set, which is opened for editing.

□ You can specify a set of variables to edit in the VAR clause and process observations conditionally for editing by supplying an *expression* in the WHERE clause.

□ The VAR and WHERE clauses are enclosed in angle brackets because they are optional.

Conventions for Output and Interactive Sessions

Many examples are presented in the context of an interactive session. In this case, statements that you enter are indicated with a right angle bracket (>). Responses from IML follow below.

Most of the programs in this book were run using the following SAS system options:

PAGESIZE=64   sets the length of the page to 64.

LINESIZE=78   sets the length of the text line to 78 characters.

Additional Documentation

The *Publications Catalog*, published twice a year, gives detailed information on the many publications available from SAS Institute. To obtain a free catalog, please send your request to the following address:

SAS Institute Inc.
Book Sales Department
SAS Campus Drive
Cary, NC 27513

In addition to *SAS/IML Software: Usage and Reference*, you will find these other books helpful when using SAS/IML software:

□ *SAS Introductory Guide, Third Edition* (order #A5685) provides information if you are unfamiliar with the SAS System or any other programming language.

□ *SAS Language: Reference, Version 6, First Edition* (order #A56076) provides detailed reference information about SAS language statements, functions, formats, and informats; the SAS Display Manager System; the SAS Text Editor; and any other element of base SAS software except for procedures.

□ *SAS Procedures Guide, Version 6, Third Edition* (order #A56080) provides detailed reference information about the procedures in base SAS software.

□ *SAS Language and Procedures: Usage, Version 6, First Edition* (order #A56075) provides task-oriented examples of the major features of base SAS software.

□ *SAS/GRAPH Software: Reference, Version 6, First Edition, Volume 1* and *Volume 2* (order #A56020) provides detailed reference information about the graphical presentation facilities available in SAS/GRAPH software.

□ *SAS/STAT User's Guide, Version 6, Fourth Edition, Volume 1* and *Volume 2* (order #A56045) provides information about the statistical procedures available in SAS/STAT software.

□ The SAS companion or other SAS documentation for your host system provides information about the host-specific features of the SAS System.

Changes and Enhancements to SAS/IML® Software

Version 6 SAS/IML software is fully compatible with Version 5 except for the graphics commands. In addition, there are a number of enhancements now available in Version 6. The rest of this chapter summarizes the enhancements available in Release 6.06 SAS/IML software.

Workspace Management

SAS/IML software has a new automatic workspace management scheme. The workspace grows dynamically as the need grows. This ensures efficient use of all the memory that the host system can provide. In addition, there is no built-in limit on how large a matrix can get. You can have a matrix with any number of rows, any number of columns, or any element size, provided there is enough memory available to store it. Overall, you should be able to run larger applications and handle larger matrices.

Programmable Windows

Programmable windowing capabilities have been added, which can be useful for developing full-screen data entry windows and displaying menus. See Chapter 11, "Windows and Display Features," for details.

Graphics

Although Version 6 graphics commands are not compatible with Version 5, they are superior in terms of flexibility, consistency, and power. See Chapter 9, "Introduction to Graphics," for details.

Graphics capabilities in Version 6 are summarized below:

☐ Graphs can now be developed incrementally. You can create a graphics segment with graphics primitives, display it, add more primitives or paste in another existing graph, display it again, and continue this process to create the final graph.

☐ All graphs are stored as segments within a SAS catalog. There are segment management routines to open (GOPEN), close (GCLOSE), delete (GDELETE), display (GSHOW), and include (GINCLUDE) segments. IML graphics catalogs and segments can be shared with other SAS products and procedures and vice versa.

☐ All graphics commands use world coordinates (except for viewport commands, which always use device coordinates). This enables you to think about a graph directly in terms of your world, without having to worry about the device coordinates. No automatic scaling is done for you, leaving the appearance of the graph totally under your control. Windows and viewports can be set or changed anytime. They can even be specified within a command to apply only to that command.

□ All graphics primitives accept attribute arguments such as color, line style, character height, character font, fill pattern, and aspect ratio. The attributes can be set globally for an IML session, for a particular graph only, or for a particular command only. All attribute values in SAS/GRAPH software are available in IML.

□ There are several new graphics commands:

GBLKVP	defines a blanking viewport.
GBLKVPD	deletes the blanking viewport.
GGRID	draws a grid.
GINCLUDE	includes a graphics segment.
GPIEXY	converts from polar to world coordinates.
GPORTPOP	pops the viewport.
GPORTSTK	stacks the viewport.
GSET	sets attributes for a graphics segment.
GSTRLEN	finds the string length.

Programming

IML now allows generation and execution of IML statements from within IML using the PUSH, QUEUE, and EXECUTE calls. These provide added flexibility to programming. You can create and execute IML statements dynamically, execute operating system or display manager commands under program control, or create your own error and interrupt handlers. See Chapter 13, "Using SAS/IML Software to Generate IML Statements," for details.

IML also allows DO loops and nested IF-THEN/ELSE statements (possibly compounded with DO-END groups) for immediate execution. In earlier releases, you could only use them in a module. Statements, in general, can be grouped together within a DO-END group outside of modules. The group is executed as a whole immediately after the END statement is issued.

The DO statement allows a DATA clause for implicit looping until the end of file, which is especially useful for data set input and output or external file input and output. See Chapter 5, "Programming," for details.

Modules

The environment for modules has many features. See Chapter 5 for details on modules. The features of modules are summarized below:

☐ User-defined modules can now be function modules; that is, they can be defined to return a value.

☐ Modules can be directly stored in their compiled form in a storage catalog. The LOAD, STORE, and REMOVE commands are enhanced to work on modules directly without converting them into equivalent character matrices.

☐ Variables defined inside a module are local to the module. The use of a GLOBAL clause allows sharing of variables between the external environment and local environment of the module.

☐ IML has more powerful module resolution, which includes comprehensive checks for

☐ temporaries (expressions or submatrices) passed as results to modules

☐ repeated arguments and conflicts with global arguments

☐ module names conflicting with IML built-in names

☐ freeing up of a module if any error occurs during its resolution

☐ automatic loading of an unresolved module from the library

☐ distinctions between function and subroutine modules.

☐ PAUSE and RESUME commands are now available which give added control over execution of user-defined modules.

☐ CALL and RUN statements can be used interchangeably. However, they have a different order of resolution, which can be advantageous when a user-defined module is given the same name as an IML built-in subroutine.

New Functions

SAS/IML software has many new functions available:

APPLY   applies an IML module to its arguments.

ARMASIM   simulates a univariate ARMA series.

BYTE   translates numbers to ordinal characters.

CHANGE   search and replace text in an array.

CHOOSE   conditionally chooses and changes elements.

CUSUM   calculates cumulative sums.

DO   produces an arithmetic series.

ECHELON   reduces a matrix to row-echelon normal form.

GENEIG	computes eigenvalues and eigenvectors of a generalized eigenproblem.
HOMOGEN	solves homogeneous linear systems.
INSERT	inserts one matrix inside another.
LCP	solves the linear complementarity problem.
LP	solves the linear programming problem.
POLYROOT	finds zeros of a real polynomial.
REMOVE	discards elements from a matrix.
SETDIF	compares elements of two matrices.
SOUND	produces a tone.
SQRSYM	converts a symmetric matrix to a square matrix.
SYMSQR	converts a square matrix to a symmetric matrix.
TYPE	determines the type of a matrix.
UNION	performs unions of sets.
UNIQUE	sorts and removes duplicates.
XSECT	intersects sets.

DATA Step Functions

All DATA step functions are now directly available in IML. DATA step functions include a variety of mathematical, trigonometric, hyperbolic, statistical, financial, character, datetime, and state-ZIP code functions.

Data Set Options

All data set options are now directly available in IML. Data set options include DROP, KEEP, RENAME, REPLACE, OBS, and FIRSTOBS and can be specified in the USE, EDIT, and CREATE statements.

Indexing SAS Data Sets

SAS data sets can be indexed on one or more variables. Indexes can be used to optimize WHERE processing, or they can be used to retrieve observations in index order.

Utility Commands

There are two new utility commands for sorting a data set and for producing summary statistics:

SORT   sorts a SAS data set.

SUMMARY   computes summary statistics for SAS data sets.

External File Input and Output

You have a full set of commands to read from or write to external text or binary files. The new commands include INFILE, FILE, INPUT, PUT, and CLOSEFILE. See Chapter 7, "File Access," for details.

RESET Statement Options

The RESET statement has several new options available:

AUTONAME   prints row and column labels for matrices.

CENTER   centers printed output.

CLIP   clips graphs across the viewport.

DEFLIB=   sets the default libname for SAS data sets.

FUZZ=   prints small numbers as 0.

LINESIZE=   sets the line size.

PAGESIZE=   sets the page size.

PRINTALL   prints final as well as intermediate results.

SPACES=   prints spacing between two adjacent matrices.

See Chapter 15, "SAS/IML Language Reference," for details on the RESET statement.

SHOW Statement Options

The SHOW statement has several new options:

ALLNAMES   shows attributes of all names including unvalued names.

FILES   shows all open external files and attributes.

MEMORY   shows largest piece of memory available.

name   shows attributes of a particular name.

PAUSE   shows all paused modules and their tracebacks.

WINDOWS shows all defined windows.

See Chapter 15 for details on the SHOW statement.

Operators

The index creation operator (:) now works on character-suffix arguments. It can also be used to generate reverse indices. The addition operator (+) now works on character operands, performing concatenation on the arguments.

Repetition Feature

A repetition feature is available for literals in braces. Also, certain special characters are now allowed in literals. See Chapter 2, "Understanding the SAS/IML Language," for more details.

Printing

The MATTRIB statement now accepts a LABEL attribute. It also accepts direct literals, expressions, and the keyword EMPTY for all its attributes. These features, combined with the new RESET options, should provide a good deal of power to generate customized printed reports. See Chapter 5 and Chapter 15 for details on the PRINT statement.

Error and Interrupt Handling

Error diagnostics and tracebacks have been standardized and simplified throughout IML. When an error occurs inside a module, it automatically pauses with a full traceback. The execution can later be resumed from the point of error. Math errors such as overflow and underflow are gracefully trapped by IML's own math error handler.

IML now has extensive provisions to recognize and honor user interrupts. Upon interrupt, execution can be suspended, resumed, or terminated.

Performance Tuning

The WORKSIZE= and SYMSIZE= options are now available in the PROC IML statement. These options can be used in conjunction with the RESET DETAILS, SHOW SPACE, and SHOW MEMORY commands to provide information for performance tuning of your IML applications. See Chapter 14, "Further Notes," for additional details.

Missing Values

All IML functions now either legitimately allow missing values or screen them out with appropriate error messages, which makes them consistent and predictable for any type of data. Division by zero now produces a warning, and the result is set to a missing value.

The MATIML Procedure

A variety of enhancements have been made to the MATIML procedure (which translates MATRIX procedure code to IML procedure code) making it reliable and fully automatic for non-macro applications. See Appendix 3, "The MATIML Procedure," for details on PROC MATIML.

Chapter **1** Introduction to
SAS/IML® Software

Introduction

SAS/IML software gives you access to a powerful and flexible programming
language (Interactive Matrix Language) in a dynamic, interactive environment.
The fundamental object of the language is a data matrix. You can use SAS/IML
software interactively (at the statement level) to see results immediately, or you
can store statements in a module and execute them later. The programming is
dynamic because necessary activities such as memory allocation and
dimensioning of matrices are done automatically.

SAS/IML software is powerful. You can access built-in operators and call
routines to perform complex tasks such as matrix inversion or eigenvector
generation. You can define your own functions and subroutines using SAS/IML
modules. You can operate on a single value or take advantage of matrix
operators to perform operations on an entire data matrix. For example, the
statement

```
x=x+1;
```

can be used to add 1 to a single value X, or to add 1 to all elements of a matrix
X.

You have access to a wide choice of data management commands. You can
read, create, and update SAS data sets from inside SAS/IML software without
ever using the DATA step. For example, reading a SAS data set to get phone
numbers for all individuals whose last name begins with "Smith" is easy:

```
read all var{phone} where(lastname=:"Smith");
```

The result is a matrix **PHONE** of phone numbers.

An Introductory Interactive Session

Here is a simple introductory session that uses SAS/IML software to estimate
the square root of a number, accurate to three decimal places. In this session,
you define a function module named APPROX to perform the calculations and
return the approximation. You then call APPROX to estimate the square root of
several numbers given in a matrix literal (enclosed in braces), and you print the
results. Throughout the session, the right angle brackets (>) indicate statements
that you submit; responses from IML follow below.

```
>   proc iml;                              /* begin IML session */

    IML Ready

>   start approx(x);                          /* begin module */
>     y=1;                                    /* initialize y */
>     do until(w<1e-3);                       /* begin do loop */
>       z=y;                                      /* set z=y */
>       y=.5#(z+x/z);                    /* estimate square root */
>       w=abs(y-z);               /* compute change in estimate */
>     end;                                       /* end do loop */
>     return(y);                      /* return approximation */
>   finish approx;                                /* end module */

    NOTE: Module APPROX defined.

>   t=approx({3,5,7,9});              /* call function APPROX  */
>   print t;                              /* print matrix */

                              T
                          1.7320508
                          2.236068
                          2.6457513
                              3

>   quit;

    Exiting IML
```

SAS/IML Software: an Overview

SAS/IML software is a programming language.
You can program easily and efficiently with the many features for arithmetic and character expressions in SAS/IML software. You have access to a wide range of built-in subroutines designed to make your programming fast, easy, and efficient. Because SAS/IML software is part of the SAS System, you can access SAS data sets or external files with an extensive set of data processing commands for data input and output, and you can edit existing SAS data sets or create new ones. SAS/IML software has a complete set of control statements, such as DO/END, START/FINISH, iterative DO, IF-THEN/ELSE, GOTO, LINK, PAUSE, and STOP, giving you all of the commands necessary for execution control and program modularization.

SAS/IML software operates on matrices.
While most programming languages deal with single data elements, the fundamental data element with SAS/IML software is the matrix, a two-dimensional (row×column) array of numeric or character values.

SAS/IML software possesses a powerful vocabulary of operators.

You can access built-in matrix operations that require calls to math-library subroutines in other languages. You have access to many operators, functions, and CALL subroutines.

SAS/IML software uses operators that apply to entire matrices.

You can add elements of the matrices **A** and **B** with the expression **A+B**. You can perform matrix multiplication with the expression **A*B** and perform elementwise multiplication with the expression **A#B**.

SAS/IML software is interactive.

You can execute a command as soon as you enter it, or you can collect commands in a module to execute later. When you execute a command, you see the results immediately. You can interact with an executing module by programming IML to pause, allowing you to enter additional statements before continuing execution.

SAS/IML software is dynamic.

You do not need to declare, dimension, and allocate storage for a data matrix. SAS/IML software does this automatically. You can change the dimension or type of a matrix at any time. You can open multiple files or access many libraries. You can reset options or replace modules at any time.

SAS/IML software processes data.

You can read all observations or read conditionally selected observations from a SAS data set into a matrix, creating either multiple vectors (one for each variable in the data set) or a matrix that contains a column for each data set variable. You can create a new SAS data set, or you can edit or append observations to an existing SAS data set.

SAS/IML software produces graphics.

You have access to a wide range of graphics commands, allowing you to visually explore relationships in data.

4

Chapter 2 Understanding the SAS/IML® Language

Defining a Matrix

The fundamental data object on which all IML commands operate is a two-dimensional (row×column) numeric or character matrix. By their very nature, matrices are useful for representing data and efficient for working with data. Matrices have the following properties:

☐ Matrices can be either numeric or character. Elements of a numeric matrix are stored in double precision. Elements of a character matrix are character strings of equal length. The length can range from 1 to 32676 characters.

☐ Matrices are referred to by valid SAS names. Names can be from 1 to 8 characters long, beginning with a letter or underscore, and continuing with letters, numbers, and underscores.

- □   Matrices have dimension defined by the number of rows and columns.

- □   Matrices can contain elements that have missing values (see "Missing Values" later in this chapter).

The dimension of a matrix is defined by the number of rows and columns it has. An $m \times n$ matrix has mn elements arranged in m rows and n columns. The following nomenclature is standard in this book:

- □ $1 \times n$ matrices are called *row vectors*.

- □ $m \times 1$ matrices are called *column vectors*.

- □ 1×1 matrices are called *scalars*.

Matrix Names and Matrix Literals

Matrix Names

A matrix is referred to by a valid SAS name. Names can be from 1 to 8 characters long, beginning with a letter or underscore and continuing with letters, numbers, and underscores. You associate a name with a matrix when you create or define the matrix. A matrix name exists independently of values. This means that at any time, you can change the values associated with a particular matrix name, change the dimension of the matrix, or even change its type (numeric or character).

Matrix Literals

A *matrix literal* is a matrix represented by its values. When you represent a matrix by a literal, you are simply specifying the values of each element of the matrix. A matrix literal can have a single element (a scalar) or have many elements arranged in a rectangular form (rows×columns). The matrix can be numeric (all elements are numeric) or character (all elements are character). The dimension of the matrix is automatically determined by the way you punctuate the values.

If there are multiple elements, use braces ({ }) to enclose the values and commas to separate the rows. Within the braces, values must be either all numeric or all character. If you use commas to create multiple rows, all rows must have the same number of elements (columns).

The values you input can be any of the following:

- □   a number, with or without decimal points, possibly in scientific notation (such as 1E-5).

- □   a character string. Character strings can be enclosed in either single quotes (') or double quotes (") but do not necessarily need quotes. Quotes are required when there are no enclosing braces or when you want to preserve case, special characters, or blanks in the string. If the string has embedded quotes, you must double them (for example, WORD='Can''t'). Special characters can be any of the following: ? = * : ( ).

- □   a period (.), representing a missing numeric value.

- □   numbers in brackets ([ ]), representing repetition factors.

Creating Matrices from Matrix Literals

Creating matrices using matrix literals is easy. You simply input the element values one at a time, usually inside of braces. Representing a matrix as a matrix literal is not the only way to create matrices. A matrix can also be created as a result of a function, a CALL statement, or an assignment statement. Below are some simple examples of matrix literals, some with a single element (scalars) and some with multiple elements. For more information on matrix literals, see Chapter 4, "Working with Matrices."

Scalar Literals

The following examples define scalars as literals. These are examples of simple assignment statements, with the matrix name on the left-hand side of the equal sign and the value on the right. Notice that you do not need to use braces when there is only one element.

```
a=12;
a=. ;
a='hi there';
a="Hello";
```

Numeric Literals

Matrix literals with multiple elements have the elements enclosed in braces. Use commas to separate the rows of a matrix. For example, the statement

```
x={1 2 3 4 5 6};
```

assigns a row vector to the matrix **X**:

```
      X
1     2     3     4     5     6
```

The statement

```
y={1,2,3,4,5};
```

assigns a column vector to the matrix **Y**:

```
Y
1
2
3
4
5
```

The statement

```
z={1 2, 3 4, 5 6};
```

assigns a 3×2 matrix literal to the matrix **Z**:

```
       Z
       1          2
       3          4
       5          6
```

The following assignment

```
w=3#z;
```

creates a matrix **W** that is three times the matrix **Z**:

```
       W
       3          6
       9         12
      15         18
```

Character Literals

You input a character matrix literal by entering character strings. If you do not use quotes, all characters are converted to uppercase. You must use either single or double quotes to preserve case or when blanks or special characters are present in the string. For character matrix literals, the length of the elements is determined from the longest element. Shorter strings are padded on the right with blanks. For example, the assignment of the literal

```
a={abc defg};
```

results in **A** being defined as a 1×2 character matrix with string length 4 (the length of the longer string).

```
       A
      ABC   DEFG
```

The assignment

```
a={'abc' 'DEFG'};
```

preserves the case of the elements, resulting in the matrix

```
       A
      abc   DEFG
```

Note that the string length is still 4.

Repetition Factors

A repetition factor can be placed in brackets before a literal element to have the element repeated. For example, the statement

```
answer={[2] 'Yes',[2] 'No'};
```

is equivalent to

```
answer={'Yes' 'Yes', 'No' 'No'};
```

and results in the matrix

```
         ANSWER
         Yes    Yes
         No     No
```

Reassigning Values

You can assign new values to elements of a matrix at any time. The statement below creates a 2×3 numeric matrix named **A**.

```
a={1 2 3, 6 5 4};
```

The statement

```
a={'Sales' 'Marketing' 'Administration'};
```

redefines the matrix **A** as a 1×3 character matrix.

Assignment Statements

Assignment statements create matrices by evaluating expressions and assigning the results to a matrix. The expressions can be composed of operators (for example, matrix multiplication) or functions (for example, matrix inversion) operating on matrices. Because of the nature of linear algebraic expressions, the resulting matrices automatically acquire appropriate characteristics and values. Assignment statements have the general form

result=*expression*;

where *result* is the name of the new matrix and *expression* is an expression that is evaluated, the results of which are assigned to the new matrix.

Functions as Expressions

Matrices can be created as a result of a function call. Scalar functions such as LOG or SQRT operate on each element of a matrix, while matrix functions, such as INV or RANK, operate on the entire matrix. For example, the statement

```
a=sqrt(b);
```

assigns the square root of each element of **B** to the corresponding element of **A**.
    The statement

```
y=inv(x);
```

calls the INV function to compute the inverse matrix of **X** and assign the results to **Y**.
    The statement

```
r=rank(x);
```

creates a matrix **R** whose elements are the ranks of the corresponding elements of **X**.

Operators within Expressions

There are three types of operators that can be used in assignment statement expressions. Be sure that the matrices on which an operator acts are conformable to the operation. For example, matrix multiplication requires that the number of columns of the left-hand matrix be equal to the number of rows of the right-hand matrix.
    The three types of operators are defined below:

prefix operators   are placed in front of an operand ($-$**A**).

infix operators   are placed between operands (**A*****B**).

postfix operators   are placed after an operand (**A**′).

    All operators can work in a one-to-many or many-to-one manner; that is, they allow you to, for example, add a scalar to a matrix or divide a matrix by a scalar. Below is an example of using operators in an assignment statement.

```
y=x#(x>0);
```

This assignment statement creates a matrix **Y** in which each negative element of the matrix **X** is replaced with zero. The statement actually has two expressions evaluated. The expression (**X**>0) is a many-to-one operation that compares each element of **X** to zero and creates a temporary matrix of results; an element of the temporary matrix is 1 when the corresponding element of **X** is positive, and 0 otherwise. The original matrix **X** is then multiplied elementwise by the temporary matrix, resulting in the matrix **Y**.
    For a complete listing and explanation of operators, see Chapter 15, "SAS/IML Language Reference," and Appendix 1.

Types of Statements

Statements in SAS/IML software can be classified into three general categories:

control statements
> direct the flow of execution. For example, the IF-THEN/ELSE statement conditionally controls statement execution.

functions and CALL statements
> perform special tasks or user-defined operations. For example, the statement CALL GSTART activates the SAS/IML graphics system.

commands
> perform special processing, such as setting options, printing, and handling input/output. For example, the command RESET PRINT turns on the automatic printing option so that matrix results are printed as you submit statements.

Control Statements

SAS/IML software has a set of statements for controlling program execution. Control statements direct the flow of execution of statements in IML. With them, you can define DO-groups and modules (also known as subroutines) and route execution of your program. Some control statements are described below.

Statements	Action
DO, END	group statements
iterative DO, END	define an iteration loop
GOTO, LINK	transfer control
IF-THEN/ELSE	routes execution conditionally
PAUSE	instructs a module to pause during execution
QUIT	ends a SAS/IML session
RESUME	instructs a module to resume execution
RETURN	returns from a LINK statement or a CALL module
RUN	executes a module
START, FINISH	define a module
STOP, ABORT	stop execution of an IML program

See Chapter 5, "Programming Statements," later in this book for more information on control statements.

Functions

The general form of a function is

result = FUNCTION(arguments);

where *arguments* can be matrix names, matrix literals, or expressions. Functions always return a single result (whereas subroutines can return multiple results or no result). If a function returns a character result, the matrix to hold the result is allocated with a string length equal to the longest element, and all shorter elements are padded with blanks.

Categories of Functions

Functions fall into the following six categories:

matrix inquiry functions
> return information about a matrix. For example, the ANY function returns a value of 1 if any of the elements of the argument matrix are nonzero.

scalar functions
> operate on each element of the matrix argument. For example, the ABS function returns a matrix whose elements are the absolute values of the corresponding elements of the argument matrix.

summary functions
> return summary statistics based on all elements of the matrix argument. For example, the SSQ function returns the sum of squares of all elements of the argument matrix.

matrix arithmetic functions
> perform matrix algebraic operations on the argument. For example, the TRACE function returns the trace of the argument matrix.

matrix reshaping functions
> manipulate the matrix argument and return a reshaped matrix. For example, the DIAG function returns a matrix whose diagonal elements are equal to the diagonal elements of a square argument matrix. All off-diagonal elements are zero.

linear algebra and statistical functions
> perform linear algebraic functions on the matrix argument. For example, the GINV function returns the matrix which is the generalized inverse of the argument matrix.

Exceptions to the SAS DATA Step

IML supports most functions supported in the SAS DATA step. These functions all accept matrix arguments, and the result has the same dimension as the argument. (See Appendix 1 for a list of these functions.) The following functions are not supported by SAS/IML software:

DIF*n*	LAG*n*
DIM	LBOUND
HBOUND	PUT
INPUT	

The following functions are implemented differently in SAS/IML software. (See Chapter 15, "SAS/IML Language Reference," for descriptions.)

MAX	SOUND
MIN	SSQ
RANK	SUBSTR
REPEAT	SUM

The random number functions, UNIFORM and NORMAL, are built-in and produce the same streams as RANUNI and RANNOR, respectively, of the DATA step. For example, to create a 10×1 vector of random numbers, use

```
x=uniform(repeat(0,10,1));
```

Also, SAS/IML software does not support the OF clause of the SAS DATA step. For example, the statement

```
a=mean(of x1-x10);
```

cannot be interpreted properly in IML. The term (X1-X10) would be interpreted as subtraction of the two matrix arguments rather than its DATA step meaning, "X1 through X10".

CALL Statements or Subroutines

CALL statements invoke a subroutine to perform calculations, operations, or a service. CALL statements are often used in place of functions when the operation returns multiple results or, in some cases, no result. The general form of the CALL statement is

CALL *SUBROUTINE(arguments)*;

where *arguments* can be matrix names, matrix literals, or expressions. If you specify several arguments, use commas to separate them. Also, when using arguments for output results, always use variable names rather than expressions or literals.

Creating Matrices with CALL Statements

Matrices are created whenever a CALL statement returns one or more result matrices. For example, the statement

```
call eigen(val,vec,t);
```

returns two matrices (vectors), **VAL** and **VEC,** containing the eigenvalues and eigenvectors, respectively, of the symmetric matrix **T.**

You can program your own subroutine using the START and FINISH statements to define a module. You can then execute the module with a CALL statement or a RUN statement. For example, the following statements define a module named MYMOD that returns matrices containing the square root and log of each element of the argument matrix:

```
start mymod(a,b,c);
   a=sqrt(c);
   b=log(c);
finish;
run mymod(s,l,x);
```

Execution of the module statements create matrices **S** and **L,** containing the square roots and logs, respectively, of the elements of **X.**

Performing Services

You can use CALL statements to perform special services, such as managing SAS data sets or accessing the graphics system. For example, the statement

```
call delete(mydata);
```

deletes the SAS data set named MYDATA.

The statements

```
call gstart;
call gopen;
call gpoint(x,y);
call gshow;
```

activate the graphics system (CALL GSTART), open a new graphics segment (CALL GOPEN), produce a scatterplot of points (CALL GPOINT), and display the graph (CALL GSHOW).

Commands

Commands are used to perform specific system actions, such as storing and loading matrices and modules, or to perform special data processing requests. Below is a list of some commands and the actions they perform.

Command	Action
FREE	frees a matrix of its values and increases available space
LOAD	loads a matrix or module from the storage library
MATTRIB	associates printing attributes with matrices
PRINT	prints a matrix or message
RESET	sets various system options
REMOVE	removes a matrix or module from library storage
SHOW	requests system information be printed
STORE	stores a matrix or module in the storage library

These commands play an important role in SAS/IML software. With them, for example, you can control printed output (with RESET PRINT, RESET NOPRINT, or MATTRIB) or get system information (with SHOW SPACE, SHOW STORAGE, or SHOW ALL).

If you are running short on available space, you can use commands to store matrices in the storage library, free the matrices of their values, and load them back later when you need them again, as shown in the example below.

Throughout this session, the right angle brackets (>) indicate statements that you submit; responses from IML follow below. First, invoke the procedure by entering PROC IML at the input prompt. Then, create matrices **A** and **B** as matrix literals.

```
> proc iml;

  IML Ready

> a={1 2 3, 4 5 6, 7 8 9};
> b={2 2 2};
```

List the names and attributes of all of your matrices with the SHOW NAMES command.

```
> show names;

  A             3 rows    3 cols num      8
  B             1 row     3 cols num      8
  Number of symbols = 2  (includes those without values)
```

Store these matrices in library storage with the STORE command, and release the space with the FREE command. To list the matrices and modules in library storage, use the SHOW STORAGE command.

```
> store a b;
> free a b;
> show storage;

  Contents of storage = SASUSER.IMLSTOR
  Matrices:
   A       B

  Modules:
```

The output from the SHOW STORAGE statement indicates that you have two matrices in storage. Because you have not stored any modules in this session, there are no modules listed in storage. Return these matrices from the storage library with the LOAD command. (See Chapter 12, "Storage Features," for details about storage.)

```
> load a b;
```

End the session with the QUIT command.

```
> quit;

  Exiting IML
```

Data Management Commands

SAS/IML software has many data management commands that allow you to manage your SAS data sets from within the SAS/IML environment. These data management commands operate on SAS data sets. There are also commands for accessing external files. Below is a list of some commands and the actions they perform.

Command	Action
APPEND	adds records to an output SAS data set
CLOSE	closes a SAS data set
CREATE	creates a new SAS data set
DELETE	deletes records in an output SAS data set
EDIT	reads from or writes to an existing SAS data set
FIND	finds records that meet some condition
LIST	lists records
PURGE	purges records marked for deletion

(*continued*)

	Command	Action
(*continued*)	READ	reads records from a SAS data set into IML variables
	SETIN	makes a SAS data set the current input data set
	SETOUT	makes a SAS data set the current output data set
	SORT	sorts a SAS data set
	USE	opens an existing SAS data set for read access

These commands can be used to perform any necessary data management functions. For example, you can read observations from a SAS data set into a target matrix with the USE or EDIT commands. You can edit a SAS data set, appending or deleting records. If you have generated data in a matrix, you can output the data to a SAS data set with the APPEND or CREATE commands. See Chapter 6, "Working with SAS Data Sets," and Chapter 7, "File Access," for more information on these commands.

Missing Values

With SAS/IML software, a numeric element can have a special value called a *missing value* that indicates that the value is unknown or unspecified. Such missing values are coded, for logical comparison purposes, in the bit pattern of very large negative numbers. A numeric matrix can have any mixture of missing and nonmissing values. A matrix with missing values should not be confused with an empty or unvalued matrix, that is, a matrix with zero rows and zero columns.

In matrix literals, a numeric missing value is specified as a single period. In data processing operations involving a SAS data set, you can append or delete missing values. All operations that move values move missing values properly.

SAS/IML software supports missing values in a limited way, however. Most matrix operators and functions do not support missing values. For example, matrix multiplication involving a matrix with missing values is not meaningful. Also, the inverse of a matrix with missing values has no meaning. Performing matrix operations such as these on matrices that have missing values can result in inconsistencies, depending on the host environment.

See Chapter 4, "Working with Matrices," and Chapter 14, "Further Notes," for more details on missing values.

In Conclusion

In this chapter, you were introduced to the fundamentals of the SAS/IML language. The basic data element, the matrix, was defined, and you learned several ways to create matrices: the matrix literal, CALL statements that return matrix results, and assignment statements.

You were introduced to the types of statements with which you can program: commands, control statements for iterative programming and module definition, functions, and CALL subroutines.

Chapter 3, "Tutorial: a Model for Linear Regression," offers an introductory tutorial that demonstrates using SAS/IML software to build and execute a module.

Chapter **3** Tutorial: a Module for Linear Regression

Introduction

SAS/IML software makes it possible for you to solve mathematical problems or implement new statistical techniques and algorithms. The language is patterned after linear algebra notation. For example, the least-squares formula familiar to statisticians

$$B = (X'X)^{-1}X'Y$$

can be easily translated into the IML statement

```
b=inv(x`*x)*x`*y;
```

This is an example of an assignment statement that uses a built-in function (INV) and operators (transpose and matrix multiplication).

If a statistical method has not been implemented directly in a SAS procedure, you may be able to program it using IML. Because the operations in IML deal with arrays of numbers rather than with one number at a time, and the most commonly used mathematical and matrix operations are built directly into the language, programs that take hundreds of lines of code in other languages often take only a few lines in IML.

Solving a System of Equations

Because IML is built around traditional matrix algebra notation, it is often possible to directly translate mathematical methods from matrix algebraic expressions into executable IML statements. For example, consider the problem of solving three simultaneous equations:

$$3x_1 - x_2 + 2x_3 = 8$$
$$2x_1 - 2x_2 + 3x_3 = 2$$
$$4x_1 + x_2 - 4x_3 = 9$$

These equations may be written in matrix form as

$$\begin{bmatrix} 3 & -1 & 2 \\ 2 & -2 & 3 \\ 4 & 1 & -4 \end{bmatrix} \begin{bmatrix} x_1 \\ x_2 \\ x_3 \end{bmatrix} = \begin{bmatrix} 8 \\ 2 \\ 9 \end{bmatrix}$$

and can be expressed symbolically as

$$\mathbf{Ax} = \mathbf{c} \quad .$$

Because **A** is nonsingular, the system has a solution given by

$$\mathbf{x} = \mathbf{A}^{-1}\mathbf{c} \quad .$$

In the following example, you will solve this system of equations using an interactive session. Submit the PROC IML statement to begin the procedure. Throughout this chapter the right angle brackets ($>$) indicate statements you submit; responses from IML follow below.

```
> proc iml;

IML Ready
```

Enter

```
> reset print;
```

The PRINT option of the RESET command causes automatic printing of results. Notice that as you submit each statement, it is executed and the results are printed. While you are learning IML or developing modules, it is a good idea to have all results printed automatically. Once you are familiar with SAS/IML software, you will not need to use automatic printing.

Next, set up the matrices **A** and **c**. Both of these matrices are input as matrix literals; that is, input the row and column values as discussed in Chapter 2, "Understanding the SAS/IML Language."

```
> a={3  -1   2,
>    2  -2   3,
>    4   1  -4};

             A        3 rows      3 cols     (numeric)

                          3          -1          2
                          2          -2          3
                          4           1         -4


> c={8, 2, 9};

             C        3 rows      1 col      (numeric)

                          8
                          2
                          9
```

Now write the solution equation, $\mathbf{x} = \mathbf{A}^{-1}\mathbf{c}$, as an IML statement. The appropriate statement is an assignment statement that uses a built-in function and an operator (INV is a built-in function that takes the inverse of a square matrix and * is the operator for matrix multiplication).

```
> x=inv(a)*c;

          X           3 rows      1 col     (numeric)

                                    3
                                    5
                                    2
```

After IML executes the statement, the first row of matrix $\mathbf{X}$ contains the x_1 value for which you are solving, the second row contains the x_2 value, and the third row contains the x_3 value.

Now end the session by entering the QUIT command.

```
>   quit;

    Exiting IML
```

A Module for Linear Regression

The previous method may be more familiar to statisticians when different notation is used. A linear model is usually written

$$\mathbf{y} = \mathbf{Xb} + \mathbf{e}$$

where $\mathbf{y}$ is the vector of responses, $\mathbf{X}$ is the design matrix, and $\mathbf{b}$ is a vector of unknown parameters estimated by minimizing the sum of squares of $\mathbf{e}$, the error or residual.

The following example is given to illustrate the programming techniques involved in performing linear regression. It is not meant to replace regression procedures such as the REG procedure, which are more efficient for regressions and offer a multitude of diagnostic options.

Suppose that you have response data $\mathbf{y}$ measured at five values of the independent variable $\mathbf{x}$ and you want to perform a quadratic regression.

Submit the PROC IML statement to begin the procedure.

```
> proc iml;

    IML Ready
```

Input the design matrix $\mathbf{X}$ and the data vector $\mathbf{y}$ as matrix literals.

```
> x={1 1 1,
>     1 2 4,
>     1 3 9,
>     1 4 16,
>     1 5 25};
```

```
X              5 rows        3 cols      (numeric)

                    1            1          1
                    1            2          4
                    1            3          9
                    1            4         16
                    1            5         25
```

```
> y={1,5,9,23,36};
```

```
Y              5 rows        1 col       (numeric)

                              1
                              5
                              9
                             23
                             36
```

Compute the least-squares estimate of **b** using the traditional formula.

```
> b=inv(x`*x)*x`*y;
```

```
B              3 rows        1 col       (numeric)

                             2.4
                            -3.2
                             2
```

The predicted values are simply the **X** matrix multiplied by the parameter estimates, and the residuals are the difference between actual and predicted **y**.

```
> yhat=x*b;
```

```
YHAT           5 rows        1 col       (numeric)

                             1.2
                             4
                            10.8
                            21.6
                            36.4
```

```
> r=y-yhat;
```

```
R              5 rows        1 col       (numeric)

                            -0.2
                             1
                            -1.8
                             1.4
                            -0.4
```

To calculate the estimate of the variance of the responses, calculate the sum of squared errors (SSE), its degrees of freedom (DFE), and the mean squared error (MSE). Note that in computing the degrees, you use the function NCOL to return the number of columns of **X**.

```
> sse=ssq(r);

            SSE            1 row       1 col      (numeric)

                                       6.4


> dfe=nrow(x)-ncol(x);

            DFE            1 row       1 col      (numeric)

                                       2


> mse=sse/dfe;

            MSE            1 row       1 col      (numeric)

                                       3.2
```

Notice that each calculation has required one simple line of code.

Now suppose you want to solve the problem repeatedly on new data sets without reentering the code. To do this, define a module (or subroutine). Modules begin with a START statement and end with a FINISH statement, with the program statements in between. The statements below define a module named REGRESS to perform linear regression.

```
> start regress;                              /* begin module */
>    xpxi=inv(t(x)*x);            /* inverse of X'X        */
>    beta=xpxi*(t(x)*y);          /* parameter estimate    */
>    yhat=x*beta;                 /* predicted values      */
>    resid=y-yhat;                 /* residuals            */
>    sse=ssq(resid);               /* SSE                  */
>    n=nrow(x);                    /* sample size          */
>    dfe=nrow(x)-ncol(x);          /* error DF             */
>    mse=sse/dfe;                  /* MSE                  */
>    cssy=ssq(y-sum(y)/n);         /* corrected total SS   */
>    rsquare=(cssy-sse)/cssy;      /* RSQUARE              */
>    print,"Regression Results",
>       sse dfe mse rsquare;
>    stdb=sqrt(vecdiag(xpxi)*mse); /* std of estimates     */
>    t=beta/stdb;                  /* parameter t-tests    */
>    prob=1-probf(t#t,1,dfe);      /* p-values             */
>    print,"Parameter Estimates",,
>       beta stdb t prob;
>    print,y yhat resid;
> finish regress;                             /* end module          */
```

Submit the module REGRESS for execution.

```
> reset noprint;
> run regress;                              /* execute module     */
```

                          Regression Results

```
        SSE       DFE       MSE    RSQUARE
        6.4         2       3.2  0.9923518
```

                         Parameter Estimates

```
      BETA      STDB         T       PROB
       2.4  3.8366652  0.6255432  0.5954801
      -3.2  2.9237940  -1.094468  0.3879690
         2  0.4780914  4.1833001  0.0526691
```

```
            Y      YHAT     RESID
            1       1.2      -0.2
            5         4         1
            9      10.8      -1.8
           23      21.6       1.4
           36      36.4      -0.4
```

At this point, you still have all of the matrices defined if you want to continue calculations. Suppose that you want to correlate the estimates. First, calculate the covariance estimate of the estimates; then, scale the covariance into a correlation matrix with values of 1 on the diagonal.

```
> reset print;                             /* turn on auto printing  */
> covb=xpxi*mse;                           /* covariance of estimates */

        COVB          3 rows     3 cols    (numeric)

                    14.72    -10.56        1.6
                   -10.56  8.5485714  -1.371429
                      1.6  -1.371429  0.2285714

> s=1/sqrt(vecdiag(covb));

        S             3 rows     1 col     (numeric)

                   0.260643
                   0.3420214
                   2.0916501
```

```
> corrb=diag(s)*covb*diag(s);              /* correlation of estimates */

            CORRB        3 rows      3 cols     (numeric)

                    1   -0.941376   0.8722784
            -0.941376           1   -0.981105
            0.8722784   -0.981105           1
```

Your module REGRESS remains available to do another regression, in this case, an orthogonalized version of the last polynomial example. In general, the columns of **X** will not be orthogonal. You can use the ORPOL function to generate orthogonal polynomials for the regression. Using them provides greater computing accuracy and reduced computing times. When using orthogonal polynomial regression, you expect the statistics of fit to be the same and the estimates to be more stable and uncorrelated.

To perform an orthogonal regression on the data, you must first create a vector containing the values of the independent variable x, which is the second column of the design matrix **X**. Then, use the ORPOL function to generate orthogonal second degree polynomials.

```
> x1={1,2,3,4,5};                          /* second column of X */

            X1           5 rows      1 col      (numeric)

                                          1
                                          2
                                          3
                                          4
                                          5


> x=orpol(x1,2);                           /* generates orthogonal polynomials */

            X            5 rows      3 cols     (numeric)

            0.4472136   -0.632456   0.5345225
            0.4472136   -0.316228   -0.267261
            0.4472136           0   -0.534522
            0.4472136   0.3162278   -0.267261
            0.4472136   0.6324555   0.5345225

> reset noprint;                           /* turns off auto printing    */
> run regress;                             /* run REGRESS */


                    Regression Results


            SSE        DFE       MSE   RSQUARE
            6.4          2       3.2 0.9923518
```

Parameter Estimates

	BETA	STDB	T	PROB
	33.093806	1.7888544	18.5	0.0029091
	27.828043	1.7888544	15.556349	0.0041068
	7.4833148	1.7888544	4.1833001	0.0526691

Y	YHAT	RESID
1	1.2	-0.2
5	4	1
9	10.8	-1.8
23	21.6	1.4
36	36.4	-0.4

```
> reset print;
> covb=xpxi*mse;
```

COVB   3 rows   3 cols   (numeric)

3.2	-2.73E-17	4.693E-16
-2.73E-17	3.2	-2.18E-15
4.693E-16	-2.18E-15	3.2

```
> s=1/sqrt(vecdiag(covb));
```

S   3 rows   1 col   (numeric)

0.559017
0.559017
0.559017

```
> corrb=diag(s)*covb*diag(s);
```

CORRB   3 rows   3 cols   (numeric)

1	-8.54E-18	1.467E-16
-8.54E-18	1	-6.8E-16
1.467E-16	-6.8E-16	1

Note that the values on the off-diagonal are printed in scientific notation; the values are close to zero but not exactly zero because of the imprecision of floating-point arithmetic. To clean up the appearance of the correlation matrix, use the FUZZ option.

```
> reset fuzz;
> corrb=diag(s)*covb*diag(s);
```

	CORRB	3 rows	3 cols	(numeric)
	1	0	0	
	0	1	0	
	0	0	1	

Plotting Regression Results

You can create some simple plots by using the PGRAF subroutine. The PGRAF subroutine produces scatterplots suitable for printing on a line printer. If you want to produce better quality graphics using color, you can use the graphics capabilities of IML (see Chapter 9, "Introduction to Graphics").

Here is how you can plot the residuals against **x**. First, create a matrix containing the pairs of points. Do this by concatenating **X1** with **RESID** using the horizontal concatenation operator ($\|$).

```
> xy=x1‖resid;
```

	XY	5 rows	2 cols	(numeric)
		1	-0.2	
		2	1	
		3	-1.8	
		4	1.4	
		5	-0.4	

Next, use a CALL statement to call the PGRAF subroutine to produce the desired plot. The arguments to PGRAF are, in order,

1. the matrix containing the pairs of points

2. a plotting symbol

3. a label for the X-axis

4. a label for the Y-axis

5. a title for the plot.

```
> call pgraf(xy,'r','x','Residuals','Plot of Residuals');
```

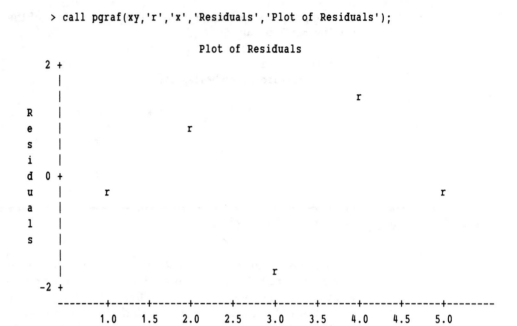

You can also plot the predicted values **yhat** against **x**. You must first create a matrix, say **XYH**, containing the points. Do this by concatenating **X1** with **YHAT**. Next, call the PGRAF subroutine to plot the points.

```
> xyh=x1‖yhat;
```

	XYH	5 rows	2 cols	(numeric)

1	1.2
2	4
3	10.8
4	21.6
5	36.4

```
> call pgraf(xyh,'*','x','Predicted','Plot of Predicted Values');
```

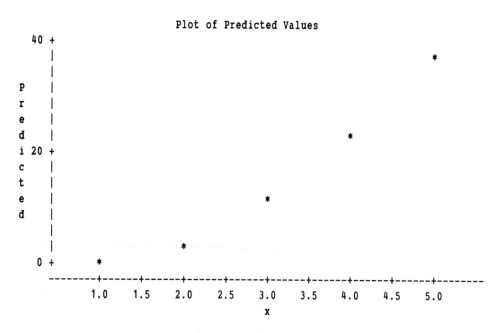

You can get a little fancier plot, denoting the observed values with a "y" and the predicted values with a "p", using the following statements. Create a matrix **NEWXY** containing the pairs of points to overlay. You need to use both the horizontal concatenation operator (‖) and the vertical concatenation operator (//). The NROW function returns the number of observations, that is, the number of rows of **X1**. The matrix **LABEL** contains the character label for each point, plotting a "y" for each observed point and a "p" for each predicted point.

```
> newxy=(x1//x1)‖(y//yhat);
```

NEWXY	10 rows	2 cols	(numeric)

1	1
2	5
3	9
4	23
5	36
1	1.2
2	4
3	10.8
4	21.6
5	36.4

```
> n=nrow(x1);
```

N	1 row	1 col	(numeric)

```
> label=repeat('y',n,1)//repeat('p',n,1);
```

```
            LABEL          10 rows       1 col      (character, size 1)

                                           y
                                           y
                                           y
                                           y
                                           y
                                           p
                                           p
                                           p
                                           p
                                           p
```

```
> call pgraf(newxy,label,'x','y','Scatter Plot with Regression Line' );
```

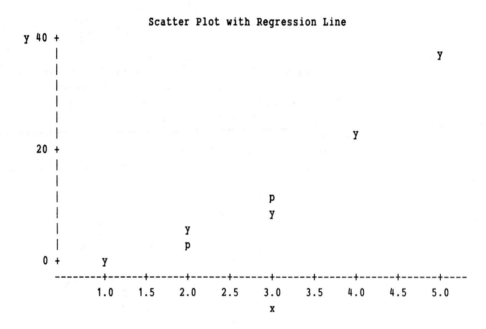

As you can see, the observed and predicted values are too close together to be able to distinguish them at all values of **x**.

In Conclusion

In this chapter, you have seen the programming techniques necessary for solving systems of equations. You have seen how to define a module for performing linear regression and obtaining covariance and correlation matrices, and how to obtain some simple diagnostic plots. Many of the ideas presented in Chapter 2, "Understanding the SAS/IML Language," such as the use of assignment statements, functions, CALL statements, and subscripting have been demonstrated.

Chapter **4** Working with Matrices

Introduction

SAS/IML software provides many ways to create matrices. You can create matrices by doing any of the following:

- entering data yourself as a matrix literal

- using assignment statements

- using matrix-generating functions

- creating submatrices from existing matrices with subscripts

- using SAS data sets (see Chapter 6, "Working with SAS Data Sets," for more information).

Once you have defined matrices, you have access to many operators and functions for working on them in matrix expressions. These operators and functions facilitate programming and make referring to submatrices efficient and simple.

Finally, you have several means available for tailoring your printed output.

Entering Data as Matrix Literals

The most basic way to create a matrix is to define a matrix literal, either numeric or character, by entering the matrix elements. A matrix literal can be a single element (called a *scalar*), a single row of data (called a *row vector*), a single column of data (called a *column vector*), or a rectangular array of data (called a *matrix*). The *dimension* of a matrix is given by its number of rows and columns. An $n \times m$ matrix has n rows and m columns.

Scalars

Scalars are matrices that have only one element. You define a scalar with the matrix name on the left-hand side of an assignment statement and its value on the right-hand side. You can use the statements below to create and print several examples of scalar literals. First, you must invoke PROC IML.

```
> proc iml;

  IML Ready

> x=12;
> y=12.34;
> z=.;
> a='Hello';
> b="Hi there";
> print x y z a b;
```

```
            X          Y        Z  A      B
           12      12.34        .  Hello  Hi there
```

Notice that when defining a character literal, you need to use either single quotes (') or double quotes ("). Using quotes preserves uppercase and lowercase distinctions and embedded blanks. It is also always correct to enclose the data values inside of braces ({ }).

Matrices with Multiple Elements

To enter a matrix having multiple elements, use braces ({ }) to enclose the data values and, if needed, commas to separate rows. Inside of the braces, all elements must be either numeric or character. You cannot have a mixture of data types within a matrix. Each row must have the same number of elements.

For example, suppose that you have one week of data on daily coffee consumption (cups per day) for your office of four people. Create a matrix

COFFEE with each person's consumption as a row of the matrix and each day represented by a column. First, submit the RESET PRINT statement so that results are printed as you submit statements.

```
> reset print;
> coffee={4 2 2 3 2,
>         3 3 1 2 1,
>         2 1 0 2 1,
>         5 4 4 3 4};
```

```
        COFFEE        4 rows      5 cols     (numeric)

            4           2           2           3           2
            3           3           1           2           1
            2           1           0           2           1
            5           4           4           3           4
```

Now create a character matrix called **NAMES** with rows containing the names of the people in your office. Note that when you do not use quotes, characters are converted to uppercase.

```
> names={Jenny, Linda, Jim, Samuel};
```

```
        NAMES        4 rows       1 col     (character, size 6)

                        JENNY
                        LINDA
                        JIM
                        SAMUEL
```

Notice that the output with the RESET PRINT statement includes the dimension, type, and when type is character, the element size of the matrix. The element size represents the length of each string and is determined from the length of the longest string.

Now, print **COFFEE** using **NAMES** as row labels with the ROWNAME= option to the PRINT statement.

```
> print coffee [rowname=names];
```

```
        COFFEE
        JENNY         4           2           2           3           2
        LINDA         3           3           1           2           1
        JIM           2           1           0           2           1
        SAMUEL        5           4           4           3           4
```

Using Assignment Statements

Assignment statements create matrices by evaluating expressions and assigning the results to a matrix. The expressions can be composed of operators (for example, the matrix addition operator (+)), functions (for example, the INV function), and subscripts. Assignment statements have the general form

result = expression;

where *result* is the name of the new matrix and *expression* is an expression that is evaluated. The resulting matrix automatically acquires the appropriate dimension, type, and value. Details on writing expressions are described in "Using Matrix Expressions" later in this chapter.

Simple Assignment Statements

Simple assignment statements involve an equation having the matrix name on the left-hand side and either an expression involving other matrices or a matrix-generating function on the right-hand side.

Suppose you want to generate some statistics for the weekly coffee data. If a cup of coffee costs 30 cents, then you can create a matrix with the daily expenses, **DAYCOST**, by multiplying the per-cup cost with the matrix **COFFEE** using the elementwise multiplication operator (#). Turn off the automatic printing so that you can tailor the output with the ROWNAME= and FORMAT= options in the PRINT statement.

```
>  reset noprint;
>  daycost=0.30#coffee;
>  print "Daily totals", daycost[rowname=names format=8.2];
```

```
       Daily totals

       DAYCOST
       JENNY      1.20    0.60    0.60    0.90    0.60
       LINDA      0.90    0.90    0.30    0.60    0.30
       JIM        0.60    0.30    0.00    0.60    0.30
       SAMUEL     1.50    1.20    1.20    0.90    1.20
```

You can calculate the weekly total cost for each person using the matrix multiplication operator (*). First create a 5×1 vector of 1's. This vector will sum the daily costs for each person when multiplied with **COFFEE**. (You will see later that there is a more efficient way to do this using subscript reduction operators.)

```
>  ones={1,1,1,1,1};
>  weektot=daycost*ones;
>  print "Week total", weektot[rowname=names format=8.2];
```

```
       Week total

       WEEKTOT
       JENNY      3.90
       LINDA      3.00
       JIM        1.80
       SAMUEL     6.00
```

Finally, you can calculate the average number of cups drunk per day by dividing the grand total of cups by days. To find the grand total, use the SUM function, which returns the sum of all elements of a matrix. Next, divide the grand total by 5, the number of days (which is the number of columns) using the division operator (/) and the NCOL function. These two matrices are created separately, but the entire calculation could be done in one statement.

```
>  grandtot=sum(coffee);
>  average=grandtot/ncol(coffee);
>  print "Total number of cups", grandtot,,"Daily average",average;
```

```
                    Total number of cups

                         GRANDTOT
                           49


                      Daily average

                         AVERAGE
                           9.8
```

Matrix-generating Functions

SAS/IML software has many built-in functions that generate useful matrices. For example, the J function creates a matrix with a given dimension and element value when you supply the number of rows, columns, and an element value for the new matrix. This function is useful to initialize a matrix to a predetermined size. Several matrix-generating functions are listed below:

BLOCK   creates a block-diagonal matrix.

DESIGNF   creates a full-rank design matrix.

I   creates an identity matrix.

J   creates a matrix of a given dimension.

SHAPE   shapes a new matrix from the argument.

The sections that follow illustrate these matrix-generating functions. Again, they are shown with automatic printing of results, activated by invoking the RESET PRINT statement.

```
reset print;
```

The BLOCK Function

The BLOCK function has the general form

BLOCK(*matrix1*,<*matrix2,...,matrix15*>)

and creates a block-diagonal matrix from the argument matrices. For example, the statements

```
>   a={1 1,1 1};
```

	A	2 rows	2 cols	(numeric)
		1	1	
		1	1	

```
>   b={2 2, 2 2};
```

	B	2 rows	2 cols	(numeric)
		2	2	
		2	2	

```
>   c=block(a,b);
```

result in the matrix

	C	4 rows	4 cols	(numeric)
1	1	0	0
1	1	0	0
0	0	2	2
0	0	2	2

The J Function

The J function has the general form

J(*nrow*<,*ncol*<,*value*>>)

and creates a matrix having *nrow* rows, *ncol* columns, and all element values equal to *value*. The *ncol* and *value* arguments are optional, but you will usually want to specify them. In many statistical applications, it is helpful to be able to create a row (or column) vector of 1's (you did so to calculate coffee totals in the last section). You can do this with the J function. For example, the statement below creates a 1×5 row vector of 1's:

```
>   one=j(1,5,1);
```

	ONE	1 row	5 cols	(numeric)
1	1	1	1	1

The I Function

The I function creates an identity matrix of a given size. It has the general form

I(*dimension*)

where *dimension* gives the number of rows. For example, the following statement creates a 3×3 identity matrix:

```
>  I3=I(3);
```

```
        I3              3 rows        3 cols     (numeric)

                    1          0          0
                    0          1          0
                    0          0          1
```

The DESIGNF Function

The DESIGNF function generates a full-rank design matrix, useful in calculating *ANOVA* tables. It has the general form

DESIGNF(*column-vector*)

For example, the following statement creates a full-rank design matrix for a one-way *ANOVA*, where the treatment factor has three levels and there are $n_1 = 3$, $n_2 = 2$, and $n_3 = 2$ observations at the factor levels:

```
>  d=designf({1,1,1,2,2,3,3});
```

```
        D               7 rows        2 cols     (numeric)

                    1          0
                    1          0
                    1          0
                    0          1
                    0          1
                   -1         -1
                   -1         -1
```

The SHAPE Function

The SHAPE function shapes a new matrix from an argument matrix. It has the general form

SHAPE(*matrix<,nrow<,ncol<,pad-value>>>*)

Although the *nrow, ncol,* and *pad-value* arguments are optional, you will usually want to specify them. The example below uses the SHAPE function to create a 3×3 matrix containing the values 99 and 33. The function cycles back and repeats values to fill in when no *pad-value* is given.

```
>  aa=shape({99 33,99 33},3,3);

             AA            3 rows        3 cols      (numeric)

                     99            33          99
                     33            99          33
                     99            33          99
```

In the next example, a *pad-value* is specified for filling in the matrix:

```
>  aa=shape({99 33,99 33},3,3,0);

             AA            3 rows        3 cols      (numeric)

                     99            33          99
                     33             0           0
                      0             0           0
```

The SHAPE function cycles through the argument matrix elements in row-major order and then fills in with 0's after the first cycle through the argument matrix.

Index Vectors

You can create a vector by using the index operator (:). Several examples of statements involving index vectors are shown below:

```
>  r=1:5;

            R            1 row        5 cols      (numeric)
                   1         2          3          4        5

>  s=10:6;

            S            1 row        5 cols      (numeric)
                  10         9          8          7        6

>  t='abc1':'abc5';

            T            1 row        5 cols      (character, size 4)
                   abc1 abc2 abc3 abc4 abc5
```

If you want an increment other than 1, use the DO function. For example, if you want a vector ranging from minus 1 to 1 by .5, use the following statement:

```
> r=do(-1,1,.5);
```

```
          R         1 row      5 cols    (numeric)
              -1       -0.5         0      0.5          1
```

Using Matrix Expressions

Matrix expressions are a sequence of names, literals, operators, and functions that perform some calculation, evaluate some condition, or manipulate values. These expressions can appear on either side of an assignment statement.

Operators

Operators used in matrix expressions fall into three general categories:

prefix operators   are placed in front of operands. For example, $-A$ uses the sign reverse prefix operator ($-$) in front of the operand **A** to reverse the sign of each element of **A**.

infix operators   are placed between operands. For example, $A+B$ uses the addition infix operator ($+$) between operands **A** and **B** to add corresponding elements of the matrices.

postfix operators   are placed after an operand. For example, **A`** uses the transpose postfix operator (`) after the operand **A** to transpose **A**.

Matrix operators are listed in Appendix 1, "SAS/IML Quick Reference," and described in detail in Chapter 15, "SAS/IML Language Reference." Table 4.1 shows the precedence of matrix operators in IML.

Table 4.1
Operator
Precedence

Priority Group	Operators
I (highest)	^ ` subscripts $-$(prefix) ## **
II	* # <> >< / @
III	+ $-$
IV	‖ // :
V	< <= > >= = ^=
VI	&
VII (lowest)	|

Compound Expressions

With SAS/IML software, you can write compound expressions involving several matrix operators and operands. For example, the following statements are valid matrix assignment statements:

```
a=x+y+z;
a=x+y*z';
a=(-x)#(y-z);
```

The rules for evaluating compound expressions are as follows:

□ Evaluation follows the order of operator precedence, as shown in Table 4.1. Group I has the highest priority; that is, Group I operators are evaluated first. Group II operators are evaluated after Group I operators, and so forth. For example, the statement

```
a=x+y*z;
```

first multiplies matrices **Y** and **Z** since the * operator (Group II) has higher precedence than the + operator (Group III). It then adds the result of this multiplication to the matrix **X**, and assigns the new matrix to **A**.

□ If neighboring operators in an expression have equal precedence, the expression is evaluated from left to right, except for the highest priority operators. For example, the statement

```
a=x/y/z;
```

first divides each element of matrix **X** by the corresponding element of matrix **Y**. Then, using the result of this division, it divides each element of the resulting matrix by the corresponding element of matrix **Z**. The operators in Group 1 in Table 4.1 are evaluated from right to left. For example, the expression

```
-x**2
```

is evaluated as

```
-(x**2)
```

When multiple prefix or postfix operators are juxtaposed, precedence is determined by their order from inside to outside.

For example, the expression

```
^-a
```

is evaluated as ^(−**A**), and the expression

```
a`[i,j]
```

is evaluated as (**A**`)[i,j].

□ All expressions enclosed in parentheses are evaluated first, using the two rules above. Thus, the IML statement

```
a=x/(y/z);
```

is evaluated by first dividing elements of **Y** by the elements of **Z**, then dividing this result into **X**.

Elementwise Binary Operators

Elementwise binary operators produce a result matrix from element-by-element operations on two argument matrices. Table 4.2 lists the elementwise binary operators.

Table 4.2
Elementwise
Binary Operators

Operator	Action
+	addition, concatenation
−	subtraction
#	elementwise multiplication
##	elementwise power
/	division
<>	element maximum
><	element minimum
|	logical OR
&	logical AND
<	less than
<=	less than or equal to
>	greater than
>=	greater than or equal to
^=	not equal to
=	equal to
MOD(m,n)	modulo (remainder)

For example, consider the two matrices **A** and **B** given below.

$$\text{Let } \mathbf{A} = \begin{bmatrix} 2 & 2 \\ 3 & 4 \end{bmatrix} \text{ and } \mathbf{B} = \begin{bmatrix} 4 & 5 \\ 1 & 0 \end{bmatrix}$$

The addition operator (+) adds corresponding matrix elements:

$$\mathbf{A}+\mathbf{B} \text{ yields } \begin{bmatrix} 6 & 7 \\ 4 & 4 \end{bmatrix}$$

The elementwise multiplication operator (#) multiplies corresponding elements:

$$\mathbf{A}\#\mathbf{B} \text{ yields } \begin{bmatrix} 8 & 10 \\ 3 & 0 \end{bmatrix}$$

The elementwise power operator (##) raises elements to powers:

$$\mathbf{A}\#\#2 \text{ yields } \begin{bmatrix} 4 & 4 \\ 9 & 16 \end{bmatrix}$$

The element maximum operator (<>) compares corresponding elements and chooses the larger:

$$\mathbf{A}<>\mathbf{B} \text{ yields } \begin{bmatrix} 4 & 5 \\ 3 & 4 \end{bmatrix}$$

The less than or equal to operator (<=) returns a 1 if an element of **A** is less than or equal to the corresponding element of **B**, and returns a 0 otherwise:

$$\mathbf{A}<=\mathbf{B} \text{ yields } \begin{bmatrix} 1 & 1 \\ 0 & 0 \end{bmatrix}$$

The modulo operator returns the remainder of each element divided by the argument.

$$\text{MOD}(\mathbf{A},3) \text{ yields } \begin{bmatrix} 2 & 2 \\ 0 & 1 \end{bmatrix}$$

All operators can also work in a one-to-many or many-to-one manner, as well as in an element-to-element manner; that is, they allow you to perform tasks such as adding a scalar to a matrix or dividing a matrix by a scalar. For example, the statement

```
x=x#(x>0);
```

replaces each negative element of the matrix **X** with 0. The expression (X>0) is a many-to-one operation that compares each element of **X** to 0 and creates a temporary matrix of results; an element in the result matrix is 1 when the expression is true and 0 when it is false. When the expression is true (the element is positive), the element is multiplied by 1. When the expression is false (the element is negative or 0), the element is multiplied by 0. To fully understand the intermediate calculations, you can use the RESET PRINTALL command to have the temporary result matrices printed.

Subscripts

Subscripts are special postfix operators placed in square brackets ([ ]) after a matrix operand. Subscript operations have the general form

operand[row,column]

where

operand	is usually a matrix name, but it can also be an expression or literal.
row	refers to an expression, either scalar or vector, for selecting one or more rows from the operand.
column	refers to an expression, either scalar or vector, for selecting one or more columns from the operand.

You can use subscripts to

□ refer to a single element of a matrix

□ refer to an entire row or column of a matrix

□ refer to any submatrix contained within a matrix

□ perform a reduction across rows or columns of a matrix.

In expressions, subscripts have the same (high) precedence as the transpose postfix operator (`). Note that when both *row* and *column* subscripts are used, they are separated by a comma.

Selecting a Single Element

You can select a single element of a matrix in two ways. You can use two subscripts (*row, column*) to refer to its location, or you can use one subscript to look for the element down the rows. For example, referring to the coffee example used earlier, to find the element corresponding to the number of cups that Linda drank on Monday, you can use either of two statements.

First, you can refer to the element by row and column location. In this case, you want the second row and first column. You can call this matrix **c21**.

```
> print coffee[rowname=names];
```

```
        COFFEE
        JENNY        4        2        2        3        2
        LINDA        3        3        1        2        1
        JIM          2        1        0        2        1
        SAMUEL       5        4        4        3        4
```

```
> c21=coffee[2,1];
> print c21;
```

```
                             C21
                              3
```

You can also look for the element down the rows. In this case, you refer to this element as the sixth element of **COFFEE** in row-major order.

```
> c6=coffee[6];
> print c6;
```

```
                             C6
                              3
```

Selecting a Row or Column

To refer to an entire row or column of a matrix, write the subscript with the row or column number, omitting the other subscript but not the comma. For example, to refer to the row of **COFFEE** that corresponds to Jim, you want the submatrix consisting of the third row and all columns:

```
> jim=coffee[3,];
> print jim;
```

```
        JIM
         2        1        0        2        1
```

If you want the data for Friday, you know that the fifth column corresponds to Friday, so you want the submatrix consisting of the fifth column and all rows:

```
> friday=coffee[,5];
> print friday;
```

```
                           FRIDAY
                             2
                             1
                             1
                             4
```

Submatrices

You refer to a submatrix by the specific rows and columns you want. Include within the brackets the rows you want, a comma, and the columns you want. For example, to create the submatrix of **COFFEE** consisting of the first and third rows and second, third, and fifth columns, submit the following statements:

```
> submat1=coffee[{1 3},{2 3 5}];
> print submat1;
```

```
        SUBMAT1
          2      2      2
          1      0      1
```

The first vector, {1 3}, selects the rows, and the second vector, {2 3 5}, selects the columns. Alternately, you can create the vectors beforehand and supply their names as arguments.

```
> rows={1 3};
> cols={2 3 5};
> submat1=coffee[rows,cols];
```

You can use index vectors generated by the index creation operator (:) in subscripts to refer to successive rows or columns. For example, to select the first three rows and last three columns of **COFFEE**, use the following statements:

```
> submat2=coffee[1:3,3:5];
> print submat2;
```

```
        SUBMAT2
          2      3      2
          1      2      1
          0      2      1
```

Note that, in each example, the number in the first subscript defines the number of rows in the new matrix; the number in the second subscript defines the number of columns.

Subscripted Assignment

You can assign values into a matrix using subscripts to refer to the element or submatrix. In this type of assignment, the subscripts appear on the left-hand side of the equal sign. For example, to change the value in the first row, second column of **COFFEE** from 2 to 4, use subscripts to refer to the appropriate element in an assignment statement:

```
> coffee[1,2]=4;
> print coffee;
```

```
        COFFEE
          4      4      2      3      2
          3      3      1      2      1
          2      1      0      2      1
          5      4      4      3      4
```

To change the values in the last column of **COFFEE** to 0s use the following statement:

```
>  coffee[,5]={0,0,0,0};
>  print coffee;
```

```
        COFFEE
          4       4       2       3       0
          3       3       1       2       0
          2       1       0       2       0
          5       4       4       3       0
```

In the next example, you first locate the positions of negative elements of a matrix and then set these elements equal to 0. This can be useful in situations where negative elements may indicate errors or be impossible values. The LOC function is useful for creating an index vector for a matrix that satisfies some condition. In the example below, the LOC function is used to find the positions of the negative elements of the matrix **T** and then to set these elements equal to 0 using subscripted assignment:

```
>  t={ 3  2 -1,
>       6 -4  3,
>       2  2  2 };
>  print t;
```

```
        T
        3       2      -1
        6      -4       3
        2       2       2
```

```
>  i=loc(t<0);
>  print i;
```

```
        I
        3       5
```

```
>  t[i]=0;
>  print t;
```

```
        T
        3       2       0
        6       0       3
        2       2       2
```

Subscripts can also contain expressions whose results are either row or column vectors. These statements can also be written

```
>  t[loc(t<0)]=0;
```

If you use a noninteger value as a subscript, only the integer portion is used. Using a subscript value less than one or greater than the dimension of the matrix results in an error.

Subscript Reduction Operators

You can use reduction operators, which return a matrix of reduced dimension, in place of values for subscripts to get reductions across all rows and columns. Table 4.3 lists the eight operators for subscript reduction in IML.

Table 4.3
Subscript
Reduction
Operators

Operator	Action
+	addition
#	multiplication
<>	maximum
><	minimum
<:>	index of maximum
>:<	index of minimum
:	mean (different from the MATRIX procedure)
##	sum of squares

For example, to get column sums of the matrix **X** (sum across the rows, which reduces the row dimension to 1), specify X[+,]. The first subscript (+) specifies summation reduction take place across the rows. Omitting the second subscript, corresponding to columns, leaves the column dimension unchanged. The elements in each column are added, and the new matrix consists of one row containing the column sums.

You can use these operators to reduce either rows or columns or both. When both rows and columns are reduced, row reduction is done first.

For example, the expression A[+,<>] results in the maximum (<>) of the column sums (+).

You can repeat reduction operators. To get the sum of the row maxima, use the expression A[,<>] [+,] .

A subscript such as A[{2 3},+] first selects the second and third rows of **A** and then finds the row sums of that matrix.

The following examples demonstrate how to use the operators for subscript reduction.

$$\text{Let } \mathbf{A} = \begin{bmatrix} 0 & 1 & 2 \\ 5 & 4 & 3 \\ 7 & 6 & 8 \end{bmatrix}$$

The following statements are true:

$$A[\{2\ 3\},+] \text{ yields } \begin{bmatrix} 12 \\ 21 \end{bmatrix} \text{(row sums for rows 2 and 3)}$$

$$A[+,<>] \text{ yields } \begin{bmatrix} 13 \end{bmatrix} \text{(maximum of column sums)}$$

$$A[<>,+] \text{ yields } \begin{bmatrix} 21 \end{bmatrix} \text{(sum of column maxima)}$$

$$A[,><][+,] \text{ yields } \begin{bmatrix} 9 \end{bmatrix} \text{(sum of row minima)}$$

$$A[,<:>] \text{ yields } \begin{bmatrix} 3 \\ 1 \\ 3 \end{bmatrix} \text{(indices of row maxima)}$$

$$A[>:<] \text{ yields } \begin{bmatrix} 1 & 1 & 1 \end{bmatrix} \text{(indices of column minima)}$$

$$A[:] \text{ yields } \begin{bmatrix} 4 \end{bmatrix} \text{(mean of all elements)}$$

Printing Matrices with Row and Column Headings

You can tailor the way your matrices are printed in several ways with the AUTONAME option, the ROWNAME= and COLNAME= options, or the MATTRIB statement.

Using the AUTONAME Option

You can use the RESET statement with the AUTONAME option to automatically print row and column headings. If your matrix has n rows and m columns, the row headings are ROW1 to ROWn and the column headings are COL1 to COLm. For example, the following statements produce the result shown below:

```
> reset autoname;
> print coffee;
```

COFFEE	COL1	COL2	COL3	COL4	COL5
ROW1	4	2	2	3	2
ROW2	3	3	1	2	1
ROW3	2	1	0	2	1
ROW4	5	4	4	3	4

Using the ROWNAME= and COLNAME= Options

You can specify your own row and column headings. The easiest way is to create vectors containing the headings and then print the matrix with the ROWNAME= and COLNAME= options. For example, the following statements produce the result shown below:

```
> names={jenny linda jim samuel};
> days={mon tue wed thu fri};
> print coffee[rowname=names colname=days];
```

COFFEE	MON	TUE	WED	THU	FRI
JENNY	4	2	2	3	2
LINDA	3	3	1	2	1
JIM	2	1	0	2	1
SAMUEL	5	4	4	3	4

Using the MATTRIB Statement

The MATTRIB statement associates printing characteristics with matrices. You can use the MATTRIB statement to print **COFFEE** with row and column headings. In addition, you can format the printed numeric output and assign a label to the matrix name. The following example shows how to tailor your printed output:

```
> mattrib coffee rowname=({jenny linda jim samuel})
>         colname=({mon tue wed thu fri})
>         label='Weekly Coffee'
>         format=2.0;
> print coffee;
```

Weekly Coffee	MON	TUE	WED	THU	FRI
JENNY	4	2	2	3	2
LINDA	3	3	1	2	1
JIM	2	1	0	2	1
SAMUEL	5	4	4	3	4

More on Missing Values

Missing values in matrices were discussed in Chapter 2, "Understanding the SAS/IML Language." You should read that chapter and Chapter 14, "Further Notes," carefully so that you are aware of the way IML treats missing values. Below are several examples that show how IML handles missing values in a matrix.

$$
\text{Let } X = \begin{bmatrix} 1 & 2 & . \\ . & 5 & 6 \\ 7 & . & 9 \end{bmatrix} \text{ and } Y = \begin{bmatrix} 4 & . & 2 \\ 2 & 1 & 3 \\ 6 & . & 5 \end{bmatrix}
$$

The following statements are true:

$$
X + Y \text{ yields } \begin{bmatrix} 5 & . & . \\ . & 6 & 9 \\ 13 & . & 14 \end{bmatrix} \text{ (matrix addition)}
$$

$$
X \# Y \text{ yields } \begin{bmatrix} 4 & . & . \\ . & 5 & 18 \\ 42 & . & 45 \end{bmatrix} \text{ (element multiplication)}
$$

$$
X[+,] \text{ yields } \begin{bmatrix} 8 & 7 & 15 \end{bmatrix} \text{ (column sums)}
$$

Chapter **5** Programming Statements

Introduction

IML is a programming language. As a programming language, it has many features that allow you to control the path of execution through the statements. The control statements in IML function similar to the corresponding statements in the SAS DATA step. This chapter presents the following control features:

- [ ] IF-THEN/ELSE statements

- [ ] DO groups

- [ ] iterative execution

- [ ] jumping (nonconsecutive execution)

- [ ] module definition and execution

- [ ] termination of execution.

IF-THEN/ELSE Statements

To perform an operation conditionally, use an IF statement to test an expression. Alternative actions appear in a THEN clause and, optionally, an ELSE statement. The general form of the IF-THEN/ELSE statement is

IF *expression* **THEN** *statement1*;
ELSE *statement2*;

The IF expression is evaluated first. If the expression is true, execution flows through the THEN alternative. If the expression is false, the ELSE statement, if present, is executed. Otherwise, the next statement is executed.

The expression to be evaluated is often a comparison, for example,

```
if max(a)<20 then p=0;
else p=1;
```

The IF statement results in the evaluation of the condition (MAX(A)<20). If the largest value found in matrix **A** is less than 20, P is set to 0. Otherwise, P is set to 1.

You can nest IF statements within the clauses of other IF or ELSE statements. Any number of nesting levels is allowed. Below is an example of nested IF statements:

```
if x=y then
   if abs(y)=z then w=-1;
   else w=0;
else w=1;
```

When the condition to be evaluated is a matrix expression, the result of the evaluation is a temporary matrix of 0s, 1s, and possibly missing values. If all values of the result matrix are nonzero and nonmissing, the condition is true; if any element in the result matrix is 0, the condition is false. This evaluation is equivalent to using the ALL function.

For example, the statement

```
if x<y then statement;
```

produces the same result as the statement

```
if all(x<y) then statement;
```

The expressions

```
if a^=b then statement;
```

and

```
if ^(a=b) then statement;
```

are valid, but the THEN clause in each case is only executed when all corresponding elements of **A** and **B** are unequal.

If you require only one element in **A** not be equal to its corresponding element in **B**, use the ANY function. For example, evaluation of the expression

```
if any(a^=b) then statement;
```

requires only one element of **A** and **B** to be unequal for the expression to be true.

DO Groups

A set of statements can be treated as a unit by putting them into a DO group, which starts with a DO statement and ends with an END statement. In this way, you can submit the entire group of statements for execution as a single unit. For some programming applications, you must use either a DO group or a module. For example, LINK and GOTO statements must be programmed inside of a DO group or a module.

The two principal uses of DO groups are

□ to group a set of statements so that they are executed as a unit

□ to group a set of statements for a conditional (IF-THEN/ELSE) clause.

DO groups have the following general form:

DO;
   additional statements
END;

You can nest DO groups to any level, just like you nest IF-THEN/ELSE statements. Below is an example of nested DO groups:

```
do;
   statements;
   do;
      statements;
      do;
         statements;
      end;
   end;
end;
```

It is good practice to indent the statements in DO groups as shown above so that their position indicates the levels of nesting.

For IF-THEN/ELSE conditionals, DO groups can be used as units for either THEN or ELSE clauses so that you can perform many statements as part of the conditional action. An example follows:

```
if x<y then
   do;
      z1=abs(x+y);
      z2=abs(x-y);
   end;
else
   do;
      z1=abs(x-y);
      z2=abs(x+y);
   end;
```

Iterative Execution

The DO statement also serves the feature of iteration. With a DO statement, you can repeatedly execute a set of statements until some condition stops the execution. A DO statement is iterative if you specify it with any of the following iteration clauses. The type of clause determines when to stop the repetition.

Clause	DO Statement
DATA	DO DATA statement
variable=*start* TO *stop* <BY *increment*>	iterative DO statement
WHILE(*expression*)	DO WHILE statement
UNTIL(*expression*)	DO UNTIL statement

A DO statement can have any combination of these four iteration clauses, but a given DO statement must be specified in the order listed above.

DO DATA Statement

The general form of the DO DATA statement is

DO DATA;

The DATA keyword specifies that iteration is to stop when an end-of-file condition occurs. The group is exited immediately upon encountering the end-of-file condition. Other DO specifications exit after tests are performed at the top or bottom of the loop. See Chapter 6, "Working with SAS Data Sets," and Chapter 7, "File Access," for more information about processing data.

You can use the DO DATA statement to read data from an external file or to process observations from a SAS data set. In the DATA step in base SAS software, the iteration is usually implied. The DO DATA statement simulates this iteration until end of file is reached.

Below is an example that reads data from an external file named MYDATA and inputs the data values into a vector. The data values are read one at a time into the dummy variable XX and collected into the vector **X** using the vertical concatenation operator (//) after each value is read.

```
infile 'mydata';          /* infile statement    */
do data;                  /* begin read loop     */
   input xx;              /* read a data value   */
   x=x//xx;               /* concatenate values  */
end;                      /* end loop            */
```

Iterative DO Statement

The general form of the iterative DO statement is

DO *variable=start* TO *stop* <BY *increment*>;

The *variable* sequence specification assigns the *start* value to the given variable. This value is then incremented by the *increment* value (or by 1 if *increment* is not specified) until it is greater than or equal to the *stop* value. (If *increment* is negative, then the iterations stop when the value is less than or equal to *stop*.)

For example, the following statement specifies a DO loop that executes by multiples of 10 until I is greater than 100:

```
do i=10 to 100 by 10;
```

DO WHILE Statement

The general form of the DO WHILE statement is

DO WHILE(*expression*);

With a WHILE clause, the expression is evaluated at the beginning of each loop, with repetition continuing until the expression is false (that is, until the value contains a 0 or missing value). Note that if the expression is false the first time it is evaluated, the loop is not executed.

For example, if the variable COUNT has an initial value of 1, the statements

```
do while(count<5);
   print count;
   count=count+1;
end;
```

print COUNT four times.

DO UNTIL Statement

The general form of the DO UNTIL statement is

DO UNTIL(*expression*);

The UNTIL clause is like the WHILE clause except that the expression is evaluated at the bottom of the loop. This means that the loop always executes at least once.

For example, if the variable COUNT has an initial value of 1, the statements

```
do until(count>5);
   print count;
   count=count+1;
end;
```

print COUNT five times.

Jumping

During normal execution, statements are executed one after another. The GOTO and LINK statements instruct IML to jump from one part of a program to another. The place to which execution jumps is identified by a *label*, which is a name followed by a colon placed before an executable statement. You can program a jump by using either the GOTO statement or the LINK statement:

GOTO *label*;
LINK *label*;

Both the GOTO and the LINK statements instruct IML to jump immediately to the labeled statement. The LINK statement, however, reminds IML where it jumped from so that execution can be returned there if a RETURN statement is encountered. The GOTO statement does not have this feature. Thus, the LINK statement provides a way of calling sections of code as if they were subroutines. The LINK statement calls the routine. The routine begins with the label and ends with a RETURN statement. LINK statements can be nested within other LINK statements to any level.

▶ *Caution* . . . . ***The GOTO and LINK statements are limited to being inside a module or DO group.*** These statements must be able to resolve the referenced label within the current unit of statements. Although matrix symbols can be shared across modules, statement labels can not. Therefore, all GOTO statement labels and LINK statement labels must be local to the DO group or module. ▲

The GOTO and LINK statements are not often used because you can usually write more understandable programs by using other features, such as DO groups for conditionals, iterative DO groups for looping, and module invocations for subroutine calls.

Here are two DO groups to show how the GOTO and LINK statements work:

```
do;                                    do;
    if x<0 then goto negative;             if x<0 then link negative;
    y=sqrt(x);                             y=sqrt(x);
    print y;                               print y;
    stop;                                  stop;
negative:                              negative:
    print "Sorry, X is negative";          print "Using Abs. value of negative X";
end;                                       x=abs(x);
                                           return;
                                       end;
```

Below is a comparable way to write the program on the left without using GOTO or LINK statements:

```
if x<0 then print "Sorry, X is negative";
else
    do;
        y=sqrt(x);
        print y;
    end;
```

Modules

Modules are used for

□ creating groups of statements that can be invoked as a unit from anywhere in the program, that is, making a subroutine or function.

□ creating a separate (symbol-table) environment, that is, defining variables that are local to the module rather than global.

A module always begins with the START statement and ends with the FINISH statement. Modules can be thought of as being either functions or subroutines. When a module returns a single parameter, it is called a function and is executed as if it were a built-in IML function; a function is invoked by its name in an assignment statement rather than with a CALL or RUN statement. Otherwise, a module is called a subroutine, and you execute the module with either the RUN statement or the CALL statement.

Defining and Executing a Module

Modules begin with a START statement, which has the general form

START <*name*> <(*arguments*)> <GLOBAL(*arguments*)>;

Modules end with a FINISH statement, which has the general form

FINISH <*name*>;

If no name appears in the START statement, the name of the module defaults to MAIN.

There are two ways you can execute a module. You can use either a RUN statement or a CALL statement. The only difference is the order of resolution.

The general forms of these statements are

RUN *name* <(*arguments*)>;
CALL *name* <(*arguments*)>;

The RUN and CALL statements must have arguments to correspond to the ones defined for the modules they invoke. A module can call other modules provided that it never recursively calls itself.

The RUN and CALL statements have orders of resolution that need to be considered only when you have given a module the same name as a built-in IML subroutine. In such cases, use the CALL statement to execute the built-in subroutine and the RUN statement to execute the user-defined module.

The RUN statement is resolved in the following order:

1. A user-defined module

2. An IML built-in function or subroutine.

The CALL statement is resolved in the following order:

1. An IML built-in subroutine

2. A user-defined module

Nesting Modules

You can nest one module within another. You must make sure that each nested module is completely contained inside of the parent module. Each module is collected independently of the others. When you nest modules, it is a good idea to indent the statements relative to the level of nesting, as shown in the example below:

```
start a;
   reset print;
   start b;
      a=a+1;
   finish b;
   run b;
finish a;
run a;
```

In this example, IML starts collecting statements for a module called A. In the middle of this module it recognizes the start of a new module called B. It saves its current work on A and collects B until encountering the first FINISH statement. It then finishes collecting A. Thus, it behaves the same as if B were collected before A, as shown below:

```
start b;
   a=a+1;
finish;
start a;
   reset print;
   run b;
finish;
run a;
```

Understanding Symbol Tables

Whenever a variable is defined outside of the module environment, its name is stored in the *global symbol table*. Whenever you are programming in immediate mode outside of a module, you are working with symbols (variables) from the global symbol table. For each module you define with arguments given in a START statement, a separate symbol table called a *local symbol table* is created for that module. All symbols (variables) used inside the module are stored in its local symbol table. There can be many local symbol tables, one for each module with arguments. A symbol can exist in the global table, one or more local tables, or in both the global table and one or more local tables. Also, depending on how a module is defined, there can be a one-to-one correspondence between variables across symbol tables (although there need not be any connection between a variable, say X, in the global table and a variable X in a local table). Values of symbols in a local table are temporary; that is, they exist only while the module is executing and are lost when the module has finished execution. Whether or not these temporary values are transferred to corresponding global variables depends on how the module is defined.

Modules with No Arguments

When you define a module with no arguments, a local symbol table is not created. All symbols (variables) are global, that is, equally accessible inside and outside the module. The symbols referenced inside the module are the same as the symbols outside the module environment. This means that variables created inside the module are also global, and any operations done on variables inside a module affect the global variables as well.

The example below shows a module with no arguments:

```
>      /* module without arguments, all symbols are global.  */
> proc iml;
> a=10;                          /* A is global      */
> b=20;                          /* B is global      */
> c=30;                          /* C is global      */
> start mod1;                    /* begin module     */
>     p=a+b;                     /* P is global      */
>     q=b-a;                     /* Q is global      */
>     c=40;                      /* C already global */
> finish;                        /* end module       */

  NOTE: Module MOD1 defined.

> run mod1;
> print a b c p q;
```

A	B	C	P	Q
10	20	40	30	10

Note that after executing the module,

□ A is still 10

□ B is still 20

□ C has been changed to 40

□ P and Q are created, added to the global symbol table, and set to 30 and 10, respectively.

Modules with Arguments

In general, the following statements are true about modules with arguments:

□ You can specify arguments as variable names.

□ If you specify several arguments, use commas to separate them.

□ If you have both output variables and input variables, it is good practice to list the output variables first.

□ When a module is invoked with either a RUN or a CALL statement, the arguments can be any name, expression, or literal. However, when using arguments for output results, use variable names rather than expressions or literals.

When a module is executed with either a RUN or a CALL statement, the value for each argument is transferred from the global symbol table to the local symbol table. For example, consider the module MOD2 defined below. The first four statements are submitted in the global environment and define variables (A,B,C, and D) whose values are stored in the global symbol table. The START statement begins definition of MOD2 and lists two variables (X and Y) as arguments. This creates a local symbol table for MOD2. All symbols used inside the module (X,Y,P,Q, and C) are in the local symbol table. There is also a one-to-one correspondence between the arguments in the RUN statement (A and

B) and the arguments in the START statement (X and Y). Also note that A, B, and D exist only in the global symbol table, whereas X, Y, P, and Q exist only in the local symbol table. The symbol C exists independently in both the local and global tables. When MOD2 is executed with the statement RUN MOD2(A,B), the value of A is transferred from the global symbol table to X in the local table. Similarly, the value of B in the global table is transferred to Y in the local table. Because C is not an argument, there is no correspondence between the value of C in the global table and the value of C in the local table. When the module finishes execution, the final values of X and Y in the local table are transferred back to A and B in the global table.

```
> proc iml;
> a=10;
> b=20;
> c=30;
> d=90;
> start mod2(x,y);                    /* begin module */
>    p=x+y;
>    q=y-x;
>    y=100;
>    c=25;
> finish mod2;                        /* end module   */

  NOTE: Module MOD2 defined.

> run mod2(a,b);
> print a b c d;
```

A	B	C	D
10	100	30	90

The PRINT statement prints the values of variables in the global symbol table. Notice that

□ A is still 10.

□ B is changed to 100 since the corresponding argument Y was changed to 100 inside the module.

□ C is still 30. Inside the module, the local symbol C was set equal to 25, but there is no correspondence between the global symbol C and the local symbol C.

□ D is still 90.

Also note that inside the module, the symbols A, B, and D do not exist. Outside the module, the symbols P, Q, X, and Y do not exist.

Defining Function Modules

Functions are special modules that return a single value. They are a special type of module because modules can, in general, return any number of values through their argument list. To write a function module, include a RETURN statement that assigns the returned value to a variable. The RETURN statement is necessary for a module to be a function. You invoke a function module in an assignment statement, as you would a standard function.

The symbol-table logic described in the preceding section also applies to function modules. Below is an example of a function module. In this module, the value of C in the local symbol table is transferred to the global symbol Z.

```
> proc iml;
> a=10;
> b=20;
> c=30;
> d=90;
> start mod3(x,y);
>    p=x+y;
>    q=y-x;
>    y=100;
>    c=40;
>    return (c);                    /* return function value */
> finish mod3;

  NOTE: Module MOD3 defined.

> z = mod3(a,b);                    /* call function         */
> print a b c d z;
```

A	B	C	D	Z
10	100	30	90	40

Note the following about this example:

□ A is still 10.

□ B is changed to 100 because Y is set to 100 inside the module, and there is a one-to-one correspondence between B and Y.

□ C is still 30. The symbol C in the global table has no connection with the symbol C in the local table.

□ Z is set to 40, which is the value of C in the local table.

Again note that inside the module, the symbols A, B, D, and Z do not exist. Outside the module, symbols P, Q, X, and Y do not exist.

In the next example, you define your own function ADD for adding two arguments:

```
> proc iml;
> reset print;
> start add(x,y);
>    sum=x+y;
>    return(sum);
> finish;
```

NOTE: Module ADD defined.

```
>   a={9 2,5 7};
```

```
                                  A
                                  9          2
                                  5          7
```

```
>   b={1 6,8 10};
```

```
                                  B
                                  1          6
                                  8         10
```

```
>   c=add(a,b);
```

```
                                  C
                                 10          8
                                 13         17
```

Function modules can also be called inside each other. For example, in the statements below, the ADD function is called twice from within the first ADD function:

```
> d=add(add(6,3),add(5,5));
> print d;
```

```
                                      D
                                     19
```

Functions are resolved in this order:

1. An IML built-in function

2. A user-defined function module

3. A SAS DATA step function.

This means that you should not use a name for a function that is already the name of an IML built-in function.

Using the GLOBAL Clause

For modules with arguments, the variables used inside the module are local and have no connection with any variables of the same name existing outside the module in the global table. However, it is possible to specify that certain variables not be placed in the local symbol table, but rather be accessed from the global table. Use the GLOBAL clause to specify variables you want shared between local and global symbol tables. Below is an example of a module using a GLOBAL clause to define the symbol C as global. This defines a one-to-one correspondence between the value of C in the global table and the value of C in the local table.

```
>  proc iml;
>  a=10;
>  b=20;
>  c=30;
>  d=90;
>  start mod4(x,y) global (c);
>      p=x+y;
>      q=y-x;
>      y=100;
>      c=40;
>      d=500;
>  finish mod4;

   NOTE: Module MOD4 defined.

>  run mod4(a,b);
>  print a b c d;
```

A	B	C	D
10	100	40	90

Note the following about this example:

□ A is still 10.

□ B is changed to 100.

□ C is changed to 40 because it was declared global. The C inside the module and outside the module were the "same."

□ D is still 90 and not 500, since D independently exists in the global and local symbol tables.

Also note that every module with arguments has its own local table, and thus it is possible to have a global and many local tables. A variable can independently exist in one or more of these tables. However, a variable can be commonly shared between the global and any number of local tables when the GLOBAL clause is used.

Nesting Modules with Arguments

For nested module calls, the concept of global and local symbol tables is somewhat different. Consider the following example:

```
> start mod1 (a,b);
>    c=a+b;
>    d=a-b;
>    run mod2 (c,d);
>    print c d;
> finish mod1;

   NOTE: Module MOD1 defined.

> start mod2 (x,y);
>    x=y-x;
>    y=x+1;
>    run mod3(x,y);
> finish mod2;

   NOTE: Module MOD2 defined.

> start mod3(w,v);
>    w=w#v;
> finish mod3;

   NOTE: Module MOD3 defined.
```

The local symbol table of MOD1 in effect becomes the global table for MOD2. The local symbol table of MOD2 is the global table for MOD3. The distinction between the global and local environments is necessary only for modules with arguments. If a module, say, A, calls another module, say, B, which has no arguments, B shares all the symbols existing in A's local symbol table.

For example, consider the following statements:

```
> x=457;
> start a;
>    print 'from a' x;
>    finish;
> start b(p);
>    print 'from b' p;
>    run a;
>    finish;
> run b(x);
```

```
                                                 P
                                   from b       457

   ERROR: Matrix X has not been set to a value.
   Error occured in module A
   called   from    module B
```

```
stmt: PRINT

Paused in module A.
```

In this example module A is called from module B. Therefore, the local
symbol table of module B becomes the global symbol table for module A.
Module A has access to all symbols available in module B. No X exists in the
local environment of module B; thus no X is available in module A as well. This
causes the error that X is unvalued.

More about Argument Passing

You can pass expressions and subscripted matrices as arguments to a module,
but you must be careful and understand the way IML evaluates the expressions
and passes results to the module. Expressions are evaluated, and the evaluated
values are stored in temporary variables. Similarly, submatrices are created
from subscripted variables and stored in temporary variables. The temporaries
are passed to the module while the original matrix remains intact. Notice that in
the example that follows, the matrix X remains intact. You might expect X to
contain the squared values of Y.

```
>  proc iml;
>  reset printall;
>  start square(a,b);
>     a=b##2;
>  finish;
>     /*  create two data matrices                     */
>  x={5 9 };

             X                1 row      2 cols     (numeric)

                                5         9

>  y={10 4};

             Y                1 row      2 cols     (numeric)

                               10         4
>     /*  pass matrices to module element-by-element    */
>  do i=1 to 2;
>     run square(x[i],y[i]);
>  end;
>     /*  RESET PRINTALL prints all intermediate results  */
```

```
I              1 row     1 col    (numeric)

                          1


#TEM1002       1 row     1 col    (numeric)

                         10


#TEM1001       1 row     1 col    (numeric)

                          5


A              1 row     1 col    (numeric)

                        100

#TEM1002       1 row     1 col    (numeric)

                          4


#TEM1001       1 row     1 col    (numeric)

                          9


A              1 row     1 col    (numeric)

                         16
```

```
>     /* show X and Y are unchanged                    */
> print x y;
```

```
          X              Y
          5       9     10      4
```

X remained unchanged because what got changed were the temporary variables
that you generally would not see. Note that IML will properly warn you of any
such instances where your results may be lost to the temporaries.

Module Storage

You can store and reload modules using the forms of the STORE and LOAD statements as they pertain to modules:

STORE MODULE=*name*;
LOAD MODULE=*name*;

You can see the names of the modules in storage with the SHOW statement:

```
show storage;
```

See Chapter 12, "Storage Features," for details on using the library storage facilities.

Stopping Execution

You can stop execution with a PAUSE, STOP, or ABORT statement. The QUIT statement is also a stopping statement, but it immediately removes you from the IML environment; the other stopping statements can be performed in the context of a program. Below are descriptions of the STOP, ABORT, and PAUSE statements.

PAUSE Statement

The general form of the PAUSE statement is

PAUSE <*message*> <***>;

The PAUSE statement

□ stops execution of the module.

□ remembers where it stopped executing.

□ prints a pause *message* that you can specify.

□ puts you in immediate mode within the module environment using the module's local symbol table. At this point you can enter more statements.

A RESUME statement allows you to continue execution at the place where the most recent PAUSE statement was executed.

You can use a STOP statement as an alternative to RESUME to remove the paused states and return to the immediate environment outside of the module.

You can specify a message in the PAUSE statement to print a message as the pause prompt. If no message is specified, IML prints the following default message:

```
paused in module XXX
```

where *XXX* is the name of the module containing the pause. To suppress the printing of any messages, use the * option:

```
pause *;
```

Below are some examples of PAUSE statements with operands:

```
pause "Please enter an assignment for X, then enter RESUME;";
```

```
msg ="Please enter an assignment for X, then enter RESUME;";
pause msg;
```

When you use the PAUSE, RESUME, STOP, or ABORT statements, you should be aware of the following details:

□ The PAUSE statement can only be issued from within a module.

□ IML diagnoses an error if you execute a RESUME statement without any pauses outstanding.

□ You can define and execute modules while paused from other modules.

□ A PAUSE statement is automatically issued if an error occurs while executing statements inside a module. This gives you an opportunity to correct the error and resume execution of the module with a RESUME statement. Alternately, you can submit a STOP or ABORT statement to exit from the module environment.

□ You cannot reenter or redefine an active (paused) module; you will get an error for recursive module execution.

□ In paused mode, you can run another module that in turn pauses; the paused environments are stacked.

□ You can put a RESUME statement inside a module. For example, suppose you are paused in module A and then run module B, which executes a RESUME statement. Execution is resumed in module A and does not return to module B.

□ IML supports stopping execution while in a paused state in both subroutine and function modules.

□ If you pause in a subroutine module that has its own symbol table, then the immediate mode during the pause uses this symbol table rather than the global one. You must use a RESUME or a STOP statement to return to the global symbol table environment.

□ You can use the PAUSE and RESUME statements, in conjunction with the PUSH, QUEUE, and EXECUTE subroutines described in Chapter 13, "Using SAS/IML Software to Generate IML Statements," to execute IML statements that you generate within a module.

STOP Statement

The general form of the STOP statement is

STOP;

The STOP statement stops execution and returns you to immediate mode, where new statements that you enter are executed. If execution was interrupted by a PAUSE statement, the STOP statement clears all pauses and returns to immediate mode of execution.

ABORT Statement

The general form of the ABORT statement is

ABORT;

The ABORT statement stops execution and exits from IML much like a QUIT statement, except that the ABORT statement is executable and programmable. For example, you may want to exit IML if a certain error occurs. You can check for the error in a module and program an ABORT statement to execute if the error occurs. The ABORT statement will not execute until the module is executed, while the QUIT statement executes immediately and ends the IML session.

In Conclusion

In this chapter you learned the basics of programming with SAS/IML software. You learned about conditional execution (IF-THEN/ELSE statements), grouping statements as a unit (DO groups), iterative execution, nonconsecutive execution, defining subroutines and functions (modules), and stopping execution. With these programming capabilities, you are able to write your own sophisticated programs and store the code as a module. You can then execute the program later with a RUN or CALL statement.

Chapter **6** Working with SAS Data Sets

Introduction

SAS/IML software has many statements for passing data from SAS data sets to matrices and from matrices to SAS data sets. You can create matrices from the variables and observations of a SAS data set in several ways. You can create a column vector for each data set variable, or you can create a matrix where columns correspond to data set variables. You can use all the observations in a data set or use a subset of them.

You can also create a SAS data set from a matrix. The columns correspond to data set variables and the rows correspond to observations. Data management commands enable you to edit, append, rename, or delete SAS data sets from within the SAS/IML environment.

When reading a SAS data set, you can read any number of observations into a matrix either sequentially, directly by record number, or conditionally according to conditions in a WHERE clause. You can also index a SAS data set. The indexing capability facilitates retrievals by the indexed variable.

Operations on SAS data sets are performed with straightforward, consistent, and powerful statements. For example, the LIST statement can perform the following tasks:

□ list the next record

□ list a specified record

□ list any number of specified records

□ list the whole file

□ list records satisfying one or more conditions

□ list specified variables or all variables.

If you want to read values into a matrix, use the READ statement instead of the LIST statement with the same operands and features as LIST. You can specify operands that control which records and variables are used indirectly, as matrices, so that you can dynamically program the records, variables, and conditional values you want.

In this chapter, you will use the SAS data set CLASS, which contains the variables NAME, SEX, AGE, HEIGHT, and WEIGHT, to learn about

□ opening a SAS data set

□ examining the contents of a SAS data set

□ printing data values with the LIST statement

□ reading observations from a SAS data set into matrices

□ editing a SAS data set

□ creating a SAS data set from a matrix

□ printing matrices with row and column headings

□ producing summary statistics

□ sorting a SAS data set

□ indexing a SAS data set

□ similarities and differences with the SAS DATA step.

Throughout this chapter, the right angle brackets (>) indicate statements you submit; responses from IML follow below.

First, invoke the IML procedure:

```
> proc iml;

IML Ready
```

Opening a SAS Data Set

Before you can access a SAS data set, you must first submit a command to open it. There are three ways to open a SAS data set:

□ To simply read from an existing data set, submit a USE statement to open it for read access. The general form of the USE statement is

USE *SAS-data-set* <VAR *operand*> <WHERE(*expression*)>;

With read access you can use the FIND, INDEX, LIST, and READ statements on the data set.

□ To read and write to an existing data set, use the EDIT statement. The general form of the EDIT statement is

EDIT *SAS-data-set* <VAR *operand*> <WHERE(*expression*)>;

This statement enables you to use both the reading statements (LIST, READ, INDEX, and FIND) and the writing statements (REPLACE, APPEND, DELETE, and PURGE).

□ To create a new data set, use the CREATE statement to open a new data set for both output and input. The general form of the CREATE statement is

CREATE *SAS-data-set* <VAR *operand*>;
CREATE *SAS-data-set* FROM *from-name*
    <[COLNAME=*column-name* ROWNAME=*row-name*]>;

Use the APPEND statement to place the matrix data into the newly created data set. If you don't use the APPEND statement, the new data set has no observations.

If you want to list observations and create matrices from the data in the SAS data set CLASS, you must first submit a statement to open CLASS. Because CLASS already exists, the USE statement is the one you want.

Making a SAS Data Set Current

IML data processing commands work on the current data set. This feature makes it unnecessary for you to specify the data set as an operand each time. There are two current data sets, one for input and one for output. IML makes a data set the current one as it is opened. You can also make a data set current by using two setting statements, SETIN and SETOUT:

□ The USE and SETIN statements make a data set current for input.

□ The SETOUT statement makes a data set current for output.

□ The CREATE and EDIT statements make a data set current for both input and output.

If you issue a USE, EDIT, or CREATE statement for a data set that is already open, the data set is made the current data set. To find out which data sets are open and which are current input and current output data sets, use the SHOW DATASETS statement.

The current observation is set by the last operation that performed input/output (I/O). If you want to set the current observation without doing any I/O, use the SETIN (or SETOUT) statement with the POINT option. After a data set is opened, the current observation is set to 0. If you attempt to list or read the current observation, the current observation is converted to 1. You can make the data set CLASS current for input and position the pointer at the tenth observation with the statement

```
> setin class point 10;
```

Displaying SAS Data Set Information

You can use SHOW statements to display information about your SAS data sets. The SHOW DATASETS statement lists all open SAS data sets and their status. The SHOW CONTENTS statement displays the variable names and types, the size, and the number of observations in the current input data set. For example, to get data set information for CLASS, issue the following statements:

```
> use class;
> show datasets;
```

```
        LIBNAME MEMNAME   OPEN MODE   STATUS
        ------- -------   ---------   ------
        WORK    .CLASS    Input       Current Input
```

```
> show contents;
```

```
        VAR NAME    TYPE   SIZE
        NAME        CHAR   8
        SEX         CHAR   8
        AGE         NUM    8
        HEIGHT      NUM    8
        WEIGHT      NUM    8
        Number of Variables:    5
        Number of Observations: 19
```

As you can see, CLASS is the only data set open. The USE statement opened it for input, and it is the current input data set. The full name for CLASS is WORK.CLASS. The libref is the default, WORK. The next section tells you how to change the libref to another name.

Referring to a SAS Data Set

The USE, EDIT, and CREATE statements take as their first operand the data set name. This name can have either one or two levels. If it is a two-level name, the first level refers to the name of the SAS data library; the second name is the data set name. If the libref is WORK, the data set is put into a directory for temporary data sets; these are automatically deleted at the end of the session. Other librefs are associated with SAS data libraries using the LIBNAME statement.

If you specify only a single name, then IML supplies a default libref. At the beginning of an IML session, the default libref is SASUSER, if SASUSER is defined as a libref, or WORK otherwise. You can reset the default libref by using the RESET DEFLIB statement. If you want to create a permanent SAS data set, you must specify a two-level name using the RESET DEFLIB statement (see Chapter 6, "SAS Files," in *SAS Language: Reference, Version 6, First Edition* for more information about permanent SAS data sets).

```
> reset deflib=name;
```

Listing Observations

You can list variables and observations in a SAS data set with the LIST statement. The general form of the LIST statement is

LIST <*range*> <VAR *operand*> <WHERE(*expression*)>;

where

range	specifies a range of observations.
operand	selects a set of variables.
expression	is an expression that is evaluated for being true or false.

The next three sections discuss how to use each of these clauses with the CLASS data set.

Specifying a Range of Observations

You can specify a range of observations with a keyword or by record number using the POINT option. You can use the *range* operand with the data management statements DELETE, FIND, LIST, READ, and REPLACE.

You can specify *range* using any of the following keywords:

ALL	all observations
CURRENT	the current observation
NEXT <*number*>	the next observation or next *number* of observations

AFTER	all observations after the current one
POINT *operand*	observations by number, where *operand* can be one of the following:

Operand	Example
a single record number	`point 5`
a literal giving several record numbers	`point {2 5 10}`
the name of a matrix containing record numbers	`point p`
an expression in parentheses	`point (p+1)`

If you want to list all observations in the CLASS data set, use the keyword ALL to indicate that the range is all observations:

```
> list all;
```

OBS	NAME	SEX	AGE	HEIGHT	WEIGHT
1	JOYCE	F	11.0000	51.3000	50.5000
2	THOMAS	M	11.0000	57.5000	85.0000
3	JAMES	M	12.0000	57.3000	83.0000
4	JANE	F	12.0000	59.8000	84.5000
5	JOHN	M	12.0000	59.0000	99.5000
6	LOUISE	F	12.0000	56.3000	77.0000
7	ROBERT	M	12.0000	64.8000	128.0000
8	ALICE	F	13.0000	56.5000	84.0000
9	BARBARA	F	13.0000	65.3000	98.0000
10	JEFFREY	M	13.0000	62.5000	84.0000
11	CAROL	F	14.0000	62.8000	102.5000
12	HENRY	M	14.0000	63.5000	102.5000
13	ALFRED	M	14.0000	69.0000	112.5000
14	JUDY	F	14.0000	64.3000	90.0000
15	JANET	F	15.0000	62.5000	112.5000
16	MARY	F	15.0000	66.5000	112.0000
17	RONALD	M	15.0000	67.0000	133.0000
18	WILLIAM	M	15.0000	66.5000	112.0000
19	PHILIP	M	16.0000	72.0000	150.0000

Without a *range* specification, the LIST statement lists only the current observation, which in this example is now the last observation because of the previous LIST statement:

```
> list;
```

OBS	NAME	SEX	AGE	HEIGHT	WEIGHT
19	PHILIP	M	16.0000	72.0000	150.0000

Use the POINT keyword with record numbers to list specific observations. You can follow the keyword POINT with a single record number or literal giving several record numbers.

```
> list point 5;
```

OBS	NAME	SEX	AGE	HEIGHT	WEIGHT
5	JOHN	M	12.0000	59.0000	99.5000

```
> list point {2 4 9};
```

OBS	NAME	SEX	AGE	HEIGHT	WEIGHT
2	THOMAS	M	11.0000	57.5000	85.0000
4	JANE	F	12.0000	59.8000	84.5000
9	BARBARA	F	13.0000	65.3000	98.0000

You can also indicate the range indirectly by creating a matrix containing the records you want listed:

```
> p={2 4 9};
> list point p;
```

OBS	NAME	SEX	AGE	HEIGHT	WEIGHT
2	THOMAS	M	11.0000	57.5000	85.0000
4	JANE	F	12.0000	59.8000	84.5000
9	BARBARA	F	13.0000	65.3000	98.0000

The *range* operand is usually listed first when you are using the access statements DELETE, FIND, LIST, READ, and REPLACE. Listed below are access statements and their default ranges.

Statement	Default Range
LIST	current
READ	current
FIND	all
REPLACE	current
APPEND	always at end
DELETE	current

Selecting a Set of Variables

You can use the VAR clause to select a set of variables. The general form of the VAR clause is

VAR *operand*

where *operand* can be specified using one of the following:

□ a literal containing variable names

□ the name of a matrix containing variable names

□ an expression in parentheses yielding variable names

□ one of the keywords listed below:

ALL   for all variables

CHAR   for all character variables

NUM   for all numeric variables.

Below are examples showing each possible way you can use the VAR clause:

```
var {time1 time5 time9};   /* a literal giving the variables */
var time;                  /* a matrix containing the names */
var('time1':'time9');              /* an expression    */
var _all_;                              /* a keyword */
```

For example, to list students' names from the CLASS data set, use the VAR clause with a literal:

```
> list point p var{name};
```

```
         OBS NAME
         ------ --------
            2 THOMAS
            4 JANE
            9 BARBARA
```

To list AGE, HEIGHT, and WEIGHT, you can use the VAR clause with a matrix giving the variables:

```
> v={age height weight};
> list point p var v;
```

```
         OBS     AGE     HEIGHT    WEIGHT
         ------ --------- --------- ---------
            2  11.0000   57.5000   85.0000
            4  12.0000   59.8000   84.5000
            9  13.0000   65.3000   98.0000
```

The VAR clause can be used with the following statements for the tasks described in the table below:

Statement	VAR Clause Function
APPEND	specifies which IML variables contain data to append to the data set
CREATE	specifies the variables to go in the data set
EDIT	limits which variables are accessed
LIST	specifies which variables to list
READ	specifies which variables to read
REPLACE	specifies which data set variable's data values to replace with corresponding IML variable data values
USE	limits which variables are accessed

Selecting Observations

The WHERE clause conditionally selects observations, within the *range* specification, according to conditions given in the *expression*. The general form of the WHERE clause is

WHERE(*variable comparison-op operand*)

where

variable	is a variable in the SAS data set.
comparison-op	is one of the following comparison operators:

$<$	less than
$<=$	less than or equal to
$=$	equal to
$>$	greater than
$>=$	greater than or equal to
$^=$	not equal to
?	contains a given string
^?	does not contain a given string
=:	begins with a given string
=*	sounds like or is spelled similar to a given string.

operand	is a literal value, a matrix name, or an expression in parentheses.

WHERE comparison arguments can be matrices. For the following operators, the WHERE clause succeeds if *all* the elements in the matrix satisfy the condition:

^=   ^ ?   <   <=   >   >=

For the following operators, the WHERE clause succeeds if *any* of the elements in the matrix satisfy the condition:

=   ?   =:   =*

Logical expressions can be specified within the WHERE clause using the AND (&) and OR (|) operators. The general form is

clause&*clause* (for an AND clause)
clause | *clause* (for an OR clause)

where *clause* can be a comparison, a parenthesized clause, or a logical expression clause that is evaluated using operator precedence.

For example, to list the names of all males in CLASS, use the following statement:

```
> list all var{name} where(sex='M');
```

```
              OBS                    NAME

              ------    ---------------------
               2                     THOMAS
               3                     JAMES
               5                     JOHN
               7                     ROBERT
              10                     JEFFREY
              12                     HENRY
              13                     ALFRED
              17                     RONALD
              18                     WILLIAM
              19                     PHILIP
```

The WHERE comparison arguments can be matrices. In the cases using the =* operator shown below, the comparison is made on each name to find a string that sounds like or is spelled similar to the given string or strings:

```
> n={name sex age};
> list all var n where(name=*{"ALFRED","CAROL","JUDY"});
```

```
        OBS        NAME      SEX          AGE

        -----   ----------------   --------   ----------
         11       CAROL      F        14.0000
         13       ALFRED     M        14.0000
         14       JUDY       F        14.0000
```

```
> list all var n where(name=*{"JON","JAN"});
```

```
        OBS NAME      SEX          AGE

        ------ --------   --------   ----------
         4 JANE       F        12.0000
         5 JOHN       M        12.0000
```

To list AGE, HEIGHT, and WEIGHT for all students in their teens, use the following statement:

```
> list all var v where(age>12);
```

```
           OBS       AGE    HEIGHT    WEIGHT

            8    13.0000   56.5000   84.0000
            9    13.0000   65.3000   98.0000
           10    13.0000   62.5000   84.0000
           11    14.0000   62.8000  102.5000
           12    14.0000   63.5000  102.5000
           13    14.0000   69.0000  112.5000
           14    14.0000   64.3000   90.0000
           15    15.0000   62.5000  112.5000
           16    15.0000   66.5000  112.0000
           17    15.0000   67.0000  133.0000
           18    15.0000   66.5000  112.0000
           19    16.0000   72.0000  150.0000
```

Note: In the WHERE clause, the expression on the left-hand side refers to values of the data set variables, and the expression on the right-hand side refers to matrix values. You cannot use comparisons involving more than one data set variable in a single comparison; for example, you cannot use either of the expressions

```
list all where(height>weight);
list all where(weight-height>0);
```

You could use the first statement if WEIGHT were a matrix name already defined rather than a variable in the SAS data set.

Reading Observations from a SAS Data Set

Transferring data from a SAS data set to a matrix is done using the READ statement. The SAS data set you want to read data from must already be open. You can open a SAS data set with either the USE or the EDIT statement. If you already have several data sets open, you can point to the one you want with the SETIN statement, making it the current input data set. The general form of the READ statement is

READ <*range*> <VAR *operand*> <WHERE(*expression*)> <INTO *name*>;

where

range	specifies a range of observations.
operand	selects a set of variables.
expression	is an expression that is evaluated for being a true or false.
name	names a target matrix for the data.

Using the READ Statement with the VAR Clause

Use the READ statement with the VAR clause to read variables from the current SAS data set into column vectors of the VAR clause. Each variable in the VAR clause becomes a column vector with the same name as the variable in the SAS data set. The number of rows is equal to the number of observations processed, depending on the range specification and the WHERE clause. For example, to read the numeric variables AGE, HEIGHT, and WEIGHT for all observations in CLASS, use the statements

```
> read all var (age height weight);
```

Now submit the SHOW NAMES statement to display all the matrices you have created so far in this chapter:

```
> show names;
```

```
            AGE            19 rows    1 col  num    8
            HEIGHT         19 rows    1 col  num    8
            N               1 row     3 cols char   4
            P               1 row     3 cols num    8
            V               1 row     3 cols char   6
            WEIGHT         19 rows    1 col  num    8
            Number of symbols = 8   (includes those without values)
```

You see that with the READ statement, you have created the three numeric vectors **AGE**, **HEIGHT**, and **WEIGHT**. (Notice that the matrices you created earlier, **N**, **P**, and **V**, are also listed.) You can select the variables you want to access with a VAR clause in the USE statement. The two previous statements can also be written

```
use class var(age height weight);
read all;
```

Using the READ Statement with the VAR and INTO Clauses

Sometimes you want to have all of the numeric variables in the same matrix so you can determine correlations. Use the READ statement with the INTO clause and the VAR clause to read the variables listed in the VAR clause into the single matrix named in the INTO clause. Each variable in the VAR clause becomes a column of the target matrix. If there are p variables in the VAR clause and n observations processed, the target matrix in the INTO clause is an $n \times p$ matrix.

The following statement creates a matrix **X** containing the numeric variables of the CLASS data set. Notice the use of the keyword _NUM_ in the VAR clause to specify all numeric variables be read.

```
> read all var _num_ into x;
> print x;
```

```
                  X
                 11    51.3    50.5
                 11    57.5      85
                 12    57.3      83
                 12    59.8    84.5
                 12      59    99.5
                 12    56.3      77
                 12    64.8     128
                 13    56.5      84
                 13    65.3      98
                 13    62.5      84
                 14    62.8   102.5
                 14    63.5   102.5
                 14      69   112.5
                 14    64.3      90
                 15    62.5   112.5
                 15    66.5     112
                 15      67     133
                 15    66.5     112
                 16      72     150
```

Using the READ Statement with the WHERE Clause

Use the WHERE clause as you did with the LIST statement, to conditionally select observations from within the specified range. If you want to create a matrix **FEMALE** containing the variables AGE, HEIGHT, and WEIGHT for females only, use the following statements:

```
> read all var _num_ into female where(sex="F");
> print female;
```

```
              FEMALE
                 11    51.3    50.5
                 12    59.8    84.5
                 12    56.3      77
                 13    56.5      84
                 13    65.3      98
                 14    62.8   102.5
                 14    64.3      90
                 15    62.5   112.5
                 15    66.5     112
```

Now try some special features of the WHERE clause to find values that begin with certain characters (the =: operator) or contain certain strings (the ? operator). To create a matrix **J** containing the students whose names begin with J, use the following statements:

```
> read all var{name} into j where(name=:"J");
> print j;
```

```
                               J
                               JOYCE
                               JAMES
                               JANE
                               JOHN
                               JEFFREY
                               JUDY
                               JANET
```

Creating a matrix **AL** of children with names containing the string AL, use the statement

```
> read all var{name} into al where(name?"AL");
> print al;
```

```
                               AL
                               ALICE
                               ALFRED
                               RONALD
```

Editing a SAS Data Set

You can edit a SAS data set using the EDIT statement. You can update values of variables, mark observations for deletion, delete the marked observations, and save the changes you make. The general form of the EDIT statement is

EDIT *SAS-data-set* <VAR *operand*> <WHERE(*expression*)>;

where

SAS-data-set	names an existing SAS data set.
operand	selects a set of variables.
expression	is an expression that is evaluated for being true or false.

Updating Observations

Suppose you have updated data and want to change some values in the CLASS data set. For instance, suppose that the student named HENRY has had a birthday since the data were added to CLASS. You can

1. make the data set CLASS current for input and output

2. read the data

3. change the appropriate data value

4. replace the changed data in the data set.

First, submit an EDIT statement to make CLASS current for input and output. Then use the FIND statement, which finds observation numbers and stores them in a matrix, to find the observation number of the data for HENRY and store it in the matrix **d**.

```
> edit class;
> find all where(name={'HENRY'}) into d;
> print d;
```

```
                 D
                 12
```

List the observation containing the data for HENRY.

```
> list point d;
```

```
    OBS NAME     SEX          AGE    HEIGHT    WEIGHT
    ------ -------- -------- --------- --------- ---------
     12 HENRY    M        14.0000   63.5000   102.5000
```

As you see, the observation number is 12. Now read the value for AGE into a matrix and update its value. Finally, replace the value in the CLASS data set and list the observation containing the data for HENRY again.

```
> age=15;
> replace;
```

```
            1 observations replaced.
```

```
> list point 12;
```

```
    OBS NAME     SEX          AGE    HEIGHT    WEIGHT
    ------ -------- -------- --------- --------- ---------
     12 HENRY    M        15.0000   63.5000   102.5000
```

Deleting Observations

Use the DELETE statement to mark an observation to be deleted. The general form of the DELETE statement is

DELETE <*range*> <WHERE(*expression*)>;

where

range   specifies a range of observations.

expression   is an expression that is evaluated for being true or false.

Below are examples of valid uses of the DELETE statement.

Code	Action
`delete;`	deletes the current observation
`delete point 12;`	deletes observation 12
`delete all where (age>12);`	deletes all observations where AGE is greater than 12

If a file accumulates a number of observations marked as deleted, you can clean out these observations and renumber the remaining observations by using the PURGE statement.

Suppose that the student named John has moved and you want to update the CLASS data set. You can remove the observation using the EDIT and DELETE statements. First, find the observation number of the data for JOHN and store it in **d** using the FIND statement. Then submit a DELETE statement to mark the record for deletion. A deleted observation is still physically in the file and still has an observation number, but it is excluded from processing. The deleted observations appear as gaps when you list the file by observation number:

```
> find all where(name={'JOHN'}) into d;
> print d;


                              D
                              5


> delete point d;


                1 observation deleted.
```

```
> list all;
```

OBS	NAME	SEX	AGE	HEIGHT	WEIGHT
1	JOYCE	F	11.0000	51.3000	50.5000
2	THOMAS	M	11.0000	57.5000	85.0000
3	JAMES	M	12.0000	57.3000	83.0000
4	JANE	F	12.0000	59.8000	84.5000
6	LOUISE	F	12.0000	56.3000	77.0000
7	ROBERT	M	12.0000	64.8000	128.0000
8	ALICE	F	13.0000	56.5000	84.0000
9	BARBARA	F	13.0000	65.3000	98.0000
10	JEFFREY	M	13.0000	62.5000	84.0000
11	CAROL	F	14.0000	62.8000	102.5000
12	HENRY	M	15.0000	63.5000	102.5000
13	ALFRED	M	14.0000	69.0000	112.5000
14	JUDY	F	14.0000	64.3000	90.0000
15	JANET	F	15.0000	62.5000	112.5000
16	MARY	F	15.0000	66.5000	112.0000
17	RONALD	M	15.0000	67.0000	133.0000
18	WILLIAM	M	15.0000	66.5000	112.0000
19	PHILIP	M	16.0000	72.0000	150.0000

Notice that there is a gap in the data where the deleted observation was (observation 5). To renumber the observations and close the gaps, submit the PURGE statement. Note that the PURGE statement deletes any indexes associated with a data set.

```
> purge;
```

Creating a SAS Data Set from a Matrix

SAS/IML software provides the ability to create a new SAS data set from a matrix. Use the CREATE and APPEND statements to create a SAS data set from a matrix, where the columns of the matrix become the data set variables and the rows of the matrix become the observations. Thus, an $n \times m$ matrix creates a SAS data set with m variables and n observations. The CREATE statement opens the new SAS data set for both input and output, and the APPEND statement writes to (outputs to) the data set.

Using the CREATE Statement with the FROM Option

You can create a SAS data set from a matrix using the CREATE statement with the FROM option. This form of the CREATE statement is·

CREATE *SAS-data-set* FROM *matrix*
        <[COLNAME=*column-name* ROWNAME=*row-name*]>;

where

SAS-data-set	names the new data set.
matrix	names the matrix containing the data.
column-name	names the variables in the data set.
row-name	adds a variable containing row titles to the data set.

Suppose you want to create a SAS data set named RATIO containing a variable with the height-to-weight ratios for each student. You first create a matrix containing the ratios from the matrices **HEIGHT** and **WEIGHT** that you have already defined. Next, use the CREATE and APPEND statements to open a new SAS data set called RATIO and append the observations, naming the data set variable HTWT instead of COL1.

```
htwt=height/weight;
create ratio from htwt[colname='htwt'];
append from htwt;
```

Now submit SHOW DATASETS and SHOW CONTENTS statements.

```
> show datasets;

  LIBNAME MEMNAME  OPEN MODE  STATUS
  ------- -------  ---------  ------
  WORK    .CLASS   Update
  WORK    .RATIO   Update     Current Input   Current Output

> show contents;

  VAR NAME   TYPE   SIZE
  HTWT       NUM      8
  Number of Variables:    1
  Number of Observations: 18

> close ratio;
```

As you can see, the new SAS data set RATIO has been created. It has 18 observations and 1 variable (recall that you deleted 1 observation earlier).

Using the CREATE Statement with the VAR Clause

You can use a VAR clause with the CREATE statement to select the variables you want to include in the new data set. In the previous example, the new data set RATIO had one variable. If you want to create a similar data set but include the second variable NAME, you use the VAR clause. You could not do this using the FROM option because the variable HTWT is numeric and NAME is character. The statements below create a new data set RATIO2 having the variables NAME and HTWT.

```
> create ratio2 var{name htwt};
> append;
> show contents;

           VAR NAME    TYPE   SIZE
           NAME        CHAR    8
           HTWT        NUM     8
           Number of Variables:    2
           Number of Observations: 18

> close ratio2;
```

Notice that now the variable NAME is in the data set.

Understanding the End-of-File Condition

If you try to read past the end of a data set or point to an observation greater than the number of observations in the data set, you create an end-of-file condition. If an end of file occurs while inside a DO DATA iteration group, IML transfers control to the next statement outside the current DO DATA group.

The example below uses a DO DATA loop while reading the CLASS data set. It reads the variable WEIGHT in one observation at a time and accumulates the weights of the students in the IML matrix SUM. When the data are read, the total class weight is stored in **SUM**.

```
setin class point 0;
sum=0;
do data;
   read next var{weight};
   sum=sum+weight;
end;
print sum;
```

Producing Summary Statistics

Summary statistics on the numeric variables of a SAS data set can be obtained with the SUMMARY statement. These statistics can be based on subgroups of the data by using the CLASS clause in the SUMMARY statement. The SAVE option in the OPT clause enables you to save the computed statistics in matrices for later perusal. For example, consider the statement below:

```
> summary var {height weight} class {sex} stat{mean std} opt{save};
```

```
      SEX    Nobs  Variable      MEAN       STD

      ----------------------------------------------
      F         9  HEIGHT      60.58889    5.01833
                   WEIGHT      90.11111   19.38391

      M         9  HEIGHT      64.45556    4.90742
                   WEIGHT     110.00000   23.84717

      All      18  HEIGHT      62.52222    5.20978
                   WEIGHT     100.05556   23.43382

      ----------------------------------------------
```

This summary statement gives the mean and standard deviation of the variables HEIGHT and WEIGHT for the two subgroups (male and female) of the data set CLASS. Since the SAVE option is set, the statistics of the variables are stored in matrices under the name of the corresponding variables, with each column corresponding to a statistic requested and each row corresponding to a subgroup. Two other vectors, SEX and _NOBS_, are created. The vector SEX contains the two distinct values of the class variable SEX used in forming the two subgroups. The vector _NOBS_ has the number of observations in each subgroup defined by the above.

Note that the combined means and standard deviations of the two subgroups are displayed but are not saved.

More than one class variable can be used, in which case a subgroup is defined by the combination of the values of the class variables.

Sorting a SAS Data Set

The observations in a SAS data set can be ordered (sorted) by specific key variables. To sort a SAS data set, close the data set if it is currently open, and issue a SORT statement for the variables by which you want the observations to be ordered. Specify an output data set name if you want to keep the original data set. For example, the statement

```
> sort class out=sorted by name;
```

creates a new SAS data set named SORTED. The new data set has the observations from the data set CLASS, ordered by the variable NAME.

The statement

```
> sort class by name;
```

sorts in place the data set CLASS by the variable NAME. However, at the completion of the SORT statement, the original data set is replaced by the sorted data set.

You can specify as many key variables as needed, and, optionally, each variable can be preceded by the keyword DESCENDING, which denotes that the variable that follows is to be sorted in descending order.

Indexing a SAS Data Set

Searching through a large data set for information about one or more specific observations may take a long time because the procedure must read each record. You can reduce this search time by first indexing the data set by a variable. The INDEX statement builds a special companion file containing the values and record numbers of the indexed variables. Once the index is built, IML may use the index for queries with WHERE clauses if it decides that indexed retrieval is more efficient. Any number of variables can be indexed, but only one index is in use at a given time. Note that purging a data set with the PURGE statement results in the loss of all associated indexes.

Once you have indexed a data set, IML can use this index whenever a search is conducted with respect to the indexed variables. The indexes are updated automatically whenever you change values in indexed variables. When an index is in use, observations cannot be randomly accessed by their physical location numbers. This means that the POINT range cannot be used when an index is in effect. However, if you purge the observations marked for deletion, or sort the data set in place, the indexes become invalid and IML automatically deletes them.

For example, if you want a list of all female students in the CLASS data set, you can first index CLASS by the variable SEX. Then use the LIST statement with a WHERE clause. Of course, the CLASS data set is small and indexing does little if anything to speed queries with the WHERE clause. If the data set had thousands of students, though, indexing could save search time.

To index the data set by the variable SEX, submit the statement

```
> index sex;

  NOTE: Variable SEX indexed.
  NOTE: Retrieval by SEX.
```

Now list all students. Notice the ordering of the special file built by indexing by the variable SEX. Retrievals by SEX will be quick.

```
> list all;
```

OBS	NAME	SEX	AGE	HEIGHT	WEIGHT
1	JOYCE	F	11.0000	51.3000	50.5000
4	JANE	F	12.0000	59.8000	84.5000
6	LOUISE	F	12.0000	56.3000	77.0000
8	ALICE	F	13.0000	56.5000	84.0000
9	BARBARA	F	13.0000	65.3000	98.0000
11	CAROL	F	14.0000	62.8000	102.5000
14	JUDY	F	14.0000	64.3000	90.0000
15	JANET	F	15.0000	62.5000	112.5000
16	MARY	F	15.0000	66.5000	112.0000
2	THOMAS	M	11.0000	57.5000	85.0000
3	JAMES	M	12.0000	57.3000	83.0000
7	ROBERT	M	12.0000	64.8000	128.0000
10	JEFFREY	M	13.0000	62.5000	84.0000
12	HENRY	M	15.0000	63.5000	102.5000
13	ALFRED	M	14.0000	69.0000	112.5000
17	RONALD	M	15.0000	67.0000	133.0000
18	WILLIAM	M	15.0000	66.5000	112.0000
19	PHILIP	M	16.0000	72.0000	150.0000

Data Set Maintenance Functions

Two functions and two subroutines are provided to perform data set maintenance:

DATASETS function
  obtains members in a data library. This function returns a character matrix containing the names of the SAS data sets in a library.

CONTENTS function
  obtains variables in a member. This function returns a character matrix containing the variable names for the SAS data set specified by *libname* and *memname*. The variable list is returned in alphabetic order.

RENAME subroutine
  renames a SAS data set member in a specified library.

DELETE subroutine
  deletes a SAS data set member in a specified library.

See Chapter 15 for details and examples of these functions and routines.

Overview of Commands

You have seen that IML has an extensive set of commands that operate on SAS data sets. The table below summarizes the data management commands you can use to perform management tasks for which you might normally use the SAS DATA step.

Table 6.1
Data Management
Commands

Command	Action
APPEND	adds observations to the end of a SAS data set
CLOSE	closes a SAS data set
CREATE	creates and opens a new SAS data set for input and output
DELETE	marks observations for deletion in a SAS data set
EDIT	opens an existing SAS data set for input and output
FIND	finds observations
INDEX	indexes variables in a SAS data set
LIST	lists observations
PURGE	purges all deleted observations from a SAS data set
READ	reads observations into IML variables
REPLACE	writes observations back into a SAS data set
RESET DEFLIB	names default libname
SAVE	saves changes and reopens a SAS data set
SETIN	selects an open SAS data set for input
SETOUT	selects an open SAS data set for output
SHOW CONTENTS	shows contents of the current input SAS data set
SHOW DATASETS	shows SAS data sets currently open
SORT	sorts a SAS data set
SUMMARY	produces summary statistics for numeric variables
USE	opens an existing SAS data set for input

Similarities and Differences with the SAS DATA Step

If you want to remain in the IML environment and mimic DATA step processing, you need to learn the basic differences between IML and the DATA step:

□ With SAS/IML software, you start with a CREATE statement instead of a DATA statement. You must explicitly set up all your variables with the right attributes before you create a data set. This means you must define character variables having the desired string length beforehand. Numeric variables are the default, so that any variable not defined as character is assumed to be numeric. In the DATA step, the variable attributes are determined from context across the whole step.

□ With SAS/IML software, you must use an APPEND statement to output an observation; in the DATA step you either use an OUTPUT statement or let the DATA step output it automatically.

□ With SAS/IML software, you iterate with a DO DATA loop. In the DATA step, the iterations are implied.

□ With SAS/IML software, you have to close the data set with a CLOSE statement unless you plan to leave the IML environment with a QUIT statement. The DATA step closes the data set automatically at the end of the step.

□ The DATA step usually executes faster than IML.

In short, the DATA step treats the problem with greater simplicity, allowing shorter programs. However, IML has more flexibility because it is both interactive and has a powerful matrix-handling capability.

In Conclusion

In this chapter, you have learned many ways to interact with SAS data sets from within the IML environment. You learned how to open and close a SAS data set, how to make it current for input and output, how to list observations by specifying a range of observations to process, a set of variables to use, and a condition for subsetting observations. You also learned summary statistics. You also know how to read observations and variables from a SAS data set into matrices as well as create a SAS data set from a matrix of values.

Chapter 7 File Access

Introduction

In this chapter you learn about external files and how to refer to an external file, whether it is a text file or a binary file. You learn how to read data from a file using the INFILE and INPUT statements and how to write data to an external file using the FILE and PUT statements.

With external files, you must know the format in which the data are stored or to be written. This is in contrast to SAS data sets, which are specialized files whose structure is already known to the SAS System.

The IML statements used to access files are very similar to the corresponding statements in the SAS DATA step. The following table summarizes the IML statements and their functions.

Statement	Function
CLOSEFILE	closes an external file
FILE	opens an external file for output
INFILE	opens an external file for input
INPUT	reads from the current input file
PUT	writes to the current output file
SHOW FILES	shows all open files, their attributes, and their status (current input and output files)

Referring to an External File

Suppose that you have data for students in a class. You have recorded the values for the variables NAME, SEX, AGE, HEIGHT, and WEIGHT for each student and have stored the data in an external text file named USER.TEXT.CLASS. If you want to read this data into IML variables, you need to indicate where the data are stored. In other words, you need to name the input file. If you want to write data from matrices to a file, you also need to name an output file.

There are two ways to refer to an input or output file: a *filepath* and a *filename*. A *filepath* is the name of the file as it is known to the operating system. A *filename* is an indirect SAS reference to the file made using the FILENAME statement. You can identify a file in either way using the FILE and INFILE statements.

For example, you can refer to the input file where the class data are stored using a literal filepath, that is, a quoted string. The statement

```
infile 'user.text.class';
```

opens the file USER.TEXT.CLASS for input. Similarly, if you want to output data to the file USER.TEXT.NEWCLASS, you need to reference the output file with the statement

```
file 'user.text.newclass';
```

You can also refer to external files using a *filename*. When using a filename as the operand, simply give the name. The name must be one already associated with a filepath by a previously issued FILENAME statement.

For example, suppose you want to reference the file with the class data using a FILENAME statement. First, you must associate the filepath with an alias (called a *fileref*), say INCLASS. Then you can refer to USER.TEXT.CLASS with the fileref INCLASS.

The following statements accomplish the same thing as the previous INFILE statement with the quoted filepath:

```
filename inclass 'user.text.class';
infile inclass;
```

You can use the same technique for output files. The following statements have the same effect as the previous file statement:

```
filename outclass 'user.text.newclass';
file outclass;
```

Three filenames have special meaning to IML: CARDS, LOG, and PRINT. These refer to the standard input and output streams for all SAS sessions, as described below:

CARDS   is a special filename for instream input data.

LOG   is a special filename for log output.

PRINT   is a special filename for standard print output.

When the filepath is specified, there is a limit of 64 characters to the operand.

Types of External Files

Most files that you work with are *text files*, which means that they can be edited and displayed without any special program. Text files under most host environments have special characters, called carriage-control characters or end-of-line characters, to separate one record from the next.

If your file does not adhere to these conventions, it is called a *binary file*. Typically, binary files do not have the usual record separators, and they may use any binary codes, including unprintable control characters. If you want to read a binary file, you must specify RECFM=N in the INFILE statement and use the byte operand (<) in the INPUT statement to specify the length of each item you want read. Treating a file as binary enables you to have direct access to a file position by byte-address using the byte operand (>) in the INPUT or PUT statement.

You write data to an external file using the FILE and PUT statements. The output file can be text or binary. If your output file is binary, you must specify RECFM=N in the FILE statement. One difference between binary and text files in output is that the PUT statement does not put the record-separator characters on the end of each record written for binary files.

Reading from an External File

After you have chosen a method to refer to the external file you want to read, you need an INFILE statement to open it for input and an INPUT statement to tell IML how to read the data.

The next several sections cover how to use an INFILE statement and how to specify an INPUT statement so that you can input data from an external file.

Using the INFILE Statement

An INFILE statement identifies an external file containing data you want to read. It opens the file for input or, if the file is already open, makes it the current input file. This means that subsequent INPUT statements read from this file until another file is made the current input file.

The following options can be used with the INFILE statement:

FLOWOVER   allows the INPUT statement to go to the next record to obtain values for the variables.

LENGTH=*variable*   names a variable containing the length of the current record, where the value is set to the number of bytes used after each INPUT statement.

MISSOVER   prevents reading from the next input record when an INPUT statement reaches the end of the current record without finding values for all variables. It assigns missing values to all values that are expected but not found.

RECFM=N   specifies that the file is to be read in as a pure binary file rather than as a file with record-separator characters. You must use the byte operands ($<$ and $>$) to get new records rather than separate INPUT statements or the new line operator (/).

STOPOVER   stops reading when an INPUT statement reaches the end of the current record without finding values for all variables in the statement. It treats going past the end of a record as an error condition, triggering an end-of-file condition. STOPOVER is the default.

The FLOWOVER, MISSOVER, and STOPOVER options control how the INPUT statement works when you try to read past the end of a record. You can specify at most one of these options. Read these options carefully so that you understand them completely.

Below is an example using the INFILE statement with a FILENAME statement to read the class data file. The MISSOVER option is used to prevent reading from the next record if values for all variables in the INPUT statement are not found.

```
filename inclass 'user.text.class';
infile inclass missover;
```

You can specify the filepath with a quoted literal also. The above statements could be written as:

```
infile 'user.text.class' missover;
```

Using the INPUT Statement

Once you have referenced the data file containing your data with an INFILE statement, you need to tell IML exactly how the data are arranged:

□   the number of variables and their names

□   each variable's type, either numeric or character

□   the format of each variable's values

□   the columns that correspond to each variable.

In other words, you must tell IML how to read the data.

The INPUT statement describes the arrangement of values in an input record. The INPUT statement reads records from a file specified in the previously executed INFILE statement, reading the values into IML variables.

There are two ways to describe a record's values in an IML INPUT statement:

□   list (or scanning) input

□   formatted input.

Here are several examples of valid INPUT statements for the class data file, depending, of course, on how the data are stored.

If the data are stored with a blank or a comma between fields, then list input can be used. For example, the INPUT statement for the class data file might look as follows:

```
infile inclass;
input name $ sex $ age height weight;
```

These statements tell IML the following:

□   There are five variables: NAME, SEX, AGE, HEIGHT, WEIGHT.

□   Data fields are separated by commas or blanks.

□   NAME and SEX are character variables, as indicated by the dollar sign ($).

□   AGE, HEIGHT, and WEIGHT are numeric variables, the default.

The data must be stored in the same order as the variables are listed on the INPUT statement.

Otherwise, you can use formatted input, which is column specific. Formatted input is the most flexible and can handle any data file. Your INPUT statement for the class data file might look as follows:

```
infile inclass;
input a1 name $char8. a10 sex $char1. a15 age 2.0
   a20 height 4.1 a25 weight 5.1;
```

These statements tell IML the following:

□ NAME is a character variable whose value begins in column 1 (indicated by @1) and occupies eight columns ($char8.).

□ SEX is a character variable whose value is in column 10 ($char1.).

□ AGE is a numeric variable whose value is found in columns 15 and 16 and has no decimal places (2.0).

□ HEIGHT is a numeric variable found in columns 20 through 23 with one decimal place implied (4.1).

□ WEIGHT is a numeric variable found in columns 25 through 29 with one decimal place implied (5.1).

The next sections discuss these two modes of input.

List Input

If your data are recorded with a comma or one or more blanks between data fields, you can use list input to read your data. If you have missing values, that is, unknown values, they must be represented by a period (.) rather than a blank field.

When IML looks for a value, it skips past blanks and tab characters. Then it scans for a delimiter to the value. The delimiter is a blank, a comma, or the end of the record. When the ampersand (&) format modifier is used, IML looks for two blanks, a comma, or the end of the record.

The general form of the INPUT statement for list input is

INPUT *variable* <$> <&> <...*variable* <$> <&>>;

where

variable   names the variable to be read by the INPUT statement.

$ indicates the preceding variable is character.

&   indicates that a character value may have a single embedded blank. Because a blank normally indicates the end of a data value, use the ampersand format modifier to indicate the end of the value with at least two blanks or a comma.

With list input, IML scans the input lines for values. Consider using list input when

□ blanks or commas separate input values

□ periods rather than blanks represent missing values.

List input is the default in several situations. Descriptions of these situations and the behavior of IML follow:

1. If no input format is specified for a variable, IML scans for a number.

2. If a single dollar sign or ampersand format modifier is specified, IML scans for a character value. The ampersand format modifier allows single embedded blanks to occur.

3. If a format is given with width unspecified or 0, IML scans for the the first blank or comma.

If the end of a record is encountered before IML finds a value, then the behavior is as described by the record overflow options in the INFILE statement discussed in "Using the INFILE Statement."

When reading with list input, the order of the variables listed in the INPUT statement must agree with the order of the values in the data file. For example, consider the following data:

```
Alice    f    10    61    97
Beth     f    11    64    105
Bill     m    12    63    110
```

You can use list input to read this data by specifying the following INPUT statement:

```
input name $ sex $ age height weight;
```

Note: This statement implies that the variables are stored in the order given. That is, each line of data contains a student's NAME, SEX, AGE, HEIGHT, and WEIGHT in that order and separated by at least one blank or by a comma.

Formatted Input

The alternative to list input is formatted input. An INPUT statement reading formatted input must have a SAS informat after each variable. An *informat* gives the data type and field width of an input value. Formatted input may be used with pointer controls and format modifiers. Note, however, that neither pointer controls nor format modifiers are necessary for formatted input.

Pointer control features

Pointer controls reset the pointer's column and line positions and tell the INPUT statement where to go to read the data value. You use pointer controls to specify the columns and lines from which you want to read:

□ *Column pointer controls* move the pointer to the column you specify.

□ *Line pointer controls* move the pointer to the next line.

□ *Line hold controls* keep the pointer on the current input line.

□ *Binary file indicator controls* indicate that the input line is from a binary file.

Column pointer controls

Column pointer controls indicate in which column an input value starts. Column pointer controls begin with either an at sign (@) or a plus sign (+).

@*n*	moves the pointer to column *n*.
@*point-variable*	moves the pointer to the column given by the current value of *point-variable*.
@*(expression)*	moves the pointer to the column given by the value of the *expression*. The *expression* must evaluate to a positive integer.
+*n*	moves the pointer *n* columns.
+*point-variable*	moves the pointer the number of columns given by the value of *point-variable*.
+*(expression)*	moves the pointer the number of columns given by the value of *expression*. The value of *expression* can be positive or negative.

Here are some examples using column pointer controls:

Example	Meaning
@12	go to column 12
@N	go to the column given by the value of N
@(N−1)	go to the column given by the value of N−1
+5	skip 5 spaces
+N	skip N spaces
+(N+1)	skip N+1 spaces

In the earlier example using formatted input, you used several pointer controls:

```
infile inclass;
input a1 name $char8. a10 sex $char1. a15 age 2.0
      a20 height 4.1 a25 weight 5.1;
```

The @1 moves the pointer to column 1, the @10 moves it to column 10, and so on. You move the pointer to the column where the data field begins and then supply an informat specifying how many columns the variable occupies.

The INPUT statement could also be written

```
input a1 name $char8. +1 sex $char1. +4 age 2. +3 height 4.1
      +1 weight 5.1;
```

In this form, you move the pointer to column 1 (@1) and read eight columns. The pointer is now at column 9. Now, move the pointer +1 columns to column 10 to read SEX. The $char1. informat says to read a character variable occupying one column. After reading the value for SEX, the pointer is at column 11, so move it to column 15 with +4 and read AGE in columns 15 and 16 (the

2. informat). The pointer is now at column 17, so move +3 columns and read HEIGHT. The same idea applies for reading WEIGHT.

Line pointer control
The line pointer control (/) directs IML to skip to the next line of input. You need a line pointer control when a record of data takes more than one line. You use the new line pointer control (/) to skip to the next line and continue reading data. In the example reading the class data, you do not need to skip a line because each line of data contains all the variables for a student.

Line hold control
The trailing at sign (@), when at the end of an INPUT statement, directs IML to hold the pointer on the current record so that you can read more data with subsequent INPUT statements. You can use it to read several records from a single line of data. Sometimes, when a record is very short, say ten columns or so, you can save space in your external file by coding several records on the same line.

Binary file indicator controls
When the external file you want to read is a binary file (RECFM=N is specified in the INFILE statement), you must tell IML how to read the values using the following binary file indicator controls:

>n	start reading the next record at the byte position n in the file.
> point-variable	start reading the next record at the byte position in the file given by point-variable.
> (expression)	start reading the next record at the byte position in the file given by expression.
<n	read the number of bytes indicated by the value of n.
< point-variable	read the number of bytes indicated by the value of point-variable.
< (expression)	read the number of bytes indicated by the value of expression.

Pattern Searching

You can have the input mechanism search for patterns of text by using the at sign (@) positional with a character operand. IML starts searching at the current position, advances until it finds the pattern, and leaves the pointer at the position immediately after the found pattern in the input record. For example, the statement

```
input a 'NAME=' name $;
```

searches for the pattern **NAME=** and then uses list input to read the value after the found pattern.

If the pattern is not found, then the pointer is left past the end of the record, and the rest of the INPUT statement follows the conventions based on the options MISSOVER, STOPOVER, and FLOWOVER described in "Using the INFILE Statement" earlier in this chapter. If you use pattern searching, you usually specify the MISSOVER option so you can control for the occurrences of the pattern not being found.

Notice that the MISSOVER feature allows you to search for a variety of items on the same record, even if some of them are not found. For example, the statements

```
infile in1 missover;
input a1 a "NAME=" name $
      a1 a "ADDR=" addr &
      a1 a "PHONE=" phone $;
```

are able to read in the ADDR variable even if **NAME=** is not found (in which case NAME is unvalued).

The pattern operand can use any characters except for the following:

% $ [ ] { } < > − ? * # @ ^ `(backquote)

Record Directives

Each INPUT statement goes to a new record except for the following special cases:

1. An at sign (@) at the end of an INPUT statement specifies that the record is to be held for future INPUT statements.

2. Binary files (RECFM=N) always hold their records until the > directive.

As discussed in the syntax of the INPUT statement, the line pointer operator (/) instructs the input mechanism to go immediately to the next record. For binary (RECFM=N) files, the > directive is used instead of the /.

Blanks

For character values, the informat determines the way blanks are interpreted. For example, the $CHARw. format reads blanks as part of the whole value, while the BZw. format turns blanks into 0s. See *SAS Language: Reference, Version 6, First Edition* for more information on informats.

Missing Values

Missing values in formatted input are represented by blanks or a single period for a numeric value and by blanks for a character value.

Matrix Use

Data values are either character or numeric. Input variables always result in scalar (one row by one column) values with type (character or numeric) and length determined by the input format.

End-of-File Condition

End of file is the condition of trying to read a record when there are no more records to read from the file. The consequences of an end of file condition are described below:

□   All the variables in the INPUT statement that encountered end of file are freed of their values. You can use the NROW or NCOL function to test if this has happened.

□   If end of file occurs while inside a DO DATA loop, execution is passed to the statement after the END statement in the loop.

For text files, the end of file is encountered first as the end of the last record. The next time input is attempted, the end of file condition is raised.

For binary files, the end of file can result in the input mechanism returning a record that is shorter than the requested length. In this case IML still attempts to process the record, using the rules described in "Using the INFILE Statement," earlier in this chapter.

The DO DATA mechanism provides a convenient mechanism for handling end of file. For example, to read the class data from the external file USER.TEXT.CLASS into a SAS data set, you need to perform the following steps:

1.   Establish a *fileref* referencing the data file.

2.   Use an INFILE statement to open the file for input.

3.   Initialize any character variables by setting the length.

4.   Create a new SAS data set with a CREATE statement. You want to list the variables you plan to input in a VAR clause.

5.   Use a DO DATA loop to read the data one line at a time.

6.   Write an INPUT statement telling IML how to read the data.

7.   Use an APPEND statement to add the new data line to the end of the new SAS data set.

8.   End the DO DATA loop.

9.   Close the new data set.

10. Close the external file with a CLOSEFILE statement.

Your code would look as follows:

```
filename inclass 'user.text.class';
infile inclass missover;
name="12345678";
sex="1";
create class  var{name sex age height weight};
do data;
   input name $ sex $ age height weight;
   append;
end;
close class;
closefile inclass;
```

Note that the APPEND statement is not executed if the INPUT statement reads past the end of file since IML escapes the loop immediately when the condition is encountered.

Differences with the SAS DATA Step

If you are familiar with the SAS DATA step, you will notice that the following features are supported differently or are not supported in IML:

□ The pound sign (#) directive supporting multiple current records is not supported.

□ Grouping parentheses are not supported.

□ The colon (:) format modifier is not supported.

□ The byte operands ($<$ and $>$) are new features supporting binary files.

□ The ampersand (&) format modifier causes IML to stop reading data if a comma is encountered. Use of the ampersand format modifier is valid with list input only.

□ RECFM=F is not supported.

Writing to an External File

If you have data in matrices and you want to write this data to an external file, you need to reference, or point to, the file as discussed in "Referring to an External File." The FILE statement opens the file for output so that you can write data to it. You need a PUT statement to direct how the data is output. These two statements are discussed in the following sections.

Using the FILE Statement

The FILE statement is used to refer to an external file. If you have values stored in matrices, you can write these values to a file. Just as with the INFILE statement, you need a fileref to point to the file you want to write to. You use a FILE statement to indicate that you want to write to rather than read from a file. For example, if you want to output to the file USER.TEXT.NEWCLASS, you can specify the file with a quoted literal filepath:

```
> file 'user.text.newclass';
```

Otherwise, you can first establish a fileref and then refer to the file by its fileref:

```
> filename outclass 'user.text.class';
> file outclass;
```

There are two options you can use in the FILE statement:

RECFM=N   specifies that the file is to be written as a pure binary file without record-separator characters.

LRECL=*operand*   specifies the size of the buffer to hold the records.

The FILE statement opens a file for output or, if the file is already open, makes it the current output file so that subsequent PUT statements write to the file. The FILE statement is similar in syntax and operation to the INFILE statement.

Using the PUT Statement

The PUT statement writes lines to the SAS log, the SAS output file, or to any external file specified in a FILE statement. The file associated with the most recently executed FILE statement is the *current output file.*

You can use the following arguments with the PUT statement:

variable   names the IML variable whose value is put to the current pointer position in the record. The variable must be scalar valued. The put variable can be followed immediately by an output format.

literal   gives a literal to be put to the current pointer position in the record. The literal can be followed immediately by an output format.

(expression)   must produce a scalar-valued result. The expression can be immediately followed by an output format.

format   names the output formats for the values.

pointer-control   moves the output pointer to a line or column.

Pointer Control Features

Most PUT statements need the added flexibility obtained with pointer controls. IML keeps track of its position on each output line with a pointer. With specifications in the PUT statement, you can control pointer movement from column to column and line to line. The pointer controls available were discussed with the INPUT statement.

Differences with the SAS DATA Step

If you are familiar with the SAS DATA step, you will notice that the following features are supported differently or are not supported:

□ The pound sign (#) directive supporting multiple current records is not supported.

□ Grouping parentheses are not supported.

□ The byte operands (< and >) are a new feature supporting binary files.

Examples

Writing a Matrix to an External File

If you have data stored in an $n \times m$ matrix and you want to output the values to an external file, you need to write out the matrix element by element.

For example, suppose you have a matrix **X** containing data you want written to the file USER.MATRIX. Suppose also that **X** contains 1s and 0s so that the format for output can be one column. You need to do the following:

1. Establish a fileref, say OUT.

2. Use a FILE statement to open the file for output.

3. Have a DO loop for the rows of the matrix.

4. Have a DO loop for the columns of the matrix.

5. Use a PUT statement to say how to write the element value.

6. End the inner DO loop.

7. Skip a line.

8. End the outer DO loop.

9. Close the file.

Your code should look as follows:

```
filename out 'user.matrix';
file out;
   do i=1 to nrow(x);
      do j=1 to ncol(x);
         put (x[i,j]) 1.0 +2 @;
      end;
      put;
   end;
closefile out;
```

The output file contains a record for each row of the matrix. For example, if your matrix was 4×4, then the file might look as follows:

```
1  1  0  1
1  0  0  1
1  1  1  0
0  1  0  1
```

Quick Printing to the PRINT File

You can use the FILE PRINT statement to route ouput to the standard print file. The following statements generate data that are output to the PRINT file:

```
> file print;
> do a=0 to 6.28 by .2;
>    x=sin(a);
>    p=(x+1)#30;
>    put @1 a 6.4 +p x 8.4;
> end;
```

The result is shown below:

```
0.0000                                        0.0000
0.2000                                          0.1987
0.4000                                            0.3894
0.6000                                              0.5646
0.8000                                                0.7174
1.0000                                                  0.8415
1.2000                                                    0.9320
1.4000                                                      0.9854
1.6000                                                      0.9996
1.8000                                                      0.9738
2.0000                                                    0.9093
2.2000                                                  0.8085
2.4000                                                0.6755
2.6000                                              0.5155
2.8000                                            0.3350
3.0000                                          0.1411
3.2000                                        -0.0584
3.4000                                      -0.2555
3.6000                                    -0.4425
3.8000                                  -0.6119
4.0000                                -0.7568
4.2000                              -0.8716
4.4000                            -0.9516
4.6000                          -0.9937
4.8000                          -0.9962
5.0000                          -0.9589
5.2000                            -0.8835
5.4000                              -0.7728
5.6000                                -0.6313
5.8000                                  -0.4646
6.0000                                    -0.2794
6.2000                                      -0.0831
```

Listing Your External Files

To list all open files and their current input or current output status, use the SHOW FILES statement.

Closing an External File

The CLOSEFILE statement closes files opened by an INFILE or a FILE statement. You specify the CLOSEFILE statement just as you do the INFILE or FILE statement. For example, the following statements open the external file USER.TEXT.CLASS for input and the close it:

```
filename in 'user.text.class';
infile in;
closefile in;
```

In Conclusion

In this chapter, you learned how to refer to, or point to, an external file with a FILENAME statement. You can use the FILENAME statement whether you want to read from or write to an external file. The file can also be referenced by a quoted literal filepath. You also learned about the difference between a text file and a binary file.

You learned how to read data from an external file with the INFILE and INPUT statements, using either list or formatted input. You learned how to write your matrices to an external file using the FILE and PUT statements. Finally, you learned how to close your files when you are finished.

Chapter 8 Applications: Statistical Examples

Introduction

SAS/IML software has many linear operators that perform high-level operations commonly needed in applying linear algebra techniques to data analysis. The similarity of the IML notation and matrix algebra notation makes translation from algorithm to program a straightforward task. The examples in this chapter show a variety of matrix operators at work.

You can use these examples to gain insight into the more complex problems you may need to solve. Some of the examples perform the same analyses as performed by procedures in SAS/STAT software and are not meant to replace them. The examples are included as learning tools.

Example 1: Correlation

This example defines modules to compute correlation coefficients between numeric variables and standardize values for a set of data.

```
    /* Module to compute correlations  */
start corr;
   n=nrow(x);                        /* number of observations */
   sum=x[+,] ;                       /* compute column sums */
   xpx=t(x)*x-t(sum)*sum/n;          /* compute sscp matrix    */
   s=diag(1/sqrt(vecdiag(xpx)));        /* scaling matrix */
   corr=s*xpx*s;                        /* correlation matrix */
   print "Correlation Matrix",,corr[rowname=nm colname=nm] ;
finish corr;

    /* Module to standardize data */
start std;
   mean=x[+,] /n;                        /* means for columns */
   x=x-repeat(mean,n,1);             /* center x to mean zero */
   ss=x[##,] ;                    /* sum of squares for columns */
   std=sqrt(ss/(n-1));               /* standard deviation estimate*/
   x=x*diag(1/std);                     /* scaling to std dev 1 */
   print ,"Standardized Data",,X[colname=nm] ;
finish std;

    /* Sample run */
x = { 1 2 3,
      3 2 1,
      4 2 1,
      0 4 1,
     24 1 0,
      1 3 8};
nm={age weight height};
run corr;
run std;
```

The results are shown below:

Correlation Matrix

```
CORR           AGE     WEIGHT     HEIGHT

AGE             1   -0.717102  -0.436558
WEIGHT  -0.717102          1   0.3508232
HEIGHT  -0.436558  0.3508232          1
```

Standardized Data

```
X        AGE      WEIGHT      HEIGHT

-0.490116  -0.322749   0.2264554
-0.272287  -0.322749  -0.452911
-0.163372  -0.322749  -0.452911
 -0.59903   1.6137431  -0.452911
2.0149206  -1.290994  -0.792594
-0.490116   0.6454972   1.924871
```

Example 2: Newton's Method for Solving Nonlinear Systems of Equations

This example solves a nonlinear system of equations by Newton's method. Let the nonlinear system be represented by

$$F(\mathbf{x}) = 0$$

where $\mathbf{x}$ is a vector and $\mathbf{F}$ is a vector-valued, possibly nonlinear function.

In order to find $\mathbf{x}$ such that F goes to 0, an initial estimate $\mathbf{x}_0$ is chosen, and Newton's iterative method for converging to the solution is used:

$$\mathbf{x}_{n+1} = \mathbf{x}_n - J^{-1}(\mathbf{x}_n) F(\mathbf{x}_n)$$

where $J(\mathbf{x})$ is the Jacobian matrix of partial derivatives of F with respect to $\mathbf{x}$.

For optimization problems, the same method is used, where $F(\mathbf{x})$ is the gradient of the objective function and $J(\mathbf{x})$ becomes the Hessian (Newton-Raphson).

In this example, the system to be solved is

$$\mathbf{x}_1 + \mathbf{x}_2 - \mathbf{x}_1{}^*\mathbf{x}_2 + 2 = 0$$

$$\mathbf{x}_1{}^*\exp(-\mathbf{x}_2) - 1 = 0 \quad .$$

The code is organized into several modules: NEWTON, FUN, and DERIV.

```
/*      Newton's Method to Solve a Nonlinear Function    */
/* The user must supply initial values, and the FUN and DERIV */
/* functions.                                           */
/* on entry: FUN evaluates the function f in terms of x  */
/* initial values are given to x                        */
/* DERIV evaluates jacobian j                           */
/* tuning variables: CONVERGE, MAXITER.                 */
/* on exit: solution in x, function value in f close to zero */
/* ITER has number of iterations.                       */

start newton;
   run fun;                 /* evaluate function at starting values */
   do iter=1 to maxiter     /* iterate until maxiter iterations */
   while(max(abs(f))>converge);              /* or convergence */
      run deriv;                  /* evaluate derivatives in j */
      delta=-solve(j,f);     /* solve for correction vector */
      x=x+delta;                  /* the new approximation */
      run fun;                    /* evaluate the function */
   end;
finish newton;

maxiter=15;                       /* default maximum iterations */
converge=.000001;                 /* default convergence criterion */


   /* User-supplied function evaluation */
start fun;
   x1=x[1] ;
   x2=x[2] ;                              /* extract the values */
   f= (x1+x2-x1*x2+2)//
   (x1*exp(-x2)-1);                       /* evaluate the function */
finish fun;

   /* User-supplied derivatives of the function */
start deriv;
   /* evaluate jacobian */
   j=((1-x2)||(1-x1) )//(exp(-x2)||(-x1*exp(-x2)));
finish deriv;

do;
   print "Solving the system: X1+X2-X1*X2+2=0, X1*EXP(-X2)-1=0" ,;
   x={.1, -2};    /* starting values */
   run newton;
   print x f;
end;
```

The results are shown below:

```
Solving the system: X1+X2-X1*X2+2=0, X1*EXP(-X2)-1=0


                            X         F
                     0.0977731 5.3523E-9
                     -2.325106 6.1501E-8
```

Example 3:   Regression

This example shows a regression module that calculates statistics not calculated by the two previous examples:

```
/*          Regression Routine           */
/* Given X, and Y, this fits Y = X B + E  */
/* by least squares.                      */

start reg;
   n=nrow(x);                         /* number of observations */
   k=ncol(x);                          /* number of variables */
   xpx=x`*x;                             /* cross-products */
   xpy=x`*y;
   xpxi=inv(xpx);                     /* inverse crossproducts */
   b=xpxi*xpy;                         /* parameter estimates */
   yhat=x*b;                            /* predicted values */
   resid=y-yhat;                            /* residuals */
   sse=resid`*resid;                 /* sum of squared errors */
   dfe=n-k;                       /* degrees of freedom error */
   mse=sse/dfe;                        /* mean squared error */
   rmse=sqrt(mse);                /* root mean squared error */
   covb=xpxi#mse;                 /* covariance of estimates */
   stdb=sqrt(vecdiag(covb));            /* standard errors */
   t=b/stdb;                        /* ttest for estimates=0 */
   probt=1-probf(t#t,1,dfe);     /* significance probability */
   print name b stdb t probt;
   s=diag(1/stdb);
   corrb=s*covb*s;                /* correlation of estimates */
   print ,"Covariance of Estimates", covb[r=name c=name] ,
          "Correlation of Estimates",corrb[r=name c=name] ;
```

```
            if nrow(tval)=0 then return;         /* is a t-value specified? */
            projx=x*xpxi*x`;                               /* hat matrix */
            vresid=(i(n)-projx)*mse;            /* covariance of residuals */
            vpred=projx#mse;                 /* covariance of predicted values */
            h=vecdiag(projx);                      /* hat leverage values */
            lowerm=yhat-tval#sqrt(h*mse); /* lower confidence limit for mean */
            upperm=yhat+tval#sqrt(h*mse);             /* upper limit for mean */
            lower=yhat-tval#sqrt(h*mse+mse);       /* lower limit for indiv */
            upper=yhat+tval#sqrt(h*mse+mse);       /* upper limit for indiv */
            print ,,"Predicted Values, Residuals, and Limits" ,,
            y yhat resid h lowerm upperm lower upper;
      finish reg;

         /* Routine to test a linear combination of the estimates  */
         /* given L, this routine tests hypothesis that LB = 0.     */

   start test;
      dfn=nrow(L);
      Lb=L*b;
      vLb=L*xpxi*L`;
      q=Lb`*inv(vLb)*Lb /dfn;
      f=q/mse;
      prob=1-probf(f,dfn,dfe);
      print ,f dfn dfe prob;
   finish test;

         /* Run it on population of U.S. for decades beginning 1790 */

   x= { 1 1 1,
        1 2 4,
        1 3 9,
        1 4 16,
        1 5 25,
        1 6 36,
        1 7 49,
        1 8 64 };

   y= {3.929,5.308,7.239,9.638,12.866,17.069,23.191,31.443};
   name={"Intercept", "Decade", "Decade**2" };
   tval=2.57; /* for 5 df at .025 level to get 95% confidence interval */
   reset fw=7;
   run reg;
   do;
      print ,"TEST Coef for Linear";
      L={0 1 0 };
      run test;
      print ,"TEST Coef for Linear,Quad";
      L={0 1 0,0 0 1};
      run test;
      print ,"TEST Linear+Quad = 0";
      L={0 1 1 };
      run test;
   end;
```

The results are shown below:

```
NAME            B     STDB       T   PROBT
Intercept 5.06934 0.96559 5.24997 0.00333
Decade    -1.1099  0.4923 -2.2546 0.07385
Decade**2 0.53964  0.0534  10.106 0.00016
```

Covariance of Estimates

COVB	Intercept	Decade	Decade**2
Intercept	0.93237	-0.4362	0.04277
Decade	-0.4362	0.24236	-0.0257
Decade**2	0.04277	-0.0257	0.00285

Correlation of Estimates

CORRB	Intercept	Decade	Decade**2
Intercept	1	-0.9177	0.8295
Decade	-0.9177	1	-0.9762
Decade**2	0.8295	-0.9762	1

Predicted Values, Residuals, and Limits

Y	YHAT	RESID	H	LOWERM	UPPERM	LOWER	UPPER
3.929	4.49904	-0.57	0.70833	3.00202	5.99606	2.17419	6.82389
5.308	5.00802	0.29998	0.27976	4.06721	5.94883	2.99581	7.02023
7.239	6.59627	0.64273	0.23214	5.73926	7.45328	4.62185	8.57069
9.638	9.26379	0.37421	0.27976	8.32298	10.2046	7.25158	11.276
12.866	13.0106	-0.1446	0.27976	12.0698	13.9514	10.9984	15.0228
17.069	17.8367	-0.7677	0.23214	16.9797	18.6937	15.8622	19.8111
23.191	23.742	-0.551	0.27976	22.8012	24.6828	21.7298	25.7542
31.443	30.7266	0.71637	0.70833	29.2296	32.2236	28.4018	33.0515

```
                            TEST Coef for Linear



                       F      DFN     DFE     PROB
                 5.08317        1       5  0.07385



                      TEST Coef for Linear,Quad



                       F      DFN     DFE     PROB
                 666.511        2       5  8.54E-7



                       TEST Linear+Quad = 0



                       F      DFN     DFE     PROB
                 1.67746        1       5  0.25184
```

Example 4: Alpha Factor Analysis

This example shows how an algorithm for computing alpha factor patterns
(Kaiser and Caffrey 1965) is transcribed into IML code.

For later reference you could store the ALPHA subroutine created below in
an IML catalog and load it when needed.

```
/*              Alpha Factor Analysis                  */
/* Ref: Kaiser et al., 1965 Psychometrika, pp. 12-13   */
/* r correlation matrix (n.s.) already set up          */
/* p number of variables                               */
/* q number of factors                                 */
/* h communalities                                     */
/* m eigenvalues                                       */
/* e eigenvectors                                      */
/* f factor pattern                                    */
/* (IQ,H2,HI,G,MM) temporary use. freed up             */
/*                                                     */
```

```
start alpha;
   p=ncol(r);
   q=0;
   h=0;                                              /* initialize */
   h2=i(p)-diag(1/vecdiag(inv(r)));                  /* smcs */
   do while(max(abs(h-h2))>.001);        /* iterate until converges */
      h=h2;
      hi=diag(sqrt(1/vecdiag(h)));
      g=hi*(r-i(p))*hi+i(p);
      call eigen(m,e,g);                  /* get eigenvalues and vecs */
      if q=0 then
      do;
         q=sum(m>1);                          /* number of factors */
         iq=1:q;
      end;                                        /* index vector */
      mm=diag(sqrt(m[iq,]));                   /* collapse eigvals */
      e=e[,iq] ;                              /* collapse eigvecs */
      h2=h*diag((e*mm) [,##]);               /* new communalities */
   end;
   hi=sqrt(h);
   h=vecdiag(h2);
   f=hi*e*mm;                               /* resulting pattern */
   free iq h2 hi g mm;                        /* free temporaries */
finish;

   /* Correlation Matrix from Harmon, Modern Factor Analysis, */
   /* 2nd edition, page 124, "Eight Physical Variables"       */


r={1.000 .846 .805 .859 .473 .398 .301 .382 ,
    .846 1.000 .881 .826 .376 .326 .277 .415 ,
    .805 .881 1.000 .801 .380 .319 .237 .345 ,
    .859 .826 .801 1.000 .436 .329 .327 .365 ,
    .473 .376 .380 .436 1.000 .762 .730 .629 ,
    .398 .326 .319 .329 .762 1.000 .583 .577 ,
    .301 .277 .237 .327 .730 .583 1.000 .539 ,
    .382 .415 .345 .365 .629 .577 .539 1.000};
nm = {Var1 Var2 Var3 Var4 Var5 Var6 Var7 Var8};
run alpha;
print ,"EIGENVALUES" , m;
print ,"COMMUNALITIES" , h[rowname=nm];
print ,"FACTOR PATTERN", f[rowname=nm];
```

The results are shown below:

EIGENVALUES

```
       M
   5.93785
   2.0622
   0.13902
   0.08211
   0.0181
  -0.0475
  -0.0915
  -0.1003
```

COMMUNALITIES

```
          H
VAR1  0.83812
VAR2  0.89057
VAR3  0.81893
VAR4  0.80673
VAR5  0.88021
VAR6  0.6392
VAR7  0.58216
VAR8  0.49981
```

FACTOR PATTERN

```
      F
VAR1  0.81339  -0.4201
VAR2  0.80284  -0.496
VAR3  0.75791  -0.4945
VAR4  0.78745  -0.432
VAR5  0.80514   0.48162
VAR6  0.68041   0.41981
VAR7  0.62062   0.44383
VAR8  0.64494   0.28959
```

Example 5: Categorical Linear Models

This example fits a linear model to a function of the response probabilities

$$\mathbf{K} \log \pi = \mathbf{X}\boldsymbol{\beta} + \mathrm{e}$$

where **K** is a matrix that compares each response category to the last. Data are from Kastenbaum and Lamphiear (1959). First, the Grizzle-Starmer-Koch (1969) approach is used to obtain generalized least-squares estimates of **β**. These form the initial values for the Newton-Raphson solution for the maximum-likelihood estimates. PROC CATMOD can also be used to analyze these binary data (see Cox 1970).

```
/* Categorical Linear Models                 */
/* by Least Squares and Maximum Likelihood   */
/*  CATLIN                                    */
/*  Input:                                    */
/*     n the s by p matrix of response counts */
/*     x the s by r design matrix             */
/*                                            */

start catlin;

   /*---find dimensions---*/
   s=nrow(n);                          /* number of populations */
   r=ncol(n);                           /* number of responses */
   q=r-1;                             /* number of function values */
   d=ncol(x);                        /* number of design parameters */
   qd=q*d;                             /* total number of parameters */

   /*---get probability estimates---*/
   rown=n[,+];                                 /* row totals */
   pr=n/(rown*repeat(1,1,r));          /* probability estimates */
   p=shape(pr[,1:q] ,0,1);           /* cut and shaped to vector */
   print "INITIAL PROBABILITY ESTIMATES" ,pr;

      /*   estimate by the GSK method   */

      /* function of probabilities */
   f=log(p)-log(pr[,r])@repeat(1,q,1);

      /* inverse covariance of f */
   si=(diag(p)-p*p`)#(diag(rown)@repeat(1,q,q));
   z=x@i(q);                           /* expanded design matrix */
   h=z`*si*z;                          /* crossproducts matrix */
   g=z`*si*f;                           /* cross with f */
   beta=solve(h,g);                    /* least squares solution */
   stderr=sqrt(vecdiag(inv(h)));         /* standard errors */
   run prob;
   print ,"GSK ESTIMATES" , beta stderr ,pi;
```

```
                /*   iterations for ML solution   */
         crit=1;
         do it=1 to 8 while(crit>.0005);       /* iterate until converge */

            /* block diagonal weighting  */
            si=(diag(pi)-pi*pi`)#(diag(rown)@repeat(1,q,q));
            g=z`*(rown@repeat(1,q,1)#(p-pi));              /* gradient */
            h=z`*si*z;                                     /* hessian */
            delta=solve(h,g);                 /* solve for correction */
            beta=beta+delta;                  /* apply the correction */
            run prob;                         /* compute prob estimates */
            crit=max(abs(delta));             /* convergence criterion */
         end;
         stderr=sqrt(vecdiag(inv(h)));               /* standard errors */
         print , "ML Estimates", beta stderr, pi;
         print , "Iterations" it "Criterion" crit;
      finish catlin;


         /*   subroutine to compute new prob estimates @ parameters   */
      start prob;
         la=exp(x*shape(beta,0,q));
         pi=la/((1+la[,+] )*repeat(1,1,q));
         pi=shape(pi,0,1);
      finish prob;



         /*---prepare frequency data and design matrix---*/
      n= { 58 11 05,
           75 19 07,
           49 14 10,
           58 17 08,
           33 18 15,
           45 22 10,
           15 13 15,
           39 22 18,
           04 12 17,
           05 15 08};     /* frequency counts*/

      x= { 1  1  1  0  0  0,
           1 -1  1  0  0  0,
           1  1  0  1  0  0,
           1 -1  0  1  0  0,
           1  1  0  0  1  0,
           1 -1  0  0  1  0,
           1  1  0  0  0  1,
           1 -1  0  0  0  1,
           1  1 -1 -1 -1 -1,
           1 -1 -1 -1 -1 -1};   /* design matrix*/

      run catlin;
```

The results are shown below:

INITIAL PROBABILITY ESTIMATES

PR

0.7837838	0.1486486	0.0675676
0.7425743	0.1881188	0.0693069
0.6712329	0.1917808	0.1369863
0.6987952	0.2048193	0.0963855
0.5	0.2727273	0.2272727
0.5844156	0.2857143	0.1298701
0.3488372	0.3023256	0.3488372
0.4936709	0.278481	0.2278481
0.1212121	0.3636364	0.5151515
0.1785714	0.5357143	0.2857143

GSK ESTIMATES

BETA	STDERR
0.9454429	0.1290925
0.4003259	0.1284867
-0.277777	0.1164699
-0.278472	0.1255916
1.4146936	0.267351
0.474136	0.294943
0.8464701	0.2362639
0.1526095	0.2633051
0.1952395	0.2214436
0.0723489	0.2366597
-0.514488	0.2171995
-0.400831	0.2285779

PI

0.7402867
0.1674472
0.7704057
0.1745023
0.6624811
0.1917744
0.7061615
0.2047033
0.516981
0.2648871
0.5697446
0.2923278
0.3988695
0.2589096
0.4667924
0.3034204
0.1320359
0.3958019
0.1651907
0.4958784

```
                    ML Estimates

            BETA      STDERR
        0.9533597   0.1286179
        0.4069338   0.1284592
        -0.279081   0.1156222
        -0.280699   0.1252816
        1.4423195   0.2669357
        0.4993123   0.2943437
        0.8411595   0.2363089
        0.1485875   0.2635159
        0.1883383   0.2202755
        0.0667313    0.236031
        -0.527163    0.216581
        -0.414965   0.2299618


                      PI
                  0.7431759
                  0.1673155
                  0.7723266
                  0.1744421
                  0.6627266
                  0.1916645
                  0.7062766
                  0.2049216
                  0.5170782
                  0.2646857
                  0.5697771
                   0.292607
                  0.3984205
                  0.2576653
                  0.4666825
                  0.3027898
                  0.1323243
                  0.3963114
                   0.165475
                  0.4972044



                    IT              CRIT
      Iterations     3 Criterion 0.0004092
```

Example 6: Regression of Subsets of Variables

The following example performs regression with variable selection similar to some of the features in the REG procedure.

```
/*        Initialization                              */
/* C,CSAVE the crossproducts matrix                   */
/* N       number of observations                     */
/* K       total number of variables to consider      */
/* L       number of variables currently in model     */
/* IN      a 0-1 vector of whether variable is in     */
/* B       print collects results (L MSE RSQ BETAS )  */

start initial;
  n=nrow(x);
  k=ncol(x);
  k1=k+1;
  ik=1:k;
  bnames={nparm mse rsquare} ||varnames;

     /* Correct by mean, adjust out intercept parameter */
  y=y-y[+,]/n;                          /* correct y by mean */
  x=x-repeat(x[+,]/n,n,1);              /* correct x by mean */
  xpy=x`*y;                                 /* crossproducts */
  ypy=y`*y;
  xpx=x`*x;
  free x y;                             /* no longer need the data */

     /* Save a copy of crossproducts matrix */
  csave=(xpx || xpy) // (xpy`|| ypy);
finish initial;


     /* Forward method */
start forward;
  print / "FORWARD SELECTION METHOD";
  free bprint;
  c=csave;
  in=repeat(0,k,1);
  l=0;                                  /* No variables are in */
  dfe=n-1;
  mse=ypy/dfe;
  sprob=0;
  do while(sprob<.15 & l<k);
     indx=loc(^in); /* where are the variables not in?*/
     cd=vecdiag(c)[indx,];      /* xpx diagonals                */
     cb=c[indx,k1];             /* adjusted xpy                 */
     tsqr=cb#cb/(cd#mse);       /* squares of t tests           */
     imax=tsqr[<:>,];           /* location of maximum in indx  */
     sprob=(1-probt(sqrt(tsqr[imax,]),dfe))*2;
     if sprob<.15 then
```

```
      do;                          /* if t-test significant      */
         ii=indx[,imax];           /* pick most significant      */
         run swp;                  /* routine to sweep           */
         run bpr;                  /* routine to collect results */
      end;
   end;
   print  bprint[colname=bnames] ;
finish forward;

   /* Backward method */
start backward;
   print / "BACKWARD ELIMINATION ";
   free bprint;
   c=csave;
   in=repeat(0,k,1);
   ii=1:k;
   run swp;
   run bpr;                        /* start with all variables in */
   sprob=1;
   do while(sprob>.15 & l>0);
      indx=loc(in);                /* where are the variables in? */
      cd=vecdiag(c)[indx,];        /* xpx diagonals               */
      cb=c[indx,k1];               /* bvalues                     */
      tsqr=cb#cb/(cd#mse);         /* squares of t tests          */
      imin=tsqr[>:<,];             /* location of minimum in indx */
      sprob=(1-probt(sqrt(tsqr[imin,]),dfe))*2;
      if sprob>.15 then
         do;                       /* if t-test nonsignificant    */
            ii=indx[,imin];        /* pick least significant       */
            run swp;               /* routine to sweep in variable */
            run bpr;               /* routine to collect results   */
         end;
   end;
   print  bprint[colname=bnames] ;
finish backward;

   /* Stepwise method */
start stepwise;
   print /"STEPWISE METHOD";
   free bprint;
   c=csave;
   in=repeat(0,k,1);
   l=0;
   dfe=n-1;
   mse=ypy/dfe;
   sprob=0;
```

```
      do while(sprob<.15 & l<k);
        indx=loc(^in);/* where are the variables not in? */
        nindx=loc(in);              /* where are the variables in?   */
        cd=vecdiag(c)[indx,];       /* xpx diagonals                 */
        cb=c[indx,k1];              /* adjusted xpy                  */
        tsqr=cb#cb/cd/mse;          /* squares of t tests            */
        imax=tsqr[<:>,];            /* location of maximum in indx   */
        sprob=(1-probt(sqrt(tsqr[imax,]),dfe))*2;
        if sprob<.15 then
        do;                         /* if t-test significant         */
           ii=indx[,imax];          /* find index into c             */
           run swp;                 /* routine to sweep              */
           run backstep;            /* check if remove any terms     */
           run bpr;                 /* routine to collect results    */
           end;
        end;
      print bprint[colname=bnames] ;
    finish stepwise;

      /* Routine to backwards-eliminate for stepwise  */
    start backstep;
      if nrow(nindx)=0 then return;
      bprob=1;
      do while(bprob>.15 & l<k);
         cd=vecdiag(c)[nindx,];             /* xpx diagonals         */
         cb=c[nindx,k1];                    /* bvalues               */
         tsqr=cb#cb/(cd#mse);               /* squares of t tests    */
         imin=tsqr[>:<,];            /* location of minimum in nindx */
         bprob=(1-probt(sqrt(tsqr[imin,]),dfe))*2;
         if bprob>.15 then
         do;
            ii=nindx[,imin];
            run swp;
            run bpr;
            end;
         end;
    finish backstep;

      /* Search all possible models */
    start all;

      /* Use method of schatzoff et al. for search technique */
      betak=repeat(0,k,k);   /* record estimates for best 1-param model*/
      msek=repeat(1e50,k,1);/* record best mse per # parms       */
      rsqk=repeat(0,k,1);    /* record best rsquare               */
      ink=repeat(0,k,k);     /* record best set per # parms       */
      limit=2##k-1;          /* number of models to examine       */
      c=csave;
      in=repeat(0,k,1);      /* start out with no variables in model   */
```

```
        do kk=1 to limit;
           run ztrail;                        /* find which one to sweep   */
           run swp;                           /* sweep it in               */
           bb=bb//(l||mse||rsq||(c[ik,k1]#in)`);
           if mse<msek[1,] then
           do;                                /* was this best for 1 parms? */
                msek[1,]=mse;                 /* record mse                */
                rsqk[1,]=rsq;                 /* record rsquare            */
                ink[,1]=in;                   /* record which parms in model */
                betak[1,]=(c[ik,k1]#in)`;     /* record estimates          */
           end;
        end;
        print / "ALL POSSIBLE MODELS" " IN SEARCH ORDER";
        print bb[colname=bnames]; free bb;
        bprint=ik`||msek||rsqk||betak;
        print ,"THE BEST MODEL FOR EACH NUMBER OF PARAMETERS";
        print bprint[colname=bnames];
           /* Mallows CP plot */
        cp=msek#(n-ik`-1)/min(msek)-(n-2#ik`);
        cp=ik`||cp;
        cpname={"nparm" "cp"};
           /* output cp out=cp colname=cpname; */
     finish all;

        /*  Subroutine to find number of trailing zeros in binary number */
        /*  on entry: kk is the number to examine                        */
        /*  on exit:  ii has the result                                  */
     start ztrail;
        ii=1;
        zz=kk;
        do while(mod(zz,2)=0);
           ii=ii+1;
           zz=zz/2;
        end;
     finish ztrail;

        /* Subroutine to sweep in a pivot                                */
        /* on entry: ii has the position(s) to pivot                     */
        /* on exit:  in, l, dfe, mse, rsq recalculated                   */
     start swp;
        if abs(c[ii,ii])<1e-9 then
        do;
           print , "FAILURE", c;
           stop;
        end;
        c=sweep(c,ii);
        in[ii,]=^in[ii,];
        l=sum(in);
        dfe=n-1-l;
        sse=c[k1,k1];
        mse=sse/dfe;
        rsq=1-sse/ypy;
     finish swp;
```

```
/* Subroutine to collect bprint results                      */
/* on entry: l,mse,rsq, and c set up to collect              */
/* on exit:  bprint has another row                          */

start bpr;
   bprint=bprint//(l||mse||rsq||(c [ik,k1]#in)`);
finish bpr;

   /*              Stepwise Methods                    */
   /* After a run to the initial routine, which sets up */
   /* the data, four different routines can be called   */
   /* to do four different model-selection methods.     */

start seq;
   run initial;        /* initialization             */
   run all;            /* all possible models        */
   run forward;        /* foreward selection method  */
   run backward;       /* backward elimination method*/
   run stepwise;       /* stepwise method            */
finish seq;


   /*                      Data on physical fitness              */
   /* these measurements were made on men involved in a physical fitness */
   /* course at n.c.state univ.  the variables are age(years), weight(kg),*/
   /* oxygen uptake rate(ml per kg body weight per minute), time to run   */
   /* 1.5 miles(minutes), heart rate while resting, heart rate while      */
   /* running (same time oxygen rate measured), and maximum heart rate    */
   /* recorded while running. certain values of maxpulse were modified    */
   /* for consistency.          data courtesy dr. a.c. linnerud           */

data=
  { 44 89.47  44.609 11.37 62 178 182 ,
    40 75.07  45.313 10.07 62 185 185 ,
    44 85.84  54.297  8.65 45 156 168 ,
    42 68.15  59.571  8.17 40 166 172 ,
    38 89.02  49.874  9.22 55 178 180 ,
    47 77.45  44.811 11.63 58 176 176 ,
    40 75.98  45.681 11.95 70 176 180 ,
    43 81.19  49.091 10.85 64 162 170 ,
    44 81.42  39.442 13.08 63 174 176 ,
    38 81.87  60.055  8.63 48 170 186 ,
    44 73.03  50.541 10.13 45 168 168 ,
    45 87.66  37.388 14.03 56 186 192 ,
    45 66.45  44.754 11.12 51 176 176 ,
    47 79.15  47.273 10.60 47 162 164 ,
    54 83.12  51.855 10.33 50 166 170 ,
    49 81.42  49.156  8.95 44 180 185 ,
    51 69.63  40.836 10.95 57 168 172 ,
    51 77.91  46.672 10.00 48 162 168 ,
    48 91.63  46.774 10.25 48 162 164 ,
    49 73.37  50.388 10.08 67 168 168 ,
```

```
57 73.37  39.407  12.63  58  174  176  ,
54 79.38  46.080  11.17  62  156  165  ,
52 76.32  45.441   9.63  48  164  166  ,
50 70.87  54.625   8.92  48  146  155  ,
51 67.25  45.118  11.08  48  172  172  ,
54 91.63  39.203  12.88  44  168  172  ,
51 73.71  45.790  10.47  59  186  188  ,
57 59.08  50.545   9.93  49  148  155  ,
49 76.32  48.673   9.40  56  186  188  ,
48 61.24  47.920  11.50  52  170  176  ,
52 82.78  47.467  10.50  53  170  172
};

x=data[,{1 2 4 5 6 7 }];
y=data[,3];
free data;
varnames={age weight runtime rstpulse runpulse maxpulse};
reset fw=8 linesize=90;
run seq;
```

The results are shown below:

ALL POSSIBLE MODELS IN SEARCH ORDER

BB	NPARM	MSE	RSQUARE	AGE	WEIGHT	RUNTIME	RSTPULSE	RUNPULSE	MAXPULSE
	1	26.63425	0.092777	-0.31136	0	0	0	0	0
	2	25.82619	0.150635	-0.37042	-0.15823	0	0	0	0
	1	28.58034	0.026488	0	-0.1041	0	0	0	0
	2	7.755636	0.744935	0	-0.02548	-3.2886	0	0	0
	3	7.226318	0.770831	-0.17388	-0.05444	-3.14039	0	0	0
	2	7.168422	0.764247	-0.15037	0	-3.20395	0	0	0
	1	7.533843	0.74338	0	0	-3.31056	0	0	0
	2	7.798261	0.743533	0	0	-3.28661	-0.00968	0	0
	3	7.336089	0.767349	-0.16755	0	-3.07925	-0.04549	0	0
	4	7.366649	0.775033	-0.19603	-0.05915	-2.9889	-0.05326	0	0
	3	8.037314	0.745111	0	-0.02569	-3.26268	-0.01041	0	0
	2	24.91487	0.180607	0	-0.09305	0	-0.27474	0	0
	3	20.28031	0.356847	-0.44698	-0.15647	0	-0.32186	0	0
	2	21.27632	0.30027	-0.38882	0	0	-0.32286	0	0
	1	24.67582	0.159485	0	0	0	-0.27921	0	0
	2	23.26003	0.235031	0	0	0	-0.20684	-0.15262	0
	3	16.81799	0.466648	-0.52338	0	0	-0.22524	-0.23769	0
	4	16.26146	0.503398	-0.56317	-0.12697	0	-0.22981	-0.2246	0
	3	23.81815	0.244651	0	-0.06381	0	-0.20843	-0.14279	0
	4	7.785151	0.762252	0	-0.01231	-3.16759	0.016669	-0.0749	0
	5	6.213174	0.817556	-0.28528	-0.05184	-2.70392	-0.02711	-0.12628	0
	4	6.166944	0.81167	-0.26213	0	-2.77733	-0.01981	-0.12874	0
	3	7.507972	0.761898	0	0	-3.17665	0.017616	-0.07658	0
	2	7.254263	0.761424	0	0	-3.14019	0	-0.07351	0
	3	5.956692	0.811094	-0.2564	0	-2.82538	0	-0.13091	0
	4	6.009033	0.816493	-0.27642	-0.04932	-2.77237	0	-0.12932	0
	3	7.510162	0.761829	0	-0.01315	-3.13261	0	-0.07189	0
	2	25.333	0.166855	0	-0.05987	0	0	-0.19797	0

NPARM	MSE	RSQUARE	AGE	WEIGHT	RUNTIME	RSTPULSE	RUNPULSE	MAXPULSE
3	18.63184	0.409126	-0.54408	-0.12049	0	0	-0.28248	0
2	18.97378	0.375995	-0.50665	0	0	0	0.239382	0
1	24.70817	0.158383	0	0	0	0	-0.2068	0
2	21.60626	0.289419	0	0	0	0	-0.6818	0.571538
3	18.21725	0.422273	-0.4214	0	0	0	-0.57966	0.361557
4	17.29877	0.47172	-0.45243	-0.14944	0	0	-0.61723	0.426862
3	21.41763	0.320779	0	-0.11815	0	0	-0.71745	0.635395
4	6.030105	0.815849	0	-0.05159	-2.9255	0	-0.39529	0.38537
5	5.176338	0.848002	-0.21962	-0.0723	-2.68252	0	-0.3734	0.304908
4	5.343462	0.836818	-0.19773	0	-2.76758	0	-0.34811	0.270513
3	5.991568	0.809988	0	0	-2.97019	0	-0.37511	0.354219
4	6.208523	0.8104	0	0	-3.00426	0.016412	-0.37778	0.353998
5	5.549941	0.837031	-0.20154	0	-2.7386	-0.01208	-0.34562	0.269064
6	5.368247	0.848672	-0.22697	-0.07418	-2.62865	-0.02153	-0.36963	0.303217
5	6.263348	0.816083	0	-0.05091	-2.95182	0.01239	-0.39704	0.384793
4	20.11235	0.385797	0	-0.1194	0	-0.19092	-0.64584	0.609632
5	15.1864	0.554066	-0.47923	-0.1527	0	-0.21555	-0.53045	0.385424
4	16.29247	0.502451	-0.44717	0	0	-0.21266	-0.49323	0.319267
3	20.37729	0.353772	0	0	0	-0.18993	-0.61019	0.545236
2	25.11456	0.174039	0	0	0	-0.25219	0	-0.07364
3	19.2347	0.390007	-0.52736	0	0	-0.26492	0	-0.20024
4	18.80875	0.425607	-0.55881	-0.12604	0	-0.27056	0	-0.17799
3	25.59719	0.188232	0	-0.07874	0	-0.25524	0	-0.05502
4	8.311496	0.746179	0	-0.02053	-3.25232	-0.00393	0	-0.02064
5	7.19584	0.788701	-0.25795	-0.04936	-2.86147	-0.04121	0	-0.08153
4	7.091611	0.783432	-0.23928	0	-2.92597	-0.0339	0	-0.08777
3	8.033673	0.745227	0	0	-3.26805	-0.00193	0	-0.02526
2	7.746932	0.745221	0	0	-3.27232	0	0	-0.02561
3	6.882626	0.78173	-0.22923	0	-3.01222	0	0	-0.09094
4	7.00018	0.786224	-0.24436	-0.04525	-2.97011	0	0	-0.08585
3	8.00441	0.746155	0	-0.02027	-3.26114	0	0	-0.02139
2	28.35356	0.067516	0	-0.07074	0	0	0	-0.12159
3	22.38148	0.290212	-0.54076	-0.11605	0	0	0	-0.24445
2	22.50135	0.259982	-0.5121	0	0	0	0	-0.2637
1	27.71259	0.056046	0	0	0	0	0	-0.13762

THE BEST MODEL FOR EACH NUMBER OF PARAMETERS

BPRINT NPARM	MSE	RSQUARE	AGE	WEIGHT	RUNTIME	RSTPULSE	RUNPULSE	MAXPULSE
1	7.533843	0.74338	0	0	-3.31056	0	0	0
2	7.168422	0.764247	-0.15037	0	-3.20395	0	0	0
3	5.956692	0.811094	-0.2564	0	-2.82538	0	-0.13091	0
4	5.343462	0.836818	-0.19773	0	-2.76758	0	-0.34811	0.270513
5	5.176338	0.848002	-0.21962	-0.0723	-2.68252	0	-0.3734	0.304908
6	5.368247	0.848672	-0.22697	-0.07418	-2.62865	-0.02153	-0.36963	0.303217

FORWARD SELECTION METHOD

BPRINT	NPARM	MSE	RSQUARE	AGE	WEIGHT	RUNTIME	RSTPULSE	RUNPULSE	MAXPULSE
	1	7.533843	0.74338	0	0	-3.31056	0	0	0
	2	7.168422	0.764247	-0.15037	0	-3.20395	0	0	0
	3	5.956692	0.811094	-0.2564	0	-2.82538	0	-0.13091	0
	4	5.343462	0.836818	-0.19773	0	-2.76758	0	-0.34811	0.270513

BACKWARD ELIMINATION

BPRINT	NPARM	MSE	RSQUARE	AGE	WEIGHT	RUNTIME	RSTPULSE	RUNPULSE	MAXPULSE
	6	5.368247	0.848672	-0.22697	-0.07418	-2.62865	-0.02153	-0.36963	0.303217
	5	5.176338	0.848002	-0.21962	-0.0723	-2.68252	0	-0.3734	0.304908
	4	5.343462	0.836818	-0.19773	0	-2.76758	0	-0.34811	0.270513

STEPWISE METHOD

BPRINT	NPARM	MSE	RSQUARE	AGE	WEIGHT	RUNTIME	RSTPULSE	RUNPULSE	MAXPULSE
	1	7.533843	0.74338	0	0	-3.31056	0	0	0
	2	7.168422	0.764247	-0.15037	0	-3.20395	0	0	0
	3	5.956692	0.811094	-0.2564	0	-2.82538	0	-0.13091	0
	4	5.343462	0.836818	-0.19773	0	-2.76758	0	-0.34811	0.270513

Example 7: Response Surface Methodology

A regression model with a complete quadratic set of regressions across several factors can be processed to yield the estimated critical values that may optimize a response. First, the regression is done for two variables according to the model

$$y = c + b_1 x_1 + b_2 x_2 + a_{11} x_1^2 + a_{12} x_1 x_2 + a_{22} x_2^2 + e \quad .$$

The estimates are then divided into a vector of linear coefficients (estimates) **b** and a matrix of quadratic coefficients **A**. The solution for critical values is

$$\mathbf{x} = -.5 \mathbf{A}^{-1} \mathbf{b} \quad .$$

The following program creates a module to perform quadratic response surface regression.

```
/*          Quadratic Response Surface Regression          */
/* This matrix routine reads in the factor variables and the    */
/* response, forms the quadratic regression model and estimates */
/* the parameters, then solves for the optimal response, prints */
/* the optimal factors and response, and then displays the      */
/* eigenvalues and eigenvectors of the matrix of quadratic      */
/* parameter estimates to determine if the solution is a maximum */
/* or minimum, or saddlepoint, and which direction has the      */
/* steepest and gentlest slopes.                                */
/*                                                              */
/* Given that d contains the factor variables,                 */
/* and y contains the response.                                */
/*                                                              */
start rsm;
   n=nrow(d);
   k=ncol(d);                                /* dimensions */
   x=j(n,1,1)||d;                            /* set up design matrix */
   do i=1 to k;
      do j=1 to i;
         x=x||d[,i] #d[,j];
      end;
   end;
   beta=solve(x`*x,x`*y);                    /* solve parameter estimates */
   print "Parameter Estimates" , beta;
   c=beta[1];                                /* intercept estimate */
   b=beta[2:(k+1)];                          /* linear estimates */
   a=j(k,k,0);
   L=k+1;                                    /* form quadratics into matrix */
   do i=1 to k;
      do j=1 to i;
         L=L+1;
         a[i,j]=beta [L,];
      end;
   end;
   a=(a+a`)*.5;                              /* symmetrize */
   xx=-.5*solve(a,b);                        /* solve for critical value */
   print , "Critical Factor Values" , xx;
      /* Compute response at critical value */
   yopt=c + b`*xx + xx`*a*xx;
   print , "Response at Critical Value" yopt;
   call eigen(eval,evec,a);
   print , "Eigenvalues and Eigenvectors", eval, evec;
   if min(eval)>0 then print , "Solution Was a Minimum";
   if max(eval)<0 then print , "Solution Was a Maximum";
finish rsm;
```

Running the module with the following sample data produces the results shown below:

```
/* Sample Problem with Two Factors */
D={-1 -1, -1 0, -1 1,
    0 -1,  0 0,  0 1,
    1 -1,  1 0,  1 1};
Y={ 71.7, 75.2, 76.3, 79.2, 81.5, 80.2, 80.1, 79.1, 75.8};

run rsm;
```

Parameter Estimates

BETA
81.222222
1.9666667
0.2166667
-3.933333
-2.225
-1.383333

Critical Factor Values

XX
0.2949376
-0.158881

YOPT
Response at Critical Value 81.495032

Eigenvalues and Eigenvectors

EVAL
-0.96621
-4.350457

EVEC
-0.351076 0.9363469
0.9363469 0.3510761

Solution Was a Maximum

Example 8: Logistic and Probit Regression for Binary Response Models

A binary response Y is fit to a linear model according to

$$\text{Prob}(Y=1) = F(\mathbf{X}\beta)$$

$$\text{Prob}(Y=0) = 1 - F(\mathbf{X}\beta)$$

where F is some smooth probability distribution function. The normal and logistic distribution functions are supported. The method is maximum likelihood via iteratively reweighted least squares (described by Charnes, Frome, and Yu 1976; Jennrich and Moore 1975; and Nelder and Wedderburn 1972). The row scaling is done by the derivative of the distribution (density). The weighting is done by $w/p(1-p)$, where w has the counts or other weights. The code below calculates logistic and probit regression for binary response models.

```
/* routine for estimating binary response models      */
/*  y is the binary response, x are regressors,        */
/*  wgt are count weights,                             */
/*  model is choice of logit probit,                   */
/*  parm has the names of the parameters               */

proc iml ;

start binest;
   b=repeat(0,ncol(x),1);
   oldb=b+1;                                /* starting values */
   do iter=1 to 20 while(max(abs(b-oldb))>1e-8);
      oldb=b;
      z=x*b;
      run f;
      loglik=sum(((y=1)#log(p) + (y=0)#log(1-p))#wgt);
      btransp=b`;
      print iter loglik btransp;
      w=wgt/(p#(1-p));
      xx=f#x;
      xpxi=inv(xx`*(w#xx));
      b=b + xpxi*(xx`*(w#(y-p)));
   end;
   p0=sum((y=1)#wgt)/sum(wgt);             /* average response */
   loglik0=sum(((y=1)#log(p0) + (y=0)#log(1-p0))#wgt);
   chisq=(2#(loglik-loglik0));
   df=ncol(x)-1;
   prob=1-probchi(chisq,df);
   print ,
      'Likelihood Ratio with Intercept-only Model' chisq df prob,;
   stderr=sqrt(vecdiag(xpxi));
   tratio=b/stderr;
   print parm b stderr tratio,,;
finish;
```

```
            /*---routine to yield distribution function and density---*/
        start f;
           if model='LOGIT' then
           do;
              p=1/(1+exp(-z));
              f=p#p#exp(-z);
           end;
           if model='PROBIT' then
           do;
              p=probnorm(z);
              f=exp(-z#z/2)/sqrt(8*atan(1));
           end;
        finish;

            /* Ingot Data From COX (1970, pp. 67-68)*/
        data={ 7 1.0 0 10, 14 1.0 0 31, 27 1.0 1 56, 51 1.0 3 13,
               7 1.7 0 17, 14 1.7 0 43, 27 1.7 4 44, 51 1.7 0 1,
               7 2.2 0 7, 14 2.2 2 33, 27 2.2 0 21, 51 2.2 0 1,
               7 2.8 0 12, 14 2.8 0 31, 27 2.8 1 22,
               7 4.0 0 9, 14 4.0 0 19, 27 4.0 1 16, 51 4.0 0 1};
        nready=data[,3];
        ntotal=data[,4];
        n=nrow(data);
        x=repeat(1,n,1)||(data[,{1 2}]);          /* intercept, heat, soak */
        x=x//x;                                              /* regressors */
        y=repeat(1,n,1)//repeat(0,n,1);                 /* binary response */
        wgt=nready//(ntotal-nready);                        /* row weights */
        parm={intercept, heat, soak};             /* names of regressors */

        model={logit};
        run binest;                                      /* run logit model */

        model={probit};
        run binest;                                     /* run probit model */
```

The results are shown below:

```
                ITER    LOGLIK   BTRANSP
                 1   -268.248        0        0        0


                ITER    LOGLIK   BTRANSP
                 2  -76.29481  -2.159406 0.0138784 0.0037327


                ITER    LOGLIK   BTRANSP
                 3  -53.38033   -3.53344 0.0363154 0.0119734


                ITER    LOGLIK   BTRANSP
                 4  -48.34609  -4.748899 0.0640013 0.0299201
```

```
ITER    LOGLIK    BTRANSP
  5 -47.69191 -5.413817 0.0790272   0.04982


ITER    LOGLIK    BTRANSP
  6 -47.67283 -5.553931 0.0819276 0.0564395


ITER    LOGLIK    BTRANSP
  7 -47.67281  -5.55916 0.0820307 0.0567708


ITER    LOGLIK    BTRANSP
  8 -47.67281 -5.559166 0.0820308 0.0567713
```

```
                                        CHISQ    DF     PROB
Likelihood Ratio with Intercept-only Model 11.64282     2  0.0029634
```

```
PARM            B    STDERR    TRATIO
INTERCEPT -5.559166 1.1196947 -4.964895
HEAT       0.0820308 0.0237345 3.4561866
SOAK       0.0567713 0.3312131 0.1714042
```

```
ITER    LOGLIK    BTRANSP
  1  -268.248         0         0         0


ITER    LOGLIK    BTRANSP
  2 -71.71043 -1.353207  0.008697 0.0023391


ITER    LOGLIK    BTRANSP
  3 -51.64122 -2.053504 0.0202739 0.0073888


ITER    LOGLIK    BTRANSP
  4 -47.88947 -2.581302  0.032626  0.018503


ITER    LOGLIK    BTRANSP
  5 -47.48924 -2.838938 0.0387625 0.0309099


ITER    LOGLIK    BTRANSP
  6 -47.47997 -2.890129 0.0398894 0.0356507
```

```
              ITER    LOGLIK   BTRANSP
               7  -47.47995  -2.89327 0.0399529 0.0362166


              ITER    LOGLIK   BTRANSP
               8  47.47995 -2.893408 0.0399553 0.0362518


              ITER    LOGLIK   BTRANSP
               9  47.47995 -2.893415 0.0399554 0.0362537


              ITER    LOGLIK   BTRANSP
              10 -47.47995 -2.893415 0.0399555 0.0362538


              ITER    LOGLIK   BTRANSP
              11 -47.47995 -2.893415 0.0399555 0.0362538
```

```
                                            CHISQ    DF      PROB
   Likelihood Ratio with Intercept-only Model 12.028543   2  0.0024436
```

```
              PARM             B    STDERR    TRATIO
              INTERCEPT -2.893415 0.5006009 -5.779884
              HEAT       0.0399555 0.0118466 3.3727357
              SOAK       0.0362538 0.1467431 0.2470561
```

Example 9:   Linear Programming

The two-phase method for linear programming can be used to solve the problem

max **c`x**
st. **Ax** <=,=,>= **b**
x>=0

A routine written in IML to solve this problem follows. The approach appends slack, surplus, and artificial variables to the model where needed. It then solves phase 1 to find a primal feasible solution. If a primal feasible solution exists and is found, the routine then goes on to phase 2 to find an optimal solution if one exists. The routine is general enough to handle minimizations as well as maximizations.

```
/*  Subroutine to solve Linear Programs                    */
/*  names:    names of the decision variables              */
/*  obj:      coefficients of the objective function       */
/*  maxormin: the value 'MAX' or 'MIN', upper or lowercase  */
/*  coef:     coefficients of the constraints              */
/*  rel:      character array of values: '<=' or '>=' or '='  */
/*  rhs:      right-hand side of constraints               */
/*  activity: returns the optimal value of decision variables */
/*                                                          */

start linprog( names, obj, maxormin, coef, rel, rhs, activity);

   bound=1.0e10;
   m=nrow(coef);
   n=ncol(coef);

      /* Convert to maximization */
   if upcase(maxormin)='MIN' then o=-1;
   else o=1;

      /* Build logical variables */
   rev=(rhs<0);
   adj=(-1*rev)+¬ rev;
   ge=((rel='>=')&¬ rev)|((rel='<=')&rev);
   eq=(rel='=');
   if max(ge)=1 then
   do;
      sr=I(m);
      logicals=-sr[,loc(ge)]||I(m);
      artobj=repeat(0,1,ncol(logicals)-m)|(eq+ge)`;
   end;
   else do;
      logicals=I(m);
      artobj=eq`;
   end;
   nl=ncol(logicals);
   nv=n+nl+2;

      /* Build coef matrix */
   a=((o*obj)||repeat(0,1,nl)||{ -1 0 })//
     (repeat(0,1,n)||-artobj||{ 0 -1 })//
     ((adj#coef)||logicals||repeat(0,m,2));

      /* rhs, lower bounds, and basis */
   b={0,0}//(adj#rhs);
   L=repeat(0,1,nv-2)||-bound||-bound;
   basis=nv-(0:nv-1);
```

```
        /* Phase 1 - primal feasibility */
call lp(rc,x,y,a,b,nv,,l,basis);
print ( { ' ',
        '**********Primal infeasible problem************',
        ' ',
        '**********Numerically unstable problem**********',
        '**********Singular basis encountered************',
        '*******Solution is numerically unstable********',
        '***Subroutine could not obtain enough memory***',
        '**********Number of iterations exceeded********'
          }[rc+1]);
if x[nv] ¬=0 then
do;
   print '**********Primal infeasible problem************';
   stop;
end;
if rc>0 then stop;

    /* phase 2 - dual feasibility */
u=repeat(.,1,nv-2)||{ . 0 };
L=repeat(0,1,nv-2)||-bound||0;
call lp(rc,x,y,a,b,nv-1,u,l,basis);

    /* Report the solution */
print ( { '************Solution is optimal**************',
        '*********Numerically unstable problem**********',
        '*************Unbounded problem****************',
        '*******Solution is numerically unstable********',
        '*********Singular basis encountered************',
        '*******Solution is numerically unstable********',
        '***Subroutine could not obtain enough memory***',
        '**********Number of iterations exceeded********'
          }[rc+1]);
value=o*x  [nv-1];
print ,'Objective Value ' value;
activity= x [1:n] ;
print ,'Decision Variables ' activity[r=names];
lhs=coef*x[1:]:;
dual=y[3:m+2];
print ,'Constraints ' lhs rel rhs dual,
      '*********************************************';

    finish;
```

Consider the following product mix example (Hadley 1963). A shop with three machines, A, B, and C, turns out products 1, 2, 3, and 4. Each product must be processed on each of the three machines (for example, lathes, drills, and milling machines). The following table shows the number of hours required by each product on each machine:

	Product			
Machine	1	2	3	4
A	1.5	1	2.4	1
B	1	5	1	3.5
C	1.5	3	3.5	1

The weekly time available on each of the machines is 2000, 8000, and 5000 hours, respectively. The products contribute 5.24, 7.30, 8.34, and 4.18 to profit, respectively. What mixture of products can be manufactured that maximizes profit? You can solve the problem as follows:

```
names={'product 1' 'product 2' 'product 3' 'product 4'};
profit={ 5.24 7.30 8.34 4.18};
tech={ 1.5 1 2.4 1 ,
       1 5 1 3.5 ,
       1.5 3 3.5 1 };
time={ 2000, 8000, 5000};
rel={ '<=', '<=', '<=' };
run linprog(names,profit,'max',tech,rel,time,products);
```

The output from this example follows:

```
*************Solution is optimal**************
```

```
                                   VALUE
                   Objective Value  12737.059
```

```
                                 ACTIVITY
           Decision Variables   product 1 294.11765
                                product 2     1500
                                product 3        0
                                product 4 58.823529
```

```
                         LHS REL      RHS      DUAL
           Constraints   2000 <=      2000 1.9535294
                         8000 <=      8000 0.2423529
                         5000 <=      5000 1.3782353
```

```
**************************************************
```

The following example shows how to find the minimum cost flow through a network using linear programming. The arcs are defined by an array of tuples; each tuple names a new arc. The elements in the arc tuples give the names of the tail and head nodes defining the arc. The following data are needed: arcs, cost for a unit of flow across the arcs, nodes, and supply and demand at each node.

A program generates the node-arc incidence matrix and calls the linear program routine for solution:

```
arcs={ 'ab' 'bd' 'ad' 'bc' 'ce' 'de' 'ae' };
cost={ 1 2 4 3 3 2 9 };
nodes={ 'a', 'b', 'c', 'd', 'e'};
supdem={ 2, 0, 0, -1, -1 };
rel=repeat('=',nrow(nodes),1);
inode=substr(arcs,1,1);
onode=substr(arcs,2,1);
free n_a_i n_a_o;
do i=1 to ncol(arcs);
   n_a_i=n_a_i || (inode[i]=nodes);
   n_a_o=n_a_o || (onode[i]=nodes);
end;
n_a=n_a_i - n_a_o;
run linprog(arcs,cost,'min',n_a,rel,supdem,x);
```

The solution is shown below:

```
*************Solution is optimal**************


                                     VALUE
                  Objective Value      8



                                   ACTIVITY
                 Decision Variables  ab         2
                                     bd         2
                                     ad         0
                                     bc         0
                                     ce         0
                                     de         1
                                     ae         0



                          LHS REL       RHS      DUAL
           Constraints      2 =           2      -2.5
                            0 =           0      -1.5
                            0 =           0      -0.5
                           -1 =          -1      -0.5
                           -1 =          -1      -2.5


    *************************************************
```

Example 10: Quadratic Programming

The quadratic program

$$\min \; c'x + x'Hx/2$$
$$\text{st. } Gx <=, =, >= b$$
$$x >= 0$$

can be solved by solving an equivalent linear complementarity problem when H is positive semidefinite. The approach is outlined in the discussion of the LCP subroutine in Chapter 15, "SAS/IML Language Reference." The following routine solves the quadratic problem.

```
/*              Routine to solve quadratic programs        */
/* names: the names of the decision variables             */
/* c:  vector of linear coefficients of the objective function */
/* H:  matrix of quadratic terms in the objective function */
/* G:  matrix of constraint coefficients                  */
/* rel: character array of values: '<=' or '>=' or '='    */
/* b:   right hand side of constraints                    */
/* activity: returns the optimal value of decision variables */

start qp( names, c, H, G, rel, b, activity);
  if min(eigval(h))<0 then
  do;
    print
       'ERROR: The minimum eigenvalue of the H matrix is negative. ';
    print '    Thus it is not positive semidefinite.          ';
    print '    QP is terminating with this error.             ';
    stop;
  end;
  nr=nrow(G);
  nc=ncol(G);

    /* Put in canonical form */
  rev=(rel='<=');
  adj=(-1 * rev) + ^rev;
    g=adj# G; b = adj # b;
    eq=( rel = '='   );
    if max(eq)=1 then
    do;
       g=g // -(diag(eq)*G)[loc(eq),];
       b=b // -(diag(eq)*b)[loc(eq)];
    end;
    m=(h || -g`) //(g || j(nrow(g),nrow(g),0));
    q=c // -b;

    /* Solve the problem */
    call lcp(rc,w,z,M,q);

    /* Report the solution */
    reset noname;
    print ( { '*************Solution is optimal***************',
            '*********No solution possible******************',
            ' ',
            ' ',
            ' ',
            '**********Solution is numerically unstable*****',
            '***********Not enough memory******************',
            '**********Number of iterations exceeded*******'}[rc+1]);
    reset name;
    activity=z[1:nc];
    objval=c`*activity + activity`*H*activity/2;
    print ,'Objective Value ' objval,
          'Decision Variables ' activity[r=names],
          '********************************************';

    finish qp;
```

As an example, consider the following problem in portfolio selection. Models used in selecting investment portfolios include assessment of the proposed portfolio's expected gain and its associated risk. One such model seeks to minimize the variance of the portfolio subject to a minimum expected gain. This can be modeled as a quadratic program in which the decision variables are the proportions to invest in each of the possible securities. The quadratic component of the objective function is the covariance of gain between the securities; the first constraint is a proportionality constraint; and the second constraint gives the minimum acceptable expected gain.

The following data are used to illustrate the model and its solution:

```
c = { 0, 0, 0, 0 };
h = { 1003.1 4.3 6.3 5.9 ,
         4.3 2.2 2.1 3.9 ,
         6.3 2.1 3.5 4.8 ,
         5.9 3.9 4.8 10 };
g = { 1    1    1    1   ,
      .17 .11 .10 .18 };
b = { 1 , .10 };
rel = { '=', '>='};
names = {'ibm', 'dec', 'dg', 'prime' };
run qp(names,c,h,g,rel,b,activity);
```

The following output shows that the minimum variance portfolio achieving the .10 expected gain is composed of DEC and DG stock in proportions of .933 and .067.

```
*************Solution is optimal***************


                            OBJVAL
              Objective Value  1.0966667


                            ACTIVITY
         Decision Variables  ibm             0
                             dec     0.9333333
                             dg      0.0666667
                             prime           0



***************************************************
```

Example 11:   Regression Quantiles

The technique of parameter estimation in linear models using the notion of regression quantiles is a generalization of the LAE (LAV) least absolute value estimation technique. For a given quantile q, the estimate $\mathbf{b}^{\bullet}$ of $\boldsymbol{\beta}$ in the model

$$\mathbf{Y} = \mathbf{X}\boldsymbol{\beta} + \boldsymbol{\varepsilon}$$

is the value of b that minimizes

$$\Sigma_{t\varepsilon T}\, q\, |y_t - x_t b| - \Sigma_{t\varepsilon S}(1 - q)\, |y_t - x_t\, b|$$

where $T = \{t \mid y_t \geq x_t b\}$ and $S = \{t \mid y_t \leq x_t\}$. For $q = 0.5$ the solution $\mathbf{b}^{\bullet}$ is identical to the estimates produced by the LAE. The following routine finds this estimate using linear programming:

```
/* Routine to find Regression Quantiles              */
/* yname:   name of dependent variable               */
/* y:       dependent variable                       */
/* xname:   names of independent variables           */
/* X:       independent variables                    */
/* b:       estimates                                */
/* predict: predicted values                         */
/* error:   difference of y and predicted            */
/* q:       quantile                                 */
/*                                                   */
/* notes: This subroutine finds the estimates b      */
/* that minimize                                     */
/*                                                   */
/* q * (y - Xb) * e - (1-q) * (Xb - y) * ¬e          */
/*                                                   */
/* where e = ( Xb <= y ). When q = .5 this is equivalent */
/* to minimizing the sum of the absolute deviations. */
/*                                                   */
/* This subroutine follows the approach given in:    */
/*                                                   */
/* Koenker, R. and G. Bassett (1978)                 */
/*                                                   */
/* Bassett, G. and R. Koenker (1982).                */
/*                                                   */
start rq(yname,y,xname,x,b,predict,error,q);
  bound=1.0e10;
  coef=x`;
  m=nrow(coef);
  n=ncol(coef);

    /* Build rhs and bounds */
  r=repeat(0,m+2,1);
  L=repeat(q-1,1,n)||repeat(0,1,m)||-bound||-bound;
  u=repeat(q,1,n)||repeat(.,1,m)||{ . . };
```

```
    /* Build coefficient matrix and basis */

a=(y`||repeat(0,1,m)|||{ -1 0 })//
  (repeat(0,1,n)||repeat(-1,1,m)|||{ 0 -1 })//
  (coef||I(m)||repeat(0,m,2));
basis=n+m+2-(0:n+m+1);

    /* Find a feasible solution */
call lp(rc,p,d,a,r,,u,L,basis);

    /* Find the optimal solution */
L=repeat(q-1,1,n)||repeat(0,1,m)||-bound||{0};
u=repeat(q,1,n)||repeat(0,1,m)|||{ . 0 };
call lp(rc,p,d,a,r,n+m+1,u,L,basis);

    /* Report the solution */
variable=xname`;
b=d[3:m+2];
predict=X*b;
error=y-predict;
wsum=sum(choose(error<0,(q-1)*error,q*error));
print ,,'Regression Quantile Estimation' ,
        'Dependent Variable: ' yname ,
        'Regression Quantile: ' q ,
        'Number of Observations: ' n ,
        'Sum of Weighted Absolute Errors: ' wsum ,
        variable b,
        X y predict error;
finish rq;
```

The following example uses data on the United States population from 1790 to 1970:

```
z = { 3.929 1790 ,
       5.308 1800 ,
       7.239 1810 ,
       9.638 1820 ,
      12.866 1830 ,
      17.069 1840 ,
      23.191 1850 ,
      31.443 1860 ,
      39.818 1870 ,
      50.155 1880 ,
      62.947 1890 ,
      75.994 1900 ,
      91.972 1910 ,
     105.710 1920 ,
     122.775 1930 ,
     131.669 1940 ,
     151.325 1950 ,
     179.323 1960 ,
     203.211 1970 };
```

```
y=z[,1];
x=repeat(1,19,1)||z[,2]||z[,2]##2;
run rq('pop',y,{'intercpt' 'year' 'yearsq'},x,b1,pred,resid,.5);
```

The output is shown below:

Regression Quantile Estimation

YNAME

Dependent Variable: pop

Q

Regression Quantile: 0.5

N

Number of Observations: 19

WSUM

Sum of Weighted Absolute Errors: 14.826429

VARIABLE	B
intercpt	21132.758
year	-23.52574
yearsq	0.006549

X		Y	PREDICT	ERROR
1	1790	3204100	3.929	5.4549176	-1.525918
1	1800	3240000	5.308	5.308	1.044E-12
1	1810	3276100	7.239	6.4708902	0.7681098
1	1820	3312400	9.638	8.9435882	0.6944118
1	1830	3348900	12.866	12.726094	0.1399059
1	1840	3385600	17.069	17.818408	-0.749408
1	1850	3422500	23.191	24.220529	-1.029529
1	1860	3459600	31.443	31.932459	-0.489459
1	1870	3496900	39.818	40.954196	-1.136196
1	1880	3534400	50.155	51.285741	-1.130741
1	1890	3572100	62.947	62.927094	0.0199059
1	1900	3610000	75.994	75.878255	0.1157451
1	1910	3648100	91.972	90.139224	1.8327765
1	1920	3686400	105.71	105.71	7.532E-13
1	1930	3724900	122.775	122.59058	0.1844157
1	1940	3763600	131.669	140.78098	-9.111976
1	1950	3802500	151.325	160.28118	-8.956176
1	1960	3841600	179.323	181.09118	-1.768184
1	1970	3880900	203.211	203.211	-3.5E-12

The L1 norm (when $q=0.5$) tends to allow the fit to be better at more points at the expense of allowing some points to fit worse, as the plot of the residuals against the least squares residuals:

```
/* Compare L1 residuals with least squares residuals */
/* Compute the least squares residuals             */
resid2=y-x*inv(x`*x)*x`*y;

/* x axis of plot */
xx=repeat(x[,2] ,3,1);

/* y axis of plot */
yy=resid//resid2//repeat(0,19,1);

/* plot character*/
id=repeat('1',19,1)//repeat('2',19,1)//repeat('-',19,1);
call pgraf(xx||yy,id,'Year','Residual',
         '1=L(1) residuals, 2=least squares residual');
```

The output generated is shown below:

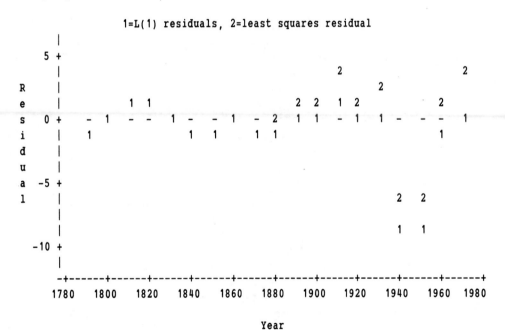

Example 12: Simulations of a Univariate ARMA Process

Simulations of time series with known ARMA structure are often needed as part of other simulations or as learning data sets for developing time series analysis skills. The following program generates a time series using the IML functions NORMAL, ARMACOV, HANKEL, PRODUCT, RATIO, TOEPLITZ, and ROOT.

```
reset noname;
start armasim(y,n,phi,theta,seed);
   /*---- ------------------------------------------------- ----*/
   /* IML Module: armasim                                       */
   /* Purpose: Simulate n data points from ARMA process         */
   /*          exact covariance method                          */
   /* Arguments:                                                 */
   /*                                                           */
   /* Input: n    : series length                               */
   /*        phi  : AR coefficients                             */
   /*        theta: MA coefficients                             */
   /*        seed : integer seed for normal deviate generator   */
   /* Output: y: realization of ARMA process                    */
   /* --- ------------------------------------------------- ----*/

   p=ncol(phi)-1;
   q=ncol(theta)-1;
   y=normal(j(1,n+q,seed));

      /* Pure MA or white noise */
   if p=0 then y=product(theta,y)[,(q+1):(n+q)];
   else do;                 /* Pure AR or ARMA */

         /* Get the autocovariance function */
      call armacov(gamma,cov,ma,phi,theta,p);
      if gamma[1]<0 then
      do;
         print 'ARMA parameters not stable.';
         print 'Execution terminating.';
         stop;
      end;

         /* Form covariance matrix */
      gamma=toeplitz(gamma);
```

```
         /* Generate covariance between initial y and */
         /* initial innovations                       */
      if q>0 then
      do;
         psi=ratio(phi,theta,q);
         psi=hankel(psi[,-((-q):(-1))]);
         m=max(1,(q-p+1));
         psi=psi[-((-q):(-m)),];
         if p>q then psi=j(p-q,q,0)//psi;
         gamma=(gamma||psi)//(psi`||i(q));
      end;

         /* Use Cholesky root to get startup values */
      gamma=root(gamma);
      startup=y[,1:(p+q)]*gamma;
      e=y[,(p+q+1):(n+q)];

         /* Generate MA part */
      if q>0 then
      do;
         e=startup[,(p+1):(p+q)]||e;
         e=product(theta,e)[,(q+1):(n+q-p)];
      end;

      y=startup[,1:p];
      phi1=phi[,-(-(p+1):(-2))]`;

         /* Use difference equation to generate */
         /* remaining values                    */
      do ii=1 to n-p;
         y=y||(e[,ii]-y[,ii:(ii+p-1)]*phi1);
      end;
   end;
   y=y`;
finish armasim; /* ARMASIM */

run armasim(y,10,{1 -0.8},{1 0.5},1234321);
print ,'Simulated Series:', y;
```

The output is shown below:

```
            Simulated Series:
               3.0764594
               1.8931735
               0.9527984
               0.0892395
              -1.811471
               -2.8063
              -2.52739
              -2.865251
              -1.332334
               0.1049046
```

Example 13:   Parameter Estimation for a Regression Model with ARMA Errors

Nonlinear estimation algorithms are required for obtaining estimates of the parameters of a regression model with innovations having an ARMA structure. The three estimation methods employed by the ARIMA procedure in SAS/ETS software are programmed in IML in the following example. The algorithms employed are slightly different from those used by PROC ARIMA, but the results obtained should be similar. This example combines the IML functions ARMALIK, PRODUCT, and RATIO to perform the estimation. Note the interactive nature of this example, illustrating how you can adjust the estimates when they venture outside of the stationary or invertible regions.

```
reset noname;
/*----------------------------------------------------------------*/
/*  name: ARMAREG Modules                                         */
/*  purpose: Perform Estimation for regression model with         */
/*  ARMA errors                                                   */
/*  usage: Before invoking the command                            */
/*                                                                */
/*  run armareg;                                                  */
/*                                                                */
/*  define the global parameters                                 */
/*                                                                */
/*  x - matrix of predictors.                                    */
/*  y - response vector.                                         */
/*  iphi - defines indices of nonzero AR parameters,             */
/*  omitting index 0 corresponding to the zero                   */
/*  order constant one.                                          */
/*  itheta - defines indices of nonzero MA parameters,           */
/*  omitting index 0 corresponding to the zero                   */
/*  order constant one.                                          */
/*  ml - estimation option: -1 if Conditional Least Squares,     */
/*  1 if Maximum Likelihood, otherwise Unconditional             */
/*  Least Squares.                                               */
/*  delta - step change in parameters (default 0.005).           */
/*  par - initial values of parms. First ncol(iphi)             */
/*  values correspond to AR parms, next ncol(itheta)            */
/*  values correspond to MA parms, and remaining are            */
/*  regression coefficients.                                    */
/*  init - undefined or zero for first call to armareg.          */
/*  maxit - maximum number of iterations. No other convergence  */
/*  criterion is used. You can invoke armareg without            */
/*  changing parameter values to continue iterations.           */
/*  nopr - undefined or zero implies no printing of             */
/*  intermediate results.                                       */
/*                                                                */
```

```
/* notes: Optimization using Gauss-Newton iterations        */
/*                                                           */
/* No checking for invertibility or stationarity during      */
/* estimation process. The parameter array par may be        */
/* modified after running armareg to place estimates         */
/* in the stationary and invertible regions, and then        */
/* armareg may be run again. If a nonstationary AR operator   */
/* is employed, a PAUSE will occur after calling ARMALIK      */
/* because of a detected singularity. Using STOP will        */
/* permit termination of ARMAREG so that the AR coefficients  */
/* may be modified.                                          */
/*                                                           */
/* T-ratios are only approximate and may be undependable,     */
/* especially for small series.                              */
/*                                                           */
/* The notation follows that of the IML function ARMALIK;     */
/* the autoregressive and moving average coefficients have    */
/* signs opposite those given by PROC ARIMA.                 */
/*---------------------------------------------------------------*/


/* Begin ARMA estimation modules */

/* Generate residuals */
start gres;
   noise=y-x*beta;
   previous=noise[:];
   if ml=-1 then do;                              /* Conditional LS */
      noise=j(nrow(y),1,previous)//noise;
      resid=product(phi,noise')[,1:nrow(noise)];
      resid=ratio(theta,resid,ncol(resid));
      resid=resid[,1:ncol(resid)]';
   end;
   else do;                                       /* Maximum likelihood */
      free l;
      call armalik(l,resid,std,noise,phi,theta);

      /* Nonstationary condition produces PAUSE */
      if nrow(l)=0 then
      do;
         print ,
         'In GRES: Parameter estimates outside stationary region.';
      end;
      else do;
         temp=l[3,]/(2#nrow(resid));
         if ml=1 then resid=resid#exp(temp);
      end;
   end;
finish gres;                                      /* finish module GRES  */
```

```
start getpar;                                        /* Get Parameters  */
   if np=0 then phi=1;
   else do;
      temp=parm[,1:np];
      phi=1||j(1,p,0);
      phi[,iphi] =temp;
   end;
   if nq=0 then theta=1;
   else do;
      temp=parm[,np+1:np+nq];
      theta=1||j(1,q,0);
      theta[,itheta] =temp;
   end;
   beta=parm[,(np+nq+1):ncol(parm)]`;
finish getpar;   /* finish module GETPAR  */

   /* Get SS Matrix - First Derivatives */
start getss;
   parm=par;
   run getpar;
   run gres;
   s=resid;
   oldsse=ssq(resid);
   do k=1 to ncol(par);
      parm=par;
      parm[,k]=parm[,k]+delta;
      run getpar;
      run gres;
      s=s||((resid-s[,1])/delta);              /* append derivatives */
   end;
   ss=s`*s;
   if nopr¬=0 then print ,'Gradient Matrix', ss;
   sssave=ss;
   do k=1 to 20;                    /* Iterate if no reduction in SSE */
      do ii=2 to ncol(ss);
         ss[ii,ii]=(1+lambda)*ss[ii,ii];
      end;
      ss=sweep(ss,2:ncol(ss));                 /* Gaussian elimination */
      delpar=ss[1,2:ncol(ss)];             /* update parm increments  */
      parm=par+delpar;
      run getpar;
      run gres;
      sse=ssq(resid);
      if sse<oldsse then
      do;                                     /* reduction, no iteration */
         lambda=lambda/10;
         k=21;
      end;
```

```
        else do;                                          /* no reduction */
                                            /* increase lambda and iterate */
            if nopr¬=0 then print ,
               'Lambda=' lambda 'SSE=' sse 'OLDSSE=' oldsse,
               'Gradient Matrix', ss ;
            lambda=10*lambda;
            ss=sssave;
            if k=20 then
            do;
               print 'In module GETSS:
                       No improvement in SSE after twenty iterations.';
               print ' Possible Ridge Problem. ';
               return;
            end;
        end;
    end;
    if nopr¬=0 then print ,'Gradient Matrix', ss;
    finish getss;                               /* Finish module GETSS  */

start armareg;                                  /* ARMAREG main module */

    /* Initialize options and parameters */
    if nrow(delta)=0 then delta=0.005;
    if nrow(maxiter)=0 then maxiter=5;
    if nrow(nopr)=0 then nopr=0;
    if nrow(ml)=0 then ml=1;
    if nrow(init)=0 then init=0;
    if init=0 then
    do;
        p=max(iphi);
        q=max(itheta);
        np=ncol(iphi);
        nq=ncol(itheta);

        /* Make indices one-based */
        do k=1 to np;
           iphi[,k]=iphi[,k]+1;
        end;
        do k=1 to nq;
           itheta[,k]=itheta[,k]+1;
        end;

        /* Create row labels for Parameter estimates */
        if p>0 then parmname = concat("AR",char(1:p,2));
        if q>0 then parmname = parmname||concat("MA",char(1:p,2));
        parmname = parmname||concat("B",char(1:ncol(x),2));

        /* Create column labels for Parameter estimates */
        pname = {"Estimate" "Std. Error" "T-Ratio"};
        init=1;
    end;
```

```
                            /* Generate starting values */
                  if nrow(par)=0 then
                  do;
                     beta=inv(x`*x)*x`*y;
                     if np+nq>0 then par=j(1,np+nq,0)||beta`;
                     else par=beta`;
                  end;
                  print ,'Parameter Starting Values',;
                  print par [colname=parmname];                   /* stderr tratio */
                  lambda=1e-6;                           /* Controls step size */
                  do iter=1 to maxiter;              /* Do maxiter iterations */
                     run getss;
                     par=par+delpar;
                     if nopr¬=0 then
                     do;
                        print ,'Parameter Update',;
                        print par [colname=parmname];            /* stderr tratio */
                        print ,'Lambda=' lambda,;
                     end;
                     if abs(par[,1] )>1 then par[,1] =-.8;
                  end;
                  sighat=sqrt(sse/(nrow(y)-ncol(par)));
                  print ,'Innovation Standard Deviation:' sighat;
                  estm=par`||(sqrt(diag(ss[2:ncol(ss),2:ncol(ss)])))
                       *j(ncol(par),1,sighat));
                  estm=estm||(estm[,1] /estm[,2]);
                  if ml=1 then print ,'Maximum Likelihood Estimation Results',;
                  else if ml=-1 then print ,
                     'Conditional Least Squares Estimation  Results',;
                  else print ,'Unconditional Least Squares Estimation Results',;
                  print estm [rowname=parmname colname=pname] ;
            finish armareg;
                  /* End of ARMA Estimation modules */

                  /* Begin estimation for Grunfeld's investment models */
                  /* See SAS/ETS User's Guide, Version 5 Edition        */

                  /* Access Grunfeld data as produced by DATA step in  */
                  /* SAS/ETS User's Guide Example                       */
            use grunfeld;
            read all var {gei} into y;
            read all var {gef gec} into x;
            x=j(nrow(x),1,1)||x;
            iphi=1;
            itheta=1;
            maxiter=10;
            delta=0.0005;
            ml=-1;
            run armareg; /* Perform CLS estimation */
```

Parameter Starting Values

AR 1	MA 1	B 1	B 2	B 3
0	0	-9.956306	0.0265512	0.1516939

Innovation Standard Deviation: 17.844166

Conditional Least Squares Estimation Results

	Estimate	Std. Error	T-Ratio
AR 1	-0.245087	0.1722625	-1.422752
MA 1	1.2646783	0.0217222	58.220574
B 1	-6.807465	0.18459	-36.87884
B 2	0.038665	0.0001244	310.91482
B 3	0.105957	0.0021439	49.421931

```
/* Transform MA coefficient to invertible region */
par[,2]=1/par[,2];
maxiter=2;
run armareg; /* Continue CLS estimation */
```

Parameter Starting Values

AR 1	MA 1	B 1	B 2	B 3
-0.245087	0.7907149	-6.807465	0.038665	0.105957

Innovation Standard Deviation: 22.60089

Conditional Least Squares Estimation Results

	Estimate	Std. Error	T-Ratio
AR 1	-0.251799	0.3380831	-0.744783
MA 1	0.7312613	0.2115922	3.455993
B 1	-11.41136	41.150426	-0.277308
B 2	0.0375525	0.0180056	2.0856074
B 3	0.1110736	0.0509154	2.1815306

```
ml=1;
maxiter=10;

  /* With CLS estimates as starting values, */
  /* perform ML estimation.                  */
run armareg;
```

Parameter Starting Values

AR 1	MA 1	B 1	B 2	B 3
-0.251799	0.7312613	-11.41136	0.0375525	0.1110736

Innovation Standard Deviation: 23.039253

Maximum Likelihood Estimation Results

	Estimate	Std. Error	T-Ratio
AR 1	-0.196224	0.3510867	-0.558904
MA 1	0.6816035	0.2712038	2.5132519
B 1	-26.47514	33.752825	-0.784383
B 2	0.0392213	0.0165545	2.3692243
B 3	0.1310306	0.0425996	3.0758634

Example 14:   Iterative Proportional Fitting

The classical use of iterative proportional fitting is to adjust frequencies to
conform to new marginal totals. Use the IPF subroutine to perform this kind of
analysis. You supply a table that contains new margins and a table that contains
old frequencies. The IPF subroutine returns a table of adjusted frequencies that
preserves any higher order interactions appearing in the initial table.

The example is a census study that estimates a population distribution
according to age and marital status (Bishop, Fienberg, and Holland 1975,
pp. 97–98). Estimates of the distribution are known for the previous year, but
only estimates of marginal totals are known for the current year. You want to
adjust the distribution of the previous year to fit the estimated marginal totals of
the current year.

```
proc iml;

  /* Stopping criteria */
mod={0.01 15};

  /* Marital status has 3 levels. age has 8 levels. */
dim={3 8};
```

```
            /* New marginal totals for age by marital status */
table={1412 0 0 ,
       1402 0 0 ,
       1174 276 0 ,
       0 1541 0 ,
       0 1681 0 ,
       0 1532 0 ,
       0 1662 0 ,
       0 5010 2634};

       /* Marginal totals are known for both */
       /* marital status and age              */
config={1 2};

       /* Use known distribution for start-up values */
initab={1306 83 0 ,
        619 765 3 ,
        263 1194 9 ,
        173 1372 28 ,
        171 1393 51 ,
        159 1372 81 ,
        208 1350 108 ,
        1116 4100 2329};

call ipf(fit,status,dim,table,config,initab,mod);

c={' SINGLE' ' MARRIED' 'WIDOWED/DIVORCED'};
r={'15 - 19' '20 - 24' '25 - 29' '30 - 34' '35 - 39' '40 - 44'
   '45 - 49' '50 OR OVER'};
print
   'POPULATION DISTRIBUTION ACCORDING TO AGE AND MARITAL STATUS',,
   'KNOWN DISTRIBUTION (PREVIOUS YEAR)',,
   initab [colname=c rowname=r format=8.0] ,,
   'ADJUSTED ESTIMATES OF DISTRIBUTION (CURRENT YEAR)',,
   fit [colname=c rowname=r format=8.2] ;
```

POPULATION DISTRIBUTION ACCORDING TO AGE AND MARITAL STATUS

KNOWN DISTRIBUTION (PREVIOUS YEAR)

	SINGLE	MARRIED	WIDOWED/DIVORCED
15 - 19	1306	83	0
20 - 24	619	765	3
25 - 29	263	1194	9
30 - 34	173	1372	28
35 - 39	171	1393	51
40 - 44	159	1372	81
45 - 49	208	1350	108
50 OR OVER	1116	4100	2329

```
        ADJUSTED ESTIMATES OF DISTRIBUTION (CURRENT YEAR)

                    SINGLE      MARRIED WIDOWED/DIVORCED

        15 - 19    1325.27       86.73           0.00
        20 - 24     615.56      783.39           3.05
        25 - 29     253.94     1187.18           8.88
        30 - 34     165.13     1348.55          27.32
        35 - 39     173.41     1454.71          52.87
        40 - 44     147.21     1308.12          76.67
        45 - 49     202.33     1352.28         107.40
        50 OR OVER 1105.16     4181.04        2357.81
```

Example 15:   Full-Screen Nonlinear Regression

This example shows how to build a menu system that allows you to do nonlinear regression from a menu. Six modules are stored on an IML storage disk. After you have stored them, use this example to try out the system. First invoke IML and set up some sample data in memory, in this case the population of the U.S. from 1790 to 1970. Then invoke the module NLIN.

```
reset storage='nlin';
load module=_all_;
uspop = {3929, 5308, 7239, 9638, 12866, 17069, 23191, 31443,
        39818, 50155, 62947, 75994, 91972, 105710, 122775, 131669,
        151325, 179323, 203211}/1000;
year=do(1790,1970,10)`;
time=year-1790;
print year time pop;
run nlin;
```

A menu appears like the one below. The entry fields are shown by underscores here, but the underscores become blanks in the real session.

```
Non-Linear Regression
Response function: _____
Predictor function: _____

Parameter Value Derivative

: _____ _____ _____
: _____ _____ _____
: _____ _____ _____
: _____ _____ _____
: _____ _____ _____
: _____ _____ _____
```

Enter an exponential model and fill in the response and predictor expression fields. For each parameter, enter the name, initial value, and derivative of the predictor with respect to the parameter.

```
Non-Linear Regression
Response function: uspop_____
Predictor function: a0*exp(a1*time)_____

Parameter Value Derivative
: a0_____ _____3.9 exp(a1*time)_____
: a1_____ _____0 time*a0*exp(a1*time)_____
: _____ _____ _____
: _____ _____ _____
: _____ _____ _____
: _____ _____ _____
```

Now press the SUBMIT key. The model compiles, the iterations start blinking on the screen, and when the model has converged, the estimates are displayed along with their standard errors, *t*-test, and significance probability.

To modify and rerun the model, submit the command

```
run nlrun;
```

Here is the code that defines and stores the modules of the system:

```
/*        Full Screen Nonlinear Regression          */
/* Six modules are defined, which constitute a system for  */
/* nonlinear regression. The interesting feature of this   */
/* system is that the problem is entered in a menu, and both */
/* iterations and final results are displayed on the same  */
/* menu.                                             */
/*                                                   */
/* Run this source to get the modules stored. Examples    */
/* of use are separate.                              */
/*                                                   */
/* Caution: this is a demonstration system only. It does not */
/* have all the necessary safeguards in it yet to         */
/* recover from user errors or rough models.         */
/* Algorithm:                                        */
/*    Gauss-Newton nonlinear regression with step-halving.   */
/* Notes: program variables all start with nd or _ to minimize */
/* the problems that would occur if user variables        */
/* interfered with the program variables.            */


/* Gauss-Newton nonlinear regression with Hartley step-halving */
```

```
/*---Routine to set up display values for new problem---*/
start nlinit;
    window nlin rows=15 columns=80 color='green'
        msgline=_msg cmndline=_cmnd
        group=title +30 'Non-Linear Regression' color='white'
        group=model / a5 'Response function:' color='white'
        +1 nddep $55. color='blue'
        / a5 'Predictor function:' color='white'
        +1 ndfun $55. color='blue'
        group=parm0 // a5 'Parameter' color='white' a15 'Value'
        a30 'Derivative'
        group=parm1 // a5 'Parameter' color='white' a15 'Value'
        group=parm2 // a5 'Parameter' color='white' a19 'Estimate'
        a33 'Std Error'
        a48 'T Ratio'
        a62 'Prob>|T|'
        group=parminit /a3 ':' color='white'
        a5 ndparm $8. color='blue'
        a15 ndbeta best12. a30 ndder $45.
        group=parmiter / a5 _parm color='white'
        a15 _beta best12. color='blue'
        group=parmest / a5 _parm color='white'
        a15 _beta best12. color='blue'
        a30 _std best12.
        a45 _t 10.4
        a60 _prob 10.4
        group=sse // a5 'Iteration =' color='white' _iter 5. color='blue'
        ' Stephalvings = ' color='white' _subit 3. color='blue'
        / a5 'Sum of Squares Error =' color='white' _sse best12.
        color='blue';
    nddep=cshape(' ',1,1,55,' ');
    ndfun=nddep;
    nd0=6;
    ndparm=repeat(' ',nd0,1);
    ndbeta=repeat(0,nd0,1);
    ndder=cshape(' ',nd0,1,55,' ');
    _msg='Enter New Nonlinear Problem';
finish nlinit;  /* Finish module NLINIT */


    /* Main routine */
start nlin;
    run nlinit;  /* initialization routine */
    run nlrun;   /* run routine */
finish nlin;
```

```
                   /* Routine to show each iteration */
            start nliter;
               display nlin.title noinput,
               nlin.model noinput,
               nlin.parm1 noinput,
               nlin.parmiter repeat noinput,
               nlin.sse noinput;
            finish nliter;


                   /* Routine for one run */
            start nlrun;
               run nlgen; /* generate the model */
               run nlest; /* estimate the model */
            finish nlrun;


                   /* Routine to generate the model */
            start nlgen;

               /* Model definition menu */
               display nlin.title, nlin.model, nlin.parm0, nlin.parminit repeat;

               /* Get number of parameters */
               t=loc(ndparm=' ');
               if nrow(t)=0 then
               do;
                  print 'no parameters';
                  stop;
               end;
               _k=t[1] -1;

                   /* Trim extra rows, and edit '*' to '#' */
               _dep=nddep; call change(_dep,'*','#',0);
               _fun=ndfun; call change(_fun,'*','#',0);
               _parm=ndparm[1:_k,];
               _beta=ndbeta[1:_k,];
               _der=ndder [1:_k,];
               call change(_der,'*','#',0);
```

```
   /* Construct nlresid module to split up parameters and */
   /* compute model                                        */
   call queue('start nlresid;');
   do i=1 to _k;
      call queue(_parm[i] ,"=_beta[",char(i,2),"] ;");
   end;
   call queue("_y = ",_dep,";",
              "_p = ",_fun,";",
              "_r = _y-_p;",
              "_sse = ssq(_r);",
              "finish;" );

      /* Construct nlderiv function */
   call queue('start nlderiv; _x = ');
   do i=1 to _k;
      call queue("(",_der[i] ,")#repeat(1,nobs,1)||");
   end;
   call queue(" nlnothin; finish;");

      /* Pause to compile the functions */
   call queue("resume;");
   pause *;
finish nlgen;  /* Finish module NLGEN   */

   /* Routine to do estimation */
start nlest;

   /*     Modified Gauss-Newton Nonlinear Regression      */
   /* _parm has parm names                                */
   /* _beta has initial values for parameters             */
   /* _k is the number of parameters                      */
   /* after nlresid:                                      */
   /* _y has response,                                    */
   /* _p has predictor after call                         */
   /* _r has residuals                                    */
   /* _sse has sse                                        */
   /* after nlderiv                                       */
   /* _x has jacobian                                     */
   /*                                                     */

   eps=1;
   _iter = 0;
   _subit = 0;
   _error = 0;
   run nlresid;        /* f, r, and sse for initial beta */
   run nliter;                      /* print iteration zero */
   nobs = nrow(_y);
   _msg = 'Iterating';
```

```
        /* Gauss-Newton iterations */
    do _iter=1 to 30 while(eps>1e-8);
       run nlderiv;                        /* subroutine for derivatives */
       _lastsse=_sse;
       _xpxi=sweep(_x`*_x);
       _delta=_xpxi*_x`*_r;                     /* correction vector */
       _old = _beta;                         /* save previous parameters */
       _beta=_beta+_delta;                      /* apply the correction */
       run nlresid;                               /* compute residual */
       run nliter;                          /* print iteration in window */
       eps=abs((_lastsse-_sse))/(_sse+1e-6); /* convergence criterion */

          /* Hartley subiterations */
       do _subit=1 to 10 while(_sse>_lastsse);
          _delta=_delta*.5;         /* halve the correction vector */
          _beta=_old+_delta;        /* apply the halved correction */
          run nlresid;                          /* find sse et al */
          run nliter;               /* print subiteration in window */
       end;
       if _subit>10 then
       do;
          _msg = "did not improve after 10 halvings";
          eps=0; /* make it fall through iter loop */
       end;
    end;

       /* print out results */
    _msg = ' ';
    if _iter>30 then
    do;
       _error=1;
       _msg = 'convergence failed';
    end;
    _iter=_iter-1;
    _dfe = nobs-_k;
    _mse = _sse/_dfe;
    _std = sqrt(vecdiag(_xpxi)#_mse);
    _t = _beta/_std;
    _prob= 1-probf(_t#_t,1,_dfe);
    display nlin.title noinput,
    nlin.model noinput,
    nlin.parm2 noinput,
    nlin.parmest repeat noinput,
    nlin.sse noinput;
 finish nlest;                               /* Finish module NLEST  */

       /* Store the modules to run later */
 reset storage='nlin';
 store module=_all_;
```

Chapter 9 Introduction to SAS/IML® Graphics

Introduction

SAS/IML software provides you with a powerful set of graphics commands (called graphics primitives) from which to create customized displays. Several basic commands are GDRAW (for drawing a line), GPOINT (for plotting points), and GPOLY (for drawing a polygon). With each primitive, you can associate a set of attributes such as color or line style.

In this chapter you learn about

□ plotting simple two-dimensional plots

□ naming and saving a graph

□ changing attributes such as color and line style

□ specifying the location and scale of your graph

□ adding axes and text.

SAS/IML graphics commands depend on the libraries and device drivers distributed with SAS/GRAPH software, and they do not work unless you have SAS/GRAPH software.

An Introductory Graph

Suppose that you have data for ACME Corporation's stock price and you want a simple PRICE×DAY graph to see the overall trend of the stock's price. The data are shown below.

Day	Price
0	43.75
5	48.00
10	59.75
15	75.5
20	59.75
25	71.50
30	70.575
35	61.125
40	79.50
45	72.375
50	67.00
55	54.125
60	58.750
65	43.625
70	47.125
75	45.50

To graph a scatterplot of these points, enter the following statements. These statements generate Figure 9.1.

```
proc iml;                          /* invoke IML        */
   call gstart;                    /* start graphics    */
   xbox={0 100 100 0};
   ybox={0 0 100 100};
   day=do(0,75,5);                 /* initialize day    */
   price={43.75,48,59.75,75.5,     /* initialize price  */
       59.75,71.5,70.575,
       61.125,79.5,72.375,67,
       54.125,58.75,43.625,
       47.125,45.50};
   call gopen;                     /* start new graph     */
   call gpoly(xbox,ybox);      /* draw a box around plot */
   call gpoint(day,price);         /* plot the points     */
   call gshow;                     /* display the graph */
```

Figure 9.1   *Scatterplot*

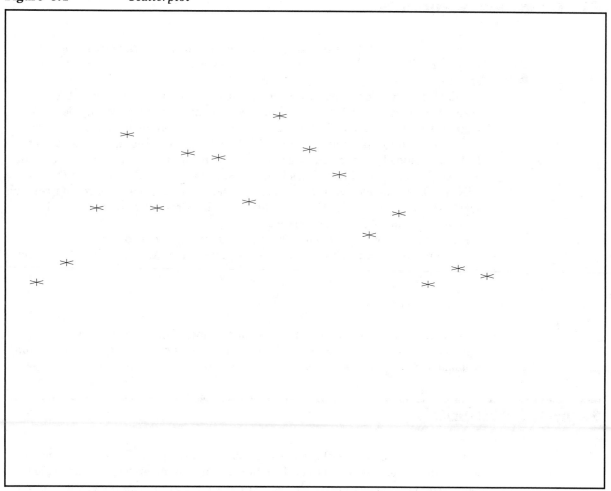

Note that the GSTART statement initializes the graphics session. It usually needs to be called only once. Next, you enter the data matrices. Then you open a graphics segment (that is, begin a new graph) with the GOPEN command. The GPOINT command draws the scatter plot of points of DAY versus PRICE. The GSHOW command displays the graph.

Notice also that, for this example, the x coordinate of the data is DAY and that $0 \leq DAY \leq 100$. The y coordinate is PRICE, which ranges from $0 \leq PRICE \leq 100$. For this example, the ranges are this way because the IML default ranges are from 0 to 100 on both the x and y axes. Later on you learn how to change the default ranges for the axes with the GWINDOW statement so that you can handle data with any range of values.

Of course, this graph is quite simple. By the end of this chapter, you will know how to add axes and titles, to scale axes, and to connect the points with lines.

IML Graphics Segments

A graph is saved in what is called a graphics segment. A *graphics segment* is simply a collection of primitives and their associated attributes that creates a graph.

Each time you create a new segment, it is named and stored in a SAS graphics catalog called WORK.GSEG. If you want to store your graphics segments in a permanent SAS catalog, do this with options to the GSTART call. You can name the segments yourself in the GOPEN statement, or you can let IML automatically generate a segment name. In this way, graphics segments that are used several times can be included in subsequent graphs by using the GINCLUDE command with the segment name. You can also manage and replay a segment using the GREPLAY procedure as well as replay it in another IML session with the GSHOW command.

To name a segment, include the name as an argument to the GOPEN statement. For example, to begin a new segment and name it STOCK1, use the statement

```
call gopen("stock1");
```

For more information about SAS catalogs and graphics, see Chapter 3, "Graphics Output," in *SAS/GRAPH Software: Reference, Version 6, First Edition, Volume 1.*

Segment Attributes

A set of attributes is initialized for each graphics segment. These attributes are color, line style, line thickness, fill pattern, font, character height, and aspect ratio. You can change any of these attributes for a graphics segment by using the GSET command. Some IML graphics commands take optional attribute arguments. The values of these arguments affect only the graphics output associated with the call.

The IML graphics subsystem uses the same conventions that SAS/GRAPH software uses in setting the default attributes. It also uses the options set in the GOPTIONS statement when applicable. The SAS/IML default values for the GSET command are given by their corresponding GOPTIONS default values. To change the default, you need to issue a GOPTIONS statement. The GOPTIONS statement can also be used to set graphics options not available through the GSET command (for example, the ROTATE option).

For more information about GOPTIONS, see Chapter 12, "The GOPTIONS Statement," in *SAS/GRAPH Software: Reference.*

Coordinate Systems

Each IML graph is associated with two independent cartesian coordinate systems, a *world coordinate system* and a *normalized coordinate system*.

Understanding World Coordinates

The *world coordinate system* is the coordinate system defined by your data. Because these coordinates help define objects in the data's two-dimensional world, these are referred to as *world coordinates*. For example, suppose you have a data set containing heights and weights, and you are interested in plotting height versus weight. Your data induces a world coordinate system in which each point (*x,y*) represents a pair of data values (*height,weight*). The world could be defined by the observed ranges of heights and weights or could be enlarged to include a range of all reasonable values for heights and weights.

Now consider a more realistic example of the stock price data for ACME Corporation. Suppose that the stock price data were actually the year end prices of ACME stock for the years 1971 through 1986, as shown below:

```
YEAR    PRICE
  71    123.75
  72    128.00
  73    139.75
  74    155.50
  75    139.75
  76    151.50
  77    150.375
  78    149.125
  79    159.50
  80    152.375
  81    147.00
  82    134.125
  83    138.75
  84    123.625
  85    127.125
  86    125.500
```

The actual range of YEAR is from 71 to 86 and the range of PRICE is from $123.625 to $159.50. These are the ranges in world coordinate space for the stock data. Of course, you could say that the range for PRICE could start at $0 and range upwards to, say, $200. Or, if you were only interested in prices during the 80's, you could say the range for PRICE is from $123.625 to $152.375. As you see, it all depends on how you want to define your world.

Figure 9.2 shows a graph of the stock data with the world defined as the actual data given. The corners of the rectangle give the actual boundaries for this data.

Figure 9.2   *World Coordinates*

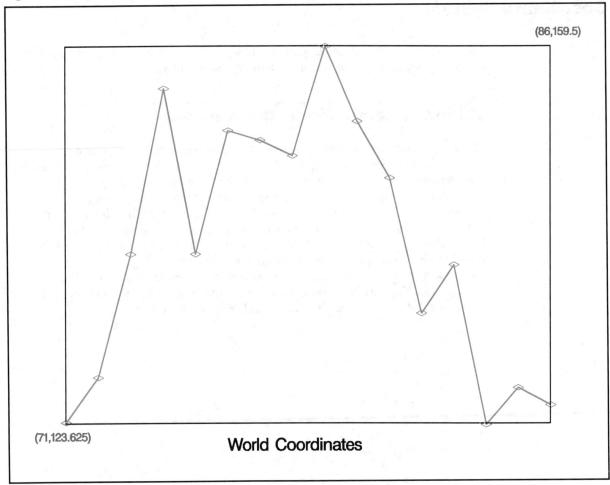

(86,159.5)

(71,123.625)

World Coordinates

Understanding Normalized Coordinates

The *normalized coordinate system* is defined relative to your display device, usually a monitor or plotter. It is always defined with points varying between (0,0) and (100,100), where (0,0) refers to the lower left corner and (100,100) refers to the upper right corner.

In summary,

□ the world coordinate system is defined relative to your data

□ the normalized coordinate system is defined relative to the display device.

Figure 9.3 shows the ACME stock data in terms of normalized coordinates. There is a natural mathematical relationship between each point in world and normalized coordinates. Figure 9.3 shows the normalized coordinate system for the ACME stock data. The normalized device coordinate system is mapped to the device display area so that (0,0), the lower left corner, corresponds to (71, 123.625) in world coordinates, and (100,100), the upper right corner, corresponds to (86,159.5) in world coordinates.

Figure 9.3   *Normalized Coordinates*

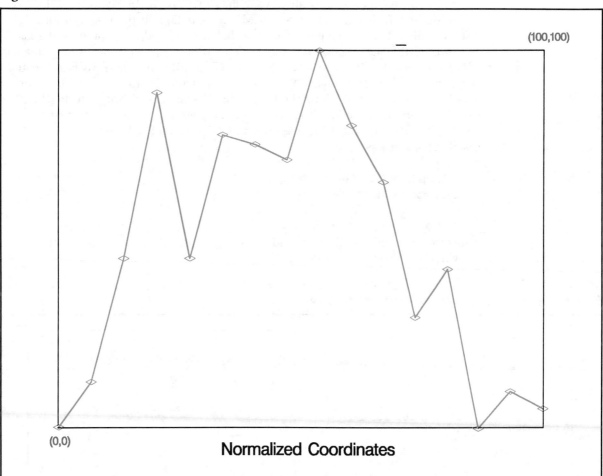

Normalized Coordinates

Windows and Viewports

A *window* defines a rectangular area in world coordinates. You define a window with a GWINDOW statement. You can define the window to be larger than, the same size as, or smaller than the actual range of data values, depending on whether you want to show all the data or only part of the data.

    A *viewport* defines in normalized coordinates a rectangular area on the display device where the image of the data appears. You define a viewport with the GPORT command. You can have your graph take up the entire display device or show it in only a portion, say the upper right part.

Mapping Windows to Viewports

A *window* and a *viewport* are related by the linear transformation that maps the window onto the viewport. A line segment in the window is mapped to a line segment in the viewport such that the relative positions are preserved.

You don't have to display all of your data in a graph. In Figure 9.4, the graph on the left displays all of the ACME stock data, and the graph on the right displays only a part of the data. Suppose that you wanted to graph only the last ten years of the stock data, say from 1977 to 1986. You would want to define a window where the YEAR axis ranges from 77 to 86, while the PRICE axis could range from 120 to 160. Figure 9.4 shows stock prices in a window defined for data from 1977 to 1986 along the horizontal direction, and from 120 to 160 along the vertical direction. The window is mapped to a viewport defined by the points (20,20) and (70,60). The appropriate GPORT and GWINDOW specifications are as follows:

```
call gwindow({77 120, 86 160});
call gport({20 20, 70 60});
```

The window, in effect, defines the portion of the graph that is to be displayed in world coordinates, and the viewport specifies the area on the device on which the image is to appear.

Figure 9.4   *Window to Viewport Mapping*

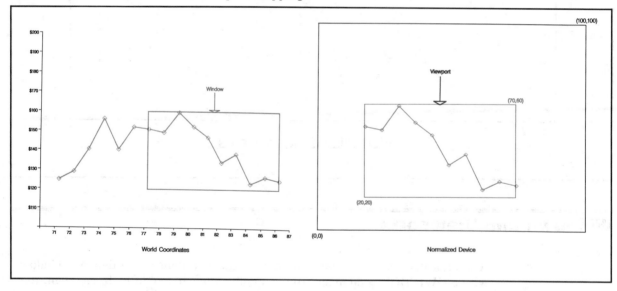

Understanding Windows

Because the default world coordinate system ranges from (0,0) to (100,100), you usually need to define a *window* in order to set the world coordinates corresponding to your data. A window specifies which part of the data in world coordinate space is to be shown. Sometimes you want all of the data shown; other times, you want to show only part of the data.

A window is defined by an array of four numbers, which define a rectangular area. You define this area by specifying the *world coordinates* of the lower left and upper right corners in the GWINDOW statement, which has the general form

CALL GWINDOW(*minimum-x minimum-y maximum-x maximum-y*);

The argument can be either a matrix or a literal. The order of the elements is important. The array of coordinates can be a 2×2, 1×4, or 4×1 matrix. These coordinates can be specified as matrix literals or as the name of a numeric matrix containing the coordinates. If you do not define a window, the default is to assume both *x* and *y* range between 0 and 100.

In summary, a window

□ defines the portion of the graph that appears in the viewport

□ is a rectangular area

□ is defined by an array of four numbers

□ is defined in world coordinates

□ scales the data relative to world coordinates.

In the previous example, the variable YEAR ranges from 1971 to 1986, while PRICE ranges from 123.625 to 159.50. Because the data do not fit nicely into the default, you want to define a window that reflects the ranges of the variables YEAR and PRICE. To draw the graph of this data to scale, you can let the YEAR axis range from 70 to 87 and the PRICE axis range from 100 to 200. Use the following statements to draw the graph, shown in Figure 9.5.

```
call gstart;
xbox={0 100 100 0};
ybox={0 0 100 100};
call gopen("stocks1");          /* begin new graph STOCKS1 */
call gset("height", 2.0);
year=do(71,86,1);                        /* initialize YEAR */
price={123.75 128.00 139.75         /* initialize PRICE */
       155.50 139.750 151.500
       150.375 149.125 159.500
       152.375 147.000 134.125
       138.750 123.625 127.125
       125.50};
call gwindow({70 100 87 200});           /* define window */
call gpoint(year,price,"diamond","green"); /* graph the points */
call gdraw(year,price,1,"green");     /* connect points */
call gshow;                           /* show the graph */
```

Figure 9.5   Stock Data

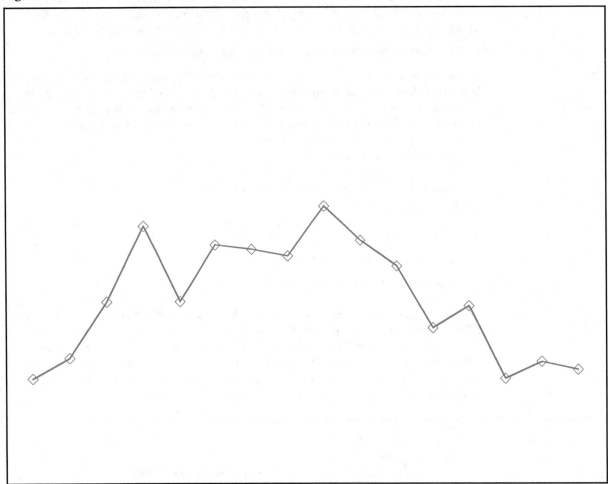

In this example, you do several things that you did not do with the previous graph:

1. You associate the name STOCKS1 with this graphics segment in the GOPEN command.

2. You define a window that reflects the actual ranges of the data with a GWINDOW command.

3. You associate a plotting symbol, the diamond, and the color green with the GPOINT command.

4. You connect the points with line segments with the GDRAW command. The GDRAW command requests that the line segments be drawn in style 1 and be green.

Understanding Viewports

A *viewport* specifies a rectangular area on the display device where the graph appears. You define this area by specifying the *normalized* coordinates, the lower left corner and the upper right corner, in the GPORT statement, which has the general form

CALL GPORT(*minimum-x minimum-y maximum-x maximum-y*);

The argument can be either a matrix or a literal. Note that both *x* and *y* must range between 0 and 100. As with the GWINDOW specification, you can give the coordinates either as a matrix literal enclosed in braces or as the name of a numeric matrix containing the coordinates. The array can be a 2×2, 1×4, or 4×1 matrix. If you do not define a viewport, the default is to span the entire display device.

In summary, a viewport

□ specifies where the image appears on the display

□ is a rectangular area

□ is specified by an array of four numbers

□ is defined in normalized coordinates

□ scales the data relative to the shape of the viewport.

To display the stock price data in a smaller area on the display device, you must define a viewport. While you are at it, add some text to the graph. You can use the graph you created and named STOCKS1 in this new graph. The statements below create the graph shown in Figure 9.6.

```
      /* module centers text strings */
start gscenter(x,y,str);
    call gstrlen(len,str);                  /* find string length */
    call gscript(x-len/2,y,str);                  /* print text */
finish gscenter;

call gopen("stocks2");                  /* open a new segment */
call gset("font","swiss");              /* set character font */
call gpoly(xbox,ybox);                  /* draw a border      */
call gwindow({70 100,87 200});          /* define a window    */
call gport({15 15,85 85});              /* define a viewport  */
call ginclude("stocks1");           /* include segment STOCKS1 */
call gxaxis({70 100},17,18, ,                /* draw x-axis    */
            ,"2.",1.5);
call gyaxis({70 100},100,11, ,                /* draw y-axis */
            ,"dollar5.",1.5);
call gset("height",2.0);               /* set character height */
call gtext(77,89,"Year");              /* print horizontal text */
call gvtext(68,200,"Price");            /* print vertical text */
call gscenter(79,210,"ACME Stock Data");       /* print title */
call gshow;
```

Figure 9.6   *Stock Data with Axes and Labels*

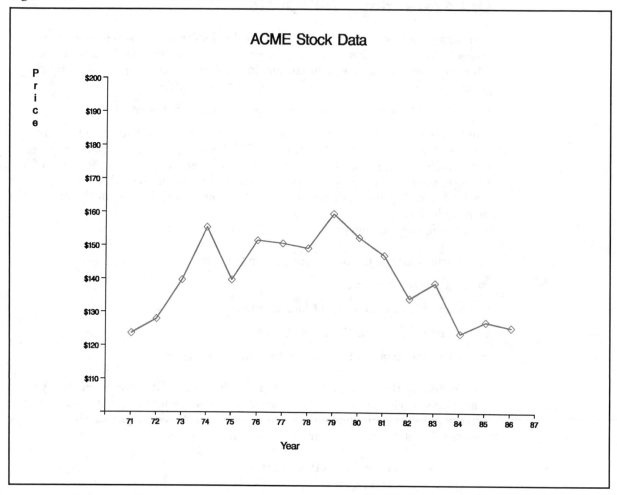

The statements that generated this graph are described below:

1. GOPEN begins a new graph and names it STOCKS2.

2. GPOLY draws a box around the display area.

3. GWINDOW defines the world coordinate space to be larger than the actual range of stock data values.

4. GPORT defines a viewport. It causes the graph to appear in the center of the display, with a border around it for text. The lower left corner has coordinates (15,15) and the upper right corner has coordinates (85,85).

5. GINCLUDE includes the graphics segment STOCKS1. This saves you from having to plot points you have already created.

6. You add axes to the graph. GXAXIS draws the *x* axis. It begins at the point (70,100) and is 17 units (years) long, divided with 18 tick marks. The axis tick marks are printed with the numeric 2.0 format, have a height of 1.5 units. GYAXIS draws the *y* axis. It also begins at (70,100) but is 100 units (dollars) long, divided with 11 tick marks. The axis tick marks are printed with the DOLLAR5.0 format and have a height of 1.5 units.

7. GSET sets the text font to be Swiss and the height of the letters to be 2.0 units. The height of the characters has been increased because the viewport definition scales character sizes relative to the viewport.

8. GTEXT prints horizontal text. It prints the text string **Year** beginning at the world coordinate point (77,89).

9. GVTEXT prints vertical text. It prints the text string **Price** beginning at the world coordinate point (68,200).

10. GSCENTER runs the module to print centered text strings.

11. GSHOW displays the graph.

Changing Windows and Viewports

Windows and viewports can be changed for the graphics segment any time that the segment is active. Using the stock price example, you can first define a window for the data during the years 1971 to 1974 and map this to the viewport defined on the upper half of the normalized device; then you can redefine the window to enclose the data for 1983 to 1986 and map this to an area in the lower half of the normalized device. Notice how the shape of the viewport affects the shape of the curve. Changing the viewport can affect the height of any printed characters as well. In this case, you can modify the HEIGHT parameter.

The statements below generate the graph in Figure 9.7:

```
      /* figure 9.7 */
reset clip;                            /* clip outside viewport */
call gopen;                            /* open a new segment     */
call gset("color","blue");
call gset("height",2.0);
call gwindow({71 120,74 175});         /* define a window        */
call gport({20 55,80 90});             /* define a viewport      */
call gpoly({71 74 74 71},{120 120 170 170});   /* draw a border */
call gscript(71.5,162,"Viewport #1 1971-74",,    /* print text */
             ,3.0,"complex","red");
call gpoint(year,price,"diamond","green");       /* draw points */
call gdraw(year,price,1,"green");              /* connect points */
call gblkvpd;
call gwindow({83 120,86 170});              /* define new window */
call gport({20 10,80 45});                /* define new viewport */
call gpoly({83 86 86 83},{120 120 170 170});   /* draw border */
call gpoint(year,price,"diamond","green");       /* draw points */
call gdraw(year,price,1,"green");             /* connect points */
call gscript(83.5,162,"Viewport #2 1983-86",,    /* print text */
             ,3.0,"complex","red");
call gshow;
```

Figure 9.7   *Multiple Viewports*

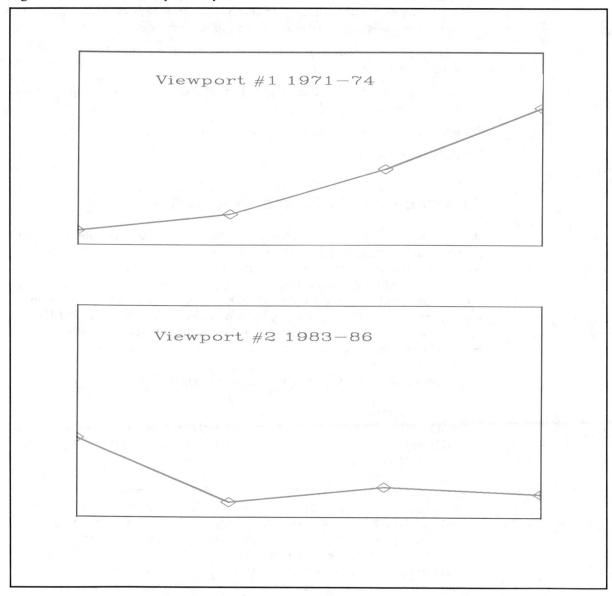

The RESET CLIP command is necessary because you are graphing only a part of the data in the window. You want to clip the data that falls outside of the window. See "Clipping Your Graphs" later in this chapter for more on clipping. In this graph, you

1. Open a new segment (GOPEN).

2. Define the first window for the first 4 years' data (GWINDOW).

3. Define a viewport in the upper part of the display device (GPORT).

4. Draw a box around the viewport (GPOLY).

5. Add text (GSCRIPT).

6. Graph the points and connect them (GPOINT and GDRAW).

7. Define the second window for the last 4 years (GWINDOW).

8. Define a viewport in the lower part of the display device (GPORT).

9. Draw a box around the viewport (GPOLY).

10. Graph the points and connect them (GPOINT and GDRAW).

11. Add text (GSCRIPT).

12. Display the graph (GSHOW).

Stacking Viewports

Viewports can be stacked; that is, a viewport can be defined relative to another viewport so that you have a viewport within a viewport.

A window or a viewport is changed globally through the IML graphics commands: the GWINDOW command for windows, and the GPORT, GPORTSTK, and GPORTPOP commands for viewports. When a window or viewport is defined, it persists across IML graphics commands until another window- or viewport-altering command is encountered. Stacking helps you define a viewport without losing the effect of a previously defined viewport. When a stacked viewport is popped, you are placed into the environment of the previous viewport.

Windows and viewports are associated with a particular segment; thus, they automatically become undefined when the segment is closed. A segment is closed whenever IML encounters a GCLOSE command or a GOPEN command. A window or a viewport can also be changed for a single graphics command. Either one can be passed as an argument to a graphics primitive, in which case any graphics output associated with the call is defined in the specified window or viewport. When a viewport is passed as an argument, it is stacked, or defined relative to the current viewport, and popped (deleted) when the graphics command is complete.

For example, suppose you want to create a legend that shows the low and peak points of the data for the ACME stock graph. Create a graphics segment showing this information:

```
call gopen("legend");
call gset('height',5);   /* enlarged to accomodate viewport later */
call gset('font','swiss');
call gscript(5,75,"Stock Peak:  159.5 in 1979");
call gscript(5,65,"Stock Low:   123.6 in 1984");
call gclose;
```

Now create a segment that highlights and labels the low and peak points of the data:

```
    /* Highlight and label the low and peak points of the stock */
call gopen("labels");
call gwindow({70 100 87 200}); /* define window */
call gpoint(84,123.625,"circle","red",4) ;
call gtext(84,120,"LOW","red");
call gpoint(79,159.5,"circle","red",4);
call gtext(79,162,"PEAK","red");
call gclose;
```

Open a new graphics segment and include the STOCK1 segment created earlier in the chapter, placing the segment in the viewport {10 10 90 90}.

```
call gopen;
call gportstk ({10 10 90 90}); /* viewport for the plot itself */
call ginclude('stocks2');
```

To place the legend in the upper right corner of this viewport, use the GPORTSTK command instead of the GPORT command to define the legend's viewport relative to the one used for the plot of the stock data:

```
call gportstk ({70 70 100 100});   /* viewport for the legend*/
call ginclude("legend");
```

Now pop the legend's viewport to get back to the viewport of the plot itself and include the segment which labels and highlights the low and peak stock points:

```
call gportpop;   /* viewport for the legend */
call ginclude ("labels");
```

Finally, display the graph.

```
call gshow;
```

Figure 9.8   *Stacking Viewports*

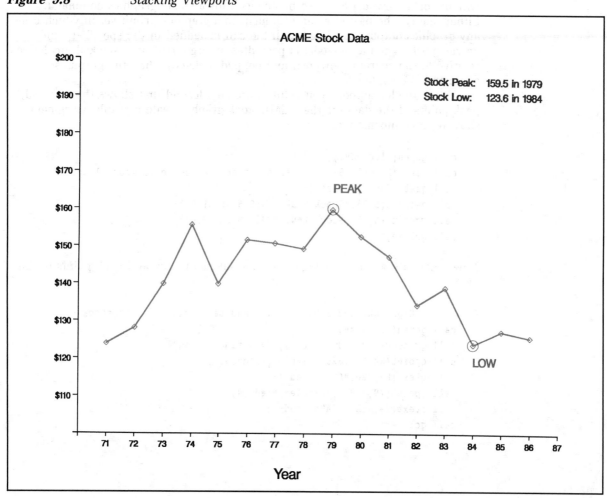

Clipping Your Graphs

The IML graphics subsystem does not automatically clip the output to the viewport. Thus, it is possible that data are graphed outside of the defined viewport. This happens when there are data points lying outside of the defined window. For instance, if you specify a window to be a subset of the world, then there will be data lying outside of the window and these points will be graphed outside of the viewport. This is usually not what you want. To clean up such graphs, you either delete the points you do not want to graph or clip the graph.

There are two ways to clip a graph. You can use the RESET CLIP command, which clips outside of a viewport. The CLIP option remains in effect until you submit a RESET NOCLIP command. You can also use the GBLKVP command, which clips either inside or outside of a viewport. Use the GBLKVP command to define a blanking area in which nothing can be drawn until the blanking area is released. Use the GBLKVPD command to release the blanking area.

Common Arguments

IML graphics commands are available in the form of call subroutines. They generally take a set of required arguments followed by a set of optional arguments. All graphics primitives take *window* and *viewport* as optional arguments. Some IML graphics commands, like GPOINT or GPIE, allow implicit repetition factors in the argument lists. The GPOINT command places as many markers as there are well defined (*x,y*) pairs. The GPIE command draws as many slices as there are well defined pies. In those cases, some of the attribute matrices can have more than one element, which are used in order. If an attribute list is exhausted before the repetition factor is completed, the last element of the list is used as the attribute for the remaining primitives.

The arguments to the IML graphics commands are positional. Thus, to skip over an optional argument from the middle of a list, a comma is needed to hold its place. For example, the command

```
call gpoint(x,y, ,"red");
```

omits the third argument from the argument list.

The following list details the arguments commonly used in IML graphics commands:

color
> is a character matrix or literal that names a valid color as specified in the GOPTIONS statement. The default color is the first color specified in the COLORS= list in the GOPTIONS statement. If no such list is given, IML uses the first default color for the graphics device. Note that *color* can be specified either as a quoted literal, such as "RED", a color number, such as 1, or the name of a matrix containing a reference to a valid color. A color number *n* refers to the *n*th color in the color list.
>
> You can change the default color with the GSET command.

font
> is a character matrix or quoted literal that specifies a valid font name. The default font is the hardware font, which can be changed by the GSET command unless a viewport is in effect.

height
> is a numeric matrix or literal that specifies the character height. The unit of height is the *gunit* of the GOPTIONS statement, when specified; otherwise, the unit is a character cell. The default height is 1 *gunit*, which you can change using the GSET command.

pattern
> is a character matrix or quoted literal that specifies the pattern to fill the interior of a closed curve. You specify a pattern by a coded character string as documented in the V= option in the PATTERN statement (see Chapter 15, "The PATTERN Statement," in *SAS/GRAPH Software: Reference*.
>
> The default pattern set by the IML graphics subsystem is "E", that is, empty. The default pattern can be changed using the GSET command.

segment-name
> is a character matrix or quoted literal that specifies a valid SAS name used to identify a graphics segment. The *segment-name* is associated with the graphics segment opened with a GOPEN command. If you do not specify *segment-name*, IML generates default names. For example, to create a graphics segment called PLOTA, enter

```
call gopen("plota");
```

> Graphics segments are not allowed to have the same name as an existing segment. If you try to create a second segment named PLOTA, (that is, when the *replace flag* is turned off), then the second segment is named PLOTA1. The *replace* flag is set by the GOPEN command for the segment that is being created. To open a new segment named PLOTA and replace an existing segment with the same name, enter

```
call gopen("plota",1);
```

> If you do not specify a *replace* argument to the GOPEN command, the default is set by the GSTART command for all subsequent segments that are created. By default, the GSTART command sets the *replace* flag to 0, so that new segments do not replace like-named segments.

style
> is a numeric matrix or literal that specifies an index corresponding to the line style documented for the SYMBOL statement in Chapter 16, "The Symbol Statement," in *SAS/GRAPH Software: Reference*. The IML graphics subsystem sets the default line style to be 1, a solid line. The default line style can be changed using the GSET command.

symbol
> is a character matrix or quoted literal that specifies either a character string corresponding to a symbol as defined for the V= option of the SYMBOL statement, or specifies the corresponding identifying symbol number. STAR is the default symbol used by the IML graphics subsystem.

SAS/IML graphics commands are described in detail in Chapter 15. See also *SAS/GRAPH Software: Reference* for additional information.

Chapter 10 Graphics Applications

Introduction

This chapter provides the details and code for three examples involving SAS/IML graphics. The first example shows a 2×2 matrix of scatterplots and a 3×3 matrix of scatterplots. A matrix of scatterplots is useful when you have several variables that you want to investigate simultaneously rather than in pairs. The second example draws a grid for representing a train schedule, with arrival and departure dates on the horizontal axis and destinations along the vertical axis. The final example plots Fisher's Iris data. This example shows how to plot several graphs on one page.

Example 1:   Scatterplot Matrix

With the viewport capability of the IML graphics subroutine, you can arrange several graphs on a page. In this example, multiple graphs are generated from three variables and are displayed in a scatterplot matrix. For each variable, one contour plot is generated with each of the other variables as the dependent variable. For the graphs on the main diagonal, a box and whisker plot is generated for each variable.

This example takes advantage of user-defined IML modules:

BOXWHSKR   computes median and quartiles.

GBXWHSKR   draws box and whisker plots.

CONTOUR   generates confidence ellipses assuming bivariate normal data.

GCONTOUR   draws the confidence ellipses for each pair of variables.

GSCATMAT   produces the $n \times n$ scatterplot matrix, where n is the number of variables.

The code for the five modules and a sample data set are given below. The modules produce Figure 10.1 and Figure 10.2.

```
/* This program generates a data set and uses iml graphics   */
/* subsystem to draw a scatterplot matrix.                   */
/*                                                           */
data factory;
  input recno prod temp a defect mon;
  cards;
    1   1.82675   71.124    1.12404   1.79845      2
    2   1.67179   70.9245   0.924523  1.05246      3
    3   2.22397   71.507    1.50696   2.36035      4
    4   2.39049   74.8912   4.89122   1.93917      5
    5   2.45503   73.5338   3.53382   2.0664       6
    6   1.68758   71.6764   1.67642   1.90495      7
    7   1.98233   72.4222   2.42221   1.65469      8
    8   1.17144   74.0884   4.08839   1.91366      9
    9   1.32697   71.7609   1.76087   1.21824     10
   10   1.86376   70.3978   0.397753  1.21775     11
   11   1.25541   74.888    4.88795   1.87875     12
   12   1.17617   73.3528   3.35277   1.15393      1
   13   2.38103   77.1762   7.17619   2.26703      2
   14   1.13669   73.0157   3.01566   1            3
   15   1.01569   70.4645   0.464485  1            4
   16   2.36641   74.1699   4.16991   1.73009      5
   17   2.27131   73.1005   3.10048   1.79657      6
   18   1.80597   72.6299   2.62986   1.8497       7
   19   2.41142   81.1973  11.1973    2.137        8
   20   1.69218   71.4521   1.45212   1.47894      9
   21   1.95271   74.8427   4.8427    1.93493     10
   22   1.28452   76.7901   6.79008   2.09208     11
   23   1.51663   83.4782  13.4782    1.81162     12
   24   1.34177   73.4237   3.42369   1.57054      1
   25   1.4309    70.7504   0.750369  1.22444      2
   26   1.84851   72.9226   2.92256   2.04468      3
   27   2.08114   78.4248   8.42476   1.78175      4
   28   1.99175   71.0635   1.06346   1.25951      5
   29   2.01235   72.2634   2.2634    1.36943      6
   30   2.38742   74.2037   4.20372   1.82846      7
   31   1.28055   71.2495   1.24953   1.8286       8
   32   2.05698   76.0557   6.05571   2.03548      9
   33   1.05429   77.721    7.72096   1.57831     10
   34   2.15398   70.8861   0.886068  2.1353      11
   35   2.46624   70.9682   0.968163  2.26856     12
   36   1.4406    73.5243   3.52429   1.72608      1
   37   1.71475   71.527    1.52703   1.72932      2
   38   1.51423   78.5824   8.5824    1.97685      3
   39   2.41538   73.7909   3.79093   2.07129      4
   40   2.28402   71.131    1.13101   2.25293      5
   41   1.70251   72.3616   2.36156   2.04926      6
```

42	1.19747	72.3894	2.3894	1	7
43	1.08089	71.1729	1.17288	1	8
44	2.21695	72.5905	2.59049	1.50915	9
45	1.52717	71.1402	1.14023	1.88717	10
46	1.5463	74.6696	4.66958	1.25725	11
47	2.34151	90	20	3.57864	12
48	1.10737	71.1989	1.19893	1.62447	1
49	2.2491	76.6415	6.64147	2.50868	2
50	1.76659	71.7038	1.70377	1.231	3
51	1.25174	76.9657	6.96572	1.99521	4
52	1.81153	73.0722	3.07225	2.15915	5
53	1.72942	71.9639	1.96392	1.86142	6
54	2.17748	78.1207	8.12068	2.54388	7
55	1.29186	77.0589	7.05886	1.82777	8
56	1.92399	72.6126	2.61256	1.32816	9
57	1.38008	70.8872	0.887228	1.37826	10
58	1.96143	73.8529	3.85289	1.87809	11
59	1.61795	74.6957	4.69565	1.65806	12
60	2.02756	75.7877	5.78773	1.72684	1
61	2.41378	75.9826	5.98255	2.76309	2
62	1.41413	71.3419	1.34194	1.75285	3
63	2.31185	72.5469	2.54685	2.27947	4
64	1.94336	71.5592	1.55922	1.96157	5
65	2.094	74.7338	4.73385	2.07885	6
66	1.19458	72.233	2.23301	1	7
67	2.13118	79.1225	9.1225	1.84193	8
68	1.48076	87.0511	17.0511	2.94927	9
69	1.98502	79.0913	9.09131	2.47104	10
70	2.25937	73.8232	3.82322	2.49798	12
71	1.18744	70.6821	0.682067	1.2848	1
72	1.20189	70.7053	0.705311	1.33293	2
73	1.69115	73.9781	3.9781	1.87517	3
74	1.0556	73.2146	3.21459	1	4
75	1.59936	71.4165	1.41653	1.29695	5
76	1.66044	70.7151	0.715145	1.22362	6
77	1.79167	74.8072	4.80722	1.86081	7
78	2.30484	71.5028	1.50285	1.60626	8
79	2.49073	71.5908	1.59084	1.80815	9
80	1.32729	70.9077	0.907698	1.12889	10
81	2.48874	83.0079	13.0079	2.59237	11
82	2.46786	84.1806	14.1806	3.35518	12
83	2.12407	73.5826	3.58261	1.98482	1
84	2.46982	76.6556	6.65559	2.48936	2
85	1.00777	70.2504	0.250364	1	3
86	1.93118	73.9276	3.92763	1.84407	4
87	1.00017	72.6359	2.63594	1.3882	5
88	1.90622	71.047	1.047	1.7595	6
89	2.43744	72.321	2.32097	1.67244	7
90	1.25712	90	20	2.63949	8
91	1.10811	71.8299	1.82987	1	9

```
92   2.25545   71.8849    1.8849   1.94247        10
93   2.47971   73.4697    3.4697   1.87842        11
94   1.93378   74.2952    4.2952   1.52478        12
95   2.17525   73.0547    3.05466  2.23563         1
96   2.18723   70.8299    0.829929 1.75177         2
97   1.69984   72.0026    2.00263  1.45564         3
98   1.12504   70.4229    0.422904 1.06042         4
99   2.41723   73.7324    3.73238  2.18307         5
;

proc iml;

   /*-- Load graphics --*/
   call gstart;


   /*--------------------*/
   /*-- Define modules --*/
   /*--------------------*/


   /*   Module : compute contours   */
   start contour(c,x,y,npoints,pvalues);

   /*   This routine computes contours for a scatter plot      */
   /*   c returns the contours as consecutive pairs of columns */
   /*   x and y are the x and y coordinates of the points      */
   /*   npoints is the number of points in a contour           */
   /*   pvalues is a column vector of contour probabilities     */
   /*   the number of contours is controlled by the ncol(pvalue) */


     xx=x||y;
     n=nrow(x);
   /* Correct for the mean */
     mean=xx[+,]/n;
     xx=xx-mean@j(n,1,1);

   /* Find principle axes of ellipses */
     xx=xx`*xx/n;
     call eigen(v,e,xx);

   /* Set contour levels */
     c=-2*log(1-pvalues);
     a=sqrt(c*v[1]); b=sqrt(c*v[2]);

   /* Parameterize the ellipse by angle */
     t=((1:npoints)-(1))#atan(1)#8/(npoints-1);
     s=sin(t);
     t=cos(t);
     s=s`*a;
     t=t`*b;
```

```
/* Form contour points */
   s=((e*(shape(s,1)//shape(t,1)))+mean`@j(1,npoints*ncol(c),1) )`;
   c=shape(s,npoints);

/* Returned as ncol pairs of columns for contours */
finish contour;
/*-- Module : draw contour curves --*/
start gcontour(t1, t2);
   run contour(t12, t1, t2, 30, {.5 .8 .9});
   window=(min(t12[,{1 3}],t1)||min(t12[,{2 4}],t2))//
          (max(t12[,{1 3}],t1)||max(t12[,{2 4}],t2));
   call gwindow(window);
   call gdraw(t12[,1],t12[,2],,'blue');
   call gdraw(t12[,3],t12[,4],,'blue');
   call gdraw(t12[,5],t12[,6],,'blue');
   call gpoint(t1,t2,,'red');
finish gcontour;



/*-- Module : find median, quartiles for box and whisker plot --*/
start boxwhskr(x, u, q2, m, q1, l);
   rx=rank(x);
   s=x;
   s[rx,]=x;
   n=nrow(x);

/*-- Median --*/
   m=floor(((n+1)/2)||((n+2)/2));
   m=(s[m,])[+,]/2;

/*-- Compute quartiles --*/
   q1=floor(((n+3)/4)||((n+6)/4));
   q1=(s[q1,])[+,]/2;
   q2=ceil(((3*n+1)/4)||((3*n-2)/4));
   q2=(s[q2,])[+,]/2;
   h=1.5*(q2-q1);          /*-- step=1.5*(interquartile range) --*/
   u=q2+h;
   l=q1-h;
   u=(u>s)[+,];            /*-- adjacent values -----------------*/
   u=s[u,];
   l=(l>s)[+,];
   l=s[l+1,];

finish boxwhskr;



/*-- Box and Whisker plot --*/
start gbxwhskr(t, ht);
   run boxwhskr(t, up, q2,med, q1, lo);
```

```
/*---Adjust screen viewport and data window  */
   y=min(t)//max(t);
   call gwindow({0, 100} || y);
   mid  = 50;
   wlen = 20;

/*-- Add whiskers */
   wstart=mid-(wlen/2);
   from=(wstart||up)//(wstart||lo);
   to=((wstart//wstart)+wlen)||from[,2];

/*-- Add box  */
   len=50;
   wstart=mid-(len/2);
   wstop=wstart+len;
   from=from//(wstart||q2)//(wstart||q1)//
        (wstart||q2)//(wstop||q2);
   to=to//(wstop||q2)//(wstop||q1)//
        (wstart||q1)//(wstop||q1);

/*---Add median line  */
   from=from//(wstart||med);
   to=to//(wstop||med);

/*---Attach whiskers to box  */
   from=from//(mid||up)//(mid||lo);
   to=to//(mid||q2)//(mid||q1);

/*-- Draw box and whiskers  */
   call gdrawl(from, to,,'red');

/*---Add minimum and maximum data points */
   call gpoint(mid, y ,3,'red');

/*---Label min, max, and mean  */
   y=med//y;
   s={'med' 'min' 'max'};
   call gset("font","swiss");
   call gset('height',13);
   call gscript(wstop+ht, y, char(y,5,2),,,,,'blue');
   call gstrlen(len, s);
   call gscript(wstart-len-ht,y,s,,,,,'blue');
   call gset('height');
finish gbxwhskr;


/*-- Module : do scatterplot matrix --*/
start gscatmat(data, vname);
   call gopen('scatter');
   nv=ncol(vname);
   if (nv=1) then nv=nrow(vname);
   cellwid=int(90/nv);
   dist=0.1*cellwid;
   width=cellwid-2*dist;
   xstart=int((90 -cellwid * nv)/2) + 5;
   xgrid=((0:nv)#cellwid + xstart)`;
```

```
    /*-- Delineate cells --*/
       cell1=xgrid;
       cell1=cell1||(cell1[nv+1]//cell1[nv+1-(0:nv-1)]);
       cell2=j(nv+1, 1, xstart);
       cell2=cell1[,1]||cell2;
       call gdrawl(cell1, cell2);
       call gdrawl(cell1[,{2 1}], cell2[,{2 1}]);
       xstart = xstart + dist;  ystart = xgrid[nv] + dist;

    /*-- Label variables ---*/
       call gset("height", 5);
       call gset("font","swiss");
       call gstrlen(len, vname);
       where=xgrid[1:nv] + (cellwid-len)/2;
       call gscript(where, 0, vname) ;
       len=len[nv-(0:nv-1)];
       where=xgrid[1:nv] + (cellwid-len)/2;
       call gscript(4,where, vname[nv - (0:nv-1)],90);

    /*-- First viewport --*/
       vp=(xstart || ystart)//((xstart || ystart) + width) ;

    /*  Since the characters are scaled to the viewport    */
    /*   (which is inversely porportional to the           */
    /*   number of variables),                             */
    /*   enlarge it proportional to the number of variables */

       ht=2*nv;
       call gset("height", ht);
       do i=1 to nv;
          do j=1 to i;
             call gportstk(vp);
             if (i=j) then run gbxwhskr(data[,i], ht);
             else run gcontour(data[,j], data[,i]);
    /*-- onto the next viewport --*/
             vp[,1] = vp[,1] + cellwid;
             call gportpop;
          end;
          vp=(xstart // xstart + width) || (vp[,2] - cellwid);
       end;
       call gwindow({0 0 100 100});
       call gshow;
finish gscatmat;
```

```
                    /*-- Placement of text is based on the character height.      */
                    /* The IML modules defined here assume percent as the unit of  */
                    /* character height for device independent control.            */
               goptions gunit=pct;

               use factory;
               vname={prod, temp, defect};
               read all var vname into xyz;
               run gscatmat(xyz, vname[1:2]);    /*-- 2 x 2 scatterplot matrix --*/
               run gscatmat(xyz, vname);         /*-- 3 x 3 scatterplot matrix --*/
               quit;

               goptions gunit=cell;              /*-- reset back to default --*/
```

Figure 10.1   *2✕2 Scatterplot Matrix*

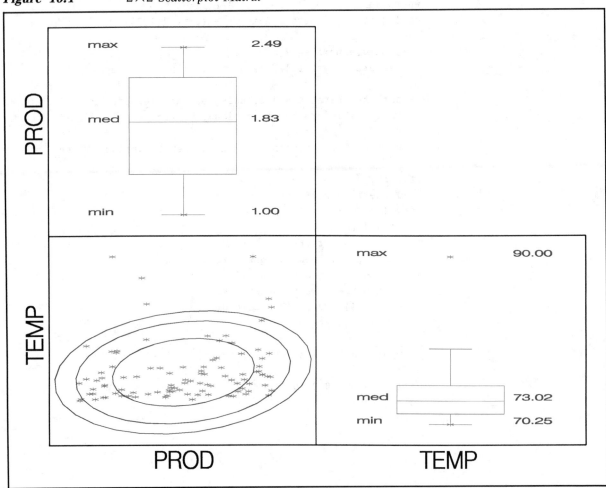

Figure 10.2   *3×3 Scatterplot Matrix*

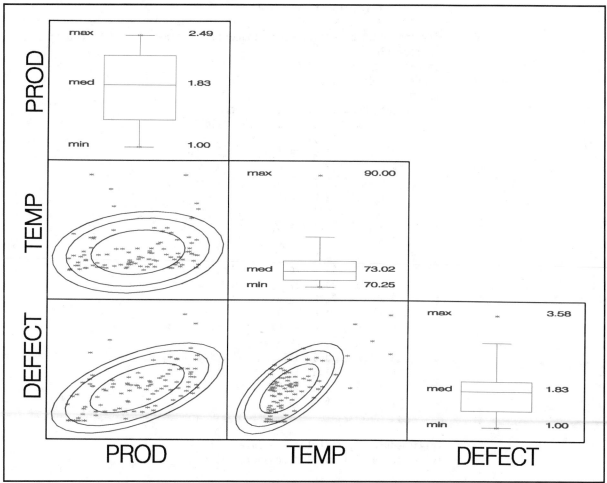

Example 2: Train Schedule

This example draws a grid on which the horizontal dimension gives the arrival/departure data and the vertical dimension gives the destination. The first section of the code defines the matrices used. The following section generates the graph. The example code given below shows some applications of the GGRID, GDRAWL, GSTRLEN, and GSCRIPT subroutines. The code below produces Figure 10.3.

```
proc iml;
    /*  Placement of text is based on the character height.    */
    /*  The graphics segment defined here assumes percent as the */
    /*  unit of character height for device independent control. */
    goptions gunit=pct;
```

```
call gstart;
/* Define several necessary matrices */
cityloc={0 27 66 110 153 180}`;
   cityname={"Paris" "Montereau" "Tonnerre" "Dijon" "Macon" "Lyons"};
   timeloc=0:30;
   timename=char(timeloc,2,0);
   /* Define a data matrix */
   schedule=
      /* origin dest start  end        comment */
         { 1     2    11.0  12.5,  /* train 1 */
           2     3    12.6  14.9,
           3     4    15.5  18.1,
           4     5    18.2  20.6,
           5     6    20.7  22.3,
           6     5    22.6  24.0,
           5     4     0.1   2.3,
           4     3     2.5   4.5,
           3     2     4.6   6.8,
           2     1     6.9   8.5,
           1     2    19.2  20.5,  /* train 2 */
           2     3    20.6  22.7,
           3     4    22.8  25.0,
           4     5     1.0   3.3,
           5     6     3.4   4.5,
           6     5     6.9   8.5,
           5     4     8.6  11.2,
           4     3    11.6  13.9,
           3     2    14.1  16.2,
           2     1    16.3  18.0
         };

   xy1=schedule[,3]||cityloc[schedule[,1]];
   xy2=schedule[,4]||cityloc[schedule[,2]];

   call gopen;
   call gwindow({-8 -35, 36 240});
   call ggrid(timeloc,cityloc,1,"red");
   call gdrawl(xy1,xy2,,"blue");

   /*-- center title -- */
   s = "Train Schedule: Paris to Lyons";
   call gstrlen(m, s,5,"titalic");
   call gscript(15-m/2,185,s,,,5,"titalic");

   /*-- find max graphics text length of cityname --*/
   call gset("height",3);
   call gset("font","italic");
   call gstrlen(len, cityname);
   m = max(len) +1.0;
   call gscript(-m, cityloc,cityname);
   call gscript(timeloc - .5,-12,timename,-90,90);
   call gshow;

quit;
goptions gunit=cell;                    /*-- reset back to default --*/
```

Figure 10.3 Train Schedule

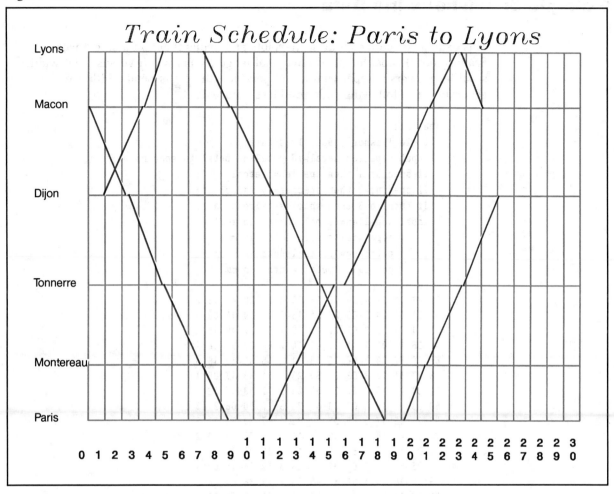

Example 3: Fisher's Iris Data

This example generates four scatterplots and prints them on a single page. Scatterplots of sepal length versus petal length, sepal width versus petal width, sepal length versus sepal width, and petal length versus petal width are generated. The following code produces Figure 10.4.

```
data iris;
    title 'Fisher (1936) Iris Data';
    input sepallen sepalwid petallen petalwid spec_no aa;
    if spec_no=1 then species='setosa    ';
    if spec_no=2 then species='versicolor';
    if spec_no=3 then species='virginica ';
    label sepallen='sepal length in mm.'
          sepalwid='sepal width  in mm.'
          petallen='petal length in mm.'
          petalwid='petal width  in mm.';
    cards;
50 33 14 02 1 64 28 56 22 3 65 28 46 15 2
67 31 56 24 3 63 28 51 15 3 46 34 14 03 1
69 31 51 23 3 62 22 45 15 2 59 32 48 18 2
46 36 10 02 1 61 30 46 14 2 60 27 51 16 2
65 30 52 20 3 56 25 39 11 2 65 30 55 18 3
58 27 51 19 3 68 32 59 23 3 51 33 17 05 1
57 28 45 13 2 62 34 54 23 3 77 38 67 22 3
63 33 47 16 2 67 33 57 25 3 76 30 66 21 3
49 25 45 17 3 55 35 13 02 1 67 30 52 23 3
70 32 47 14 2 64 32 45 15 2 61 28 40 13 2
48 31 16 02 1 59 30 51 18 3 55 24 38 11 2
63 25 50 19 3 64 32 53 23 3 52 34 14 02 1
49 36 14 01 1 54 30 45 15 2 79 38 64 20 3
44 32 13 02 1 67 33 57 21 3 50 35 16 06 1
58 26 40 12 2 44 30 13 02 1 77 28 67 20 3
63 27 49 18 3 47 32 16 02 1 55 26 44 12 2
50 23 33 10 2 72 32 60 18 3 48 30 14 03 1
51 38 16 02 1 61 30 49 18 3 48 34 19 02 1
50 30 16 02 1 50 32 12 02 1 61 26 56 14 3
64 28 56 21 3 43 30 11 01 1 58 40 12 02 1
51 38 19 04 1 67 31 44 14 2 62 28 48 18 3
49 30 14 02 1 51 35 14 02 1 56 30 45 15 2
58 27 41 10 2 50 34 16 04 1 46 32 14 02 1
60 29 45 15 2 57 26 35 10 2 57 44 15 04 1
50 36 14 02 1 77 30 61 23 3 63 34 56 24 3
58 27 51 19 3 57 29 42 13 2 72 30 58 16 3
54 34 15 04 1 52 41 15 01 1 71 30 59 21 3
64 31 55 18 3 60 30 48 18 3 63 29 56 18 3
49 24 33 10 2 56 27 42 13 2 57 30 42 12 2
55 42 14 02 1 49 31 15 02 1 77 26 69 23 3
60 22 50 15 3 54 39 17 04 1 66 29 46 13 2
52 27 39 14 2 60 34 45 16 2 50 34 15 02 1
44 29 14 02 1 50 20 35 10 2 55 24 37 10 2
58 27 39 12 2 47 32 13 02 1 46 31 15 02 1
69 32 57 23 3 62 29 43 13 2 74 28 61 19 3
```

```
59 30 42 15 2 51 34 15 02 1 50 35 13 03 1
56 28 49 20 3 60 22 40 10 2 73 29 63 18 3
67 25 58 18 3 49 31 15 01 1 67 31 47 15 2
63 23 44 13 2 54 37 15 02 1 56 30 41 13 2
63 25 49 15 2 61 28 47 12 2 64 29 43 13 2
51 25 30 11 2 57 28 41 13 2 65 30 58 22 3
69 31 54 21 3 54 39 13 04 1 51 35 14 03 1
72 36 61 25 3 65 32 51 20 3 61 29 47 14 2
56 29 36 13 2 69 31 49 15 2 64 27 53 19 3
68 30 55 21 3 55 25 40 13 2 48 34 16 02 1
48 30 14 01 1 45 23 13 03 1 57 25 50 20 3
57 38 17 03 1 51 38 15 03 1 55 23 40 13 2
66 30 44 14 2 68 28 48 14 2 54 34 17 02 1
51 37 15 04 1 52 35 15 02 1 58 28 51 24 3
67 30 50 17 2 63 33 60 25 3 53 37 15 02 1
;

proc iml;

use iris; read all;

    /*-------------------------------------------------------  */
    /*  Create 5 graphs, PETAL, SEPAL, SPWID, SPLEN, and ALL4 */
    /*   After the graphs are created, to see any one, type   */
    /*            CALL GSHOW("name");                         */
    /*   where name is the name of any one of the 5 graphs    */
    /* -------------------------------------------------------  */


call gstart;                      /*-- always start with GSTART --*/

    /*-- Spec_no will be used as marker index, change 3 to 4 */
    /*-- 1 is + , 2 is x, 3 is *, 4 is a square -------------*/
do i=1 to 150;
    if (spec_no[i] = 3) then spec_no[i] = 4;
end;

    /*-- Creates 4 x-y plots stored in 4 different segments */

    /*-- Creates a segment called petal, petallen by petalwid --*/
call gopen("petal");
    wp = { -10 -5, 90 30};
    call gwindow(wp);
    call gxaxis({0 0}, 75, 6,,,'5.1');
    call gyaxis({0 0}, 25, 5,,,'5.1');
    call gpoint(petallen, petalwid, spec_no, 'blue');
    labs = "Petallen vs Petalwid";
    call gstrlen(len, labs,2, 'swiss');
    call gscript(40-len/2,-4,labs,,,2,'swiss');
```

```
                    /*-- Creates a segment called sepal, sepallen by sepalwid --*/
                call gopen("sepal");
                    ws = {35 15 85 55};
                    call gwindow(ws);
                    call gxaxis({40 20}, 40, 9, , ,'5.1');
                    call gyaxis({40 20}, 28, 7, , ,'5.1');
                    call gpoint(sepallen, sepalwid, spec_no, 'blue');
                    labs = "Sepallen vs Sepalwid";
                    call gstrlen(len, labs,2, 'swiss');
                    call gscript(60-len/2,16,labs,,,2,'swiss');


                    /*-- Creates a segment called spwid, petalwid by sepalwid --*/
                call gopen("spwid");
                    wspwid = { 15 -5 55 30};
                    call gwindow(wspwid);
                    call gxaxis({20 0}, 28, 7,,,'5.1');
                    call gyaxis({20 0}, 25, 5,,,'5.1');
                    call gpoint(sepalwid, petalwid, spec_no, 'green');
                    labs = "Sepalwid vs Petalwid";
                    call gstrlen(len, labs,2,'swiss');
                    call gscript(35-len/2,-4,labs,,,2,'swiss');


                    /*-- Creates a segment called splen, petallen by sepallen --*/
                call gopen("splen");
                    wsplen = {35 -15 85 90};
                    call gwindow(wsplen);
                    call gxaxis({40 0}, 40, 9,,,'5.1');
                    call gyaxis({40 0}, 75, 6,,,'5.1');
                    call gpoint(sepallen, petallen, spec_no, 'red');
                    labs = "Sepallen vs Petallen";
                    call gstrlen(len, labs,2,'swiss');
                    call gscript(60-len/2,-14,labs,,,2,'swiss');


                    /*-- Create a new segment */
                call gopen("all4");
                    call gport({50 0, 100 50});   /* change viewport, lower right ----*/
                    call ginclude("sepal");       /* include sepal in this graph -----*/
                    call gport({0 50, 50 100});   /* change the viewport, upper left */
                    call ginclude("petal");       /* include petal ------------------*/
                    call gport({0 0, 50 50});     /* change the viewport, lower left */
                    call ginclude("spwid");       /* include spwid ------------------*/
                    call gport({50 50, 100 100});/* change the viewport, upper right */
                    call ginclude("splen");       /* include splen ------------------*/

                call gshow("Petal");
```

Figure 10.4   *Petal Length versus Petal Width*

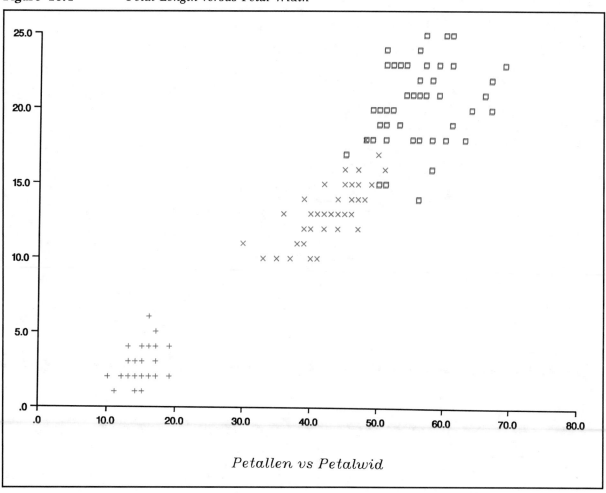

Petallen vs Petalwid

Chapter 11 Window and Display Features

Introduction

The dynamic nature of IML gives you the ability to create windows on your display for full-screen data entry or menuing. Using the WINDOW statement, you can define a window, its fields, and its attributes. Using the DISPLAY statement, you can display a window and await data entry.

These statements are similar in form and function to the corresponding statements in the SAS DATA step. The specification of fields in the WINDOW or DISPLAY statements is similar to the specifications used in the INPUT and PUT statements. Using these statements you can write applications that behave similarly to other full-screen facilities in the SAS System, such as the BUILD procedure in SAS/AF software and the FSEDIT procedure in SAS/FSP software.

Creating a Display Window for Data Entry

Suppose your application is a data entry system for a mailing list. You want to create a data set called MAILLIST by prompting the user with a window that displays all the entry fields. You want the data entry window to look as follows:

```
+--MAILLIST------------------------------------------------+
| Command==>                                               |
|                                                          |
|                                                          |
| NAME:                                                    |
| ADDRESS:                                                 |
| CITY: STATE: ZIP:                                        |
| PHONE:                                                   |
|                                                          |
+----------------------------------------------------------+
```

The process for creating a display window for this application consists of

1. initializing the variables

2. creating a SAS data set

3. defining a module for collecting data that

   a. defines a window

   b. defines the data fields

   c. defines a loop for collecting data

   d. provides an exit from the loop

4. executing the data-collecting routine.

The whole system can be implemented with the following code to define modules INITIAL and MAILGET:

```
/*   module to initialize the variables                   */
/*                                                        */
start initial;
  name='         ';
  addr='                      ';
  city='                      ';
  state='  ';
  zip='     ';
  phone='          ';
finish initial;
```

This defines a module named INITIAL that initializes the variables you want to collect. The initialization sets the string length for the character fields. You need to do this prior to creating your data set.

Now define a module for collecting the data:

```
/* module to collect data                                    */
/*                                                           */
start mailget;
  /* define the window                                       */
  window maillist cmndline=cmnd msgline=msg
    group=addr
    #2 " NAME: " name
    #3 " ADDRESS:" addr
    #4 " CITY: " city +2 "STATE: " state +2 "ZIP: " zip
    #5 " PHONE: " phone;
  /*                                                          */
  /* collect addresses until the user enters exit            */
  /*                                                          */
  do until(cmnd="EXIT");
    run initial;
    msg="ENTER SUBMIT TO APPEND OBSERVATION, EXIT TO END";
  /*                                                          */
  /* loop until user types submit or exit                    */
  /*                                                          */
    do until(cmnd="SUBMIT"|cmnd="EXIT");
      display maillist.addr;
    end;
    if cmnd="SUBMIT" then append;
  end;
  window close=maillist;
finish mailget;
  /* initialize variables                                    */
run initial;
  /* create the new data set                                 */
create maillist var{name addr city state zip phone};
  /* collect data                                            */
run mailget;
  /* close the new data set                                  */
close maillist;
```

In the module MAILGET, the WINDOW statement creates a window named MAILLIST with a group of fields (the group is named ADDR) presenting data fields for data entry. The program sends messages to the window through the MSGLINE= variable MSG. The program receives commands you enter through the CMNDLINE= variable CMND.

You can enter data into the fields after each prompt field. After you are finished with the entry, press a key defined as SUBMIT, or type SUBMIT in the command field. The data are appended to the data set MAILLIST. When data entry is complete, enter EXIT in the command field. If you enter a command other than SUBMIT, EXIT, or a valid display manager command in the command field, you get the message on the message line:

ENTER SUBMIT TO APPEND OBSERVATION, EXIT TO END.

Using the WINDOW Statement

You use the WINDOW statement to define a window, its fields, and its attributes. The general form of the WINDOW statement is

WINDOW <CLOSE=>*window-name* <*window-options*>
<GROUP=*group-name-1 field-specs*
<...GROUP=*group-name-n field-specs*>>;

The following options can be used with the WINDOW statement:

CLOSE=	is used only when you want to close the window.
window-name	is a valid SAS name for the window. This name is displayed in the upper left border of the window.
window-options	control the size, position, and other attributes of the window. You can change the attributes interactively with window commands such as WGROW, WDEF, WSHRINK, and COLOR. These options are described in the next section.
GROUP=*group-name*	starts a repeating sequence of groups of fields defined for the window. The *group-name* is a valid SAS variable name used to identify a group of fields in a DISPLAY statement that occurs later in the program.
field-specs	is a sequence of field specifications made up of positionals, field operands, formats, and options. These are described in "Field Specifications" later in this chapter.

Window Options

Window-options control the attributes of the window. The following options are valid in the WINDOW statement:

CMNDLINE=*name*
   names a character variable in which the command line entered by the user is stored.

COLOR=*operand*
   specifies the background color for the window. The *operand* can be either a quoted character literal or the name of a character variable containing the color. The valid values are BLACK, GREEN, MAGENTA, RED, CYAN, GRAY, and BLUE. BLACK is the default.

COLUMNS=*operand*
   specifies the starting number of columns of the window. The *operand* can be either a literal number, a variable name, or an expression in parentheses. The default is 78.

ICOLUMN=*operand*

    specifies the initial column position of the window on the display screen. The *operand* can be either a literal number or a variable name. The default is column 1.

IROW=*operand*

    specifies the initial row position of the window on the display screen. The *operand* can be either a literal number or a variable name. The default is row 1.

MSGLINE=*operand*

    specifies the message to be displayed on the standard message line when the window is made active. The *operand* is a quoted character literal or the name of a character variable containing the message.

ROWS=*operand*

    determines the starting number of rows of the window. The *operand* is either a literal number, the name of a variable containing the number, or an expression in parentheses yielding the number. The default is 23 rows.

Field Specifications

Both the WINDOW and DISPLAY statements allow field specifications. Field specifications have the general form

<positionals> field-operand <format> <field-options>

Positionals

The *positionals* are directives specifying the position on the screen in which to begin the field. There are four kinds of positionals, any number of which are allowed for each field operand. Positionals are the following:

# *operand*	specifies the row position; that is, it moves the current position to column 1 of the specified line. The *operand* is either a number, a variable name, or an expression in parentheses. The expression must evaluate to a positive number.
/	instructs IML to go to column 1 of the next row.
@ *operand*	specifies the column position. The *operand* is either a number, a variable name, or an expression in parentheses. The @ directive should come after the pound sign (#) positional, if it is specified.
+ *operand*	instructs IML to skip columns. The *operand* is either a number, a variable name, or an expression in parentheses.

Field Operands

The *field-operand* specifies what goes in the field. It is either a character literal in quotes or the name of a character variable.

Formats

The *format* is the format used for display, for the value, and also as the informat applied to entered values. If no format is specified, the standard numeric or character format is used.

Field Options

The *field-options* specify the attributes of the field as follows:

PROTECT=YES P=YES	specifies that the field is protected; that is, you cannot enter values in the field. If the field operand is a literal, it is already protected.
COLOR=*operand*	specifies the color of the field. The *operand* can be either a literal character value in quotes, a variable name, or an expression in parentheses. The colors available are WHITE, BLACK, GREEN, MAGENTA, RED, YELLOW, CYAN, GRAY, and BLUE. BLUE is the default. Note that the color specification is different from that of the corresponding DATA step value because it is an operand rather than a name without quotes.

Using the DISPLAY Statement

After you have opened a window with the WINDOW statement, you can use the DISPLAY statement to display the fields in the window.

The DISPLAY statement specifies a list of groups to be displayed. Each group is separated from the next by a comma. The general form of the DISPLAY statement is

DISPLAY <*group-spec-1 group-options*,<*...,group-spec-n group-options*>>;

Group Specifications

The group specification names a group, either a compound name of the form *windowname.groupname* or a *windowname* followed by a group defined by fields and enclosed in parentheses. For example, you can specify *windowname.groupname* or *windowname(field-specs)*, where *field-specs* are as defined earlier for the WINDOW statement.

In the example, you used the statement

```
display maillist.addr;
```

to display the window MAILLIST and the group ADDR.

Group Options

The *group-options* can be any of the following:

BELL
> rings the bell, sounds the alarm, or beeps the speaker at your workstation when the window is displayed.

NOINPUT
> requests that the group be displayed with all the fields protected so that no data entry can be done.

REPEAT
> specifies that the group be repeated for each element of the matrices specified as *field-operands*. See "Repeating Fields" later in this chapter.

Details about Windows

The following sections discuss some of the ideas behind windows.

The Number and Position of Windows

You can have any number of windows. They can overlap each other or be disjoint. Each window behaves independently from the others. You can specify the starting size, position, and color of the window when you create it. Each window responds to display manager commands so that it can be moved, sized, popped, or changed in color dynamically by the user.

You can list all active windows in a session by using the SHOW WINDOWS command. This makes it easy to keep track of multiple windows.

Windows and the Display Surface

A window is really a viewport into a display. The display can be larger or smaller than the window. If the display is larger than the window, you can use scrolling commands to move the surface under the window (or equivalently, move the window over the display surface). The scrolling commands are as follows:

RIGHT $<n>$	scrolls right.
LEFT $<n>$	scrolls left.
FORWARD $<n>$	scrolls forward (down).
BACKWARD $<n>$	scrolls backward (up).
TOP	scrolls to the top of the display surface.
BOTTOM	scrolls to the bottom of the display surface.

The argument *n* is an optional numeric argument that indicates the number of positions to scroll. The default is 5.

Only one window is active at a time. You can move, zoom, enlarge, shrink, or recolor inactive windows, but you cannot scroll or enter data.

Each display starts with the same standard lines: first a command line for entering commands, then a message line for displaying messages (such as error messages).

The remainder of the display is up to you to design. You can put fields in any positive row and column position of the display surface, even if it is off the displayed viewport.

Deciding Where to Define Fields

You have a choice of whether to define your fields in the WINDOW statement, the DISPLAY statement, or both. Defining field groups in the WINDOW statement saves work if you access the window from many different DISPLAY statements. Specifying field groups in the DISPLAY statement provides more flexibility.

Groups of Fields

All fields must be part of field groups. The group is just a mechanism to treat multiple fields together as a unit in the DISPLAY statement. There is only one rule about the field positions of different groups: active fields must not overlap. Overlapping is acceptable among fields as long as they are not simultaneously active. Active fields are the ones that are specified together in the current DISPLAY statement.

You name groups specified in the WINDOW statement. You specify groups in the DISPLAY statement just by putting them in parentheses; they are not named.

Field Attributes

There are two types of fields you can define:

□ Protected fields are for constants on the screen

□ Unprotected fields accept data entry.

If the field consists of a character string in quotes, it is protected. If the field is a variable name, it is not protected unless you specify PROTECT=YES as a field option. If you want all fields protected, specify the NOINPUT group option in the DISPLAY statement.

Display Execution

When you execute a DISPLAY statement, the SAS System displays the window with all current values of the variables. You can then enter data into the unprotected fields. All the basic editing keys (cursor controls, delete, end, insert, and so forth) work, as well as display manager commands to scroll or otherwise manage the window. Control does not return to the IML code until you enter a command on the command line that is not recognized as a display manager command. Typically, a SUBMIT command is used since most users define a function key for this command. Before control is returned to you, IML moves all modified field values from the screen back into IML variables using standard or specified informat routines. If you have specified the CMNDLINE= option in the

WINDOW statement, the current command line is passed back to the specified variable.

The window remains visible with the last values entered until the next DISPLAY statement or until the window is closed by a WINDOW statement with the CLOSE= option.

Only one window is active at a time. Every window may be subject to display manager commands, but only the window specified in the current DISPLAY statement transfers data to IML.

Each window is composed dynamically every time it is displayed. If you position fields by variables, you can make them move to different parts of the screen simply by programming the values of the variables.

The DISPLAY statement even allows general expressions in parentheses as positional or field operands. The WINDOW statement only allows literal constants or variable names as operands. If a field operand is an expression in parentheses, then it is always a protected field. You cannot use the statement

```
display w(log(X));
```

and expect it to return the log function of the data entered. You would need the following code to do that:

```
lx=log(x);
display w(lx);
```

Field Formatting and Inputting

The length of a field on the screen is specified in the format after the field operand, if you give one. If a format is not given, IML uses standard character or numeric formats and informats. Numeric informats allow scientific notation and missing values (represented with periods). The default length for character variables is the size of the variable element. The default size for numeric fields is as given with the FW= option (see the discussion of the RESET statement in Chapter 15, "SAS/IML Language Reference"). If you specify a named format (such as DATE7.), IML attempts to use it for both the output format and input informat. If IML cannot find an input informat of that name, it uses the standard informats.

Display-only Windows

If a window consists only of protected fields, it is merely displayed; that is, it does not wait for user input. These display-only windows can be displayed rapidly.

Opening Windows

The WINDOW statement is executable. When a WINDOW statement is executed, it checks to see if the specific window has already been opened. If it has not been opened, then the WINDOW statement opens it; otherwise, the WINDOW statement does nothing.

Closing Windows

To close a window, use the CLOSE= option in the WINDOW statement. In the example given earlier, you closed MAILLIST with the statement

```
window close=maillist;
```

Repeating Fields

If you specify an operand for a field that is a multi-element matrix, the routines deal with the first value of the matrix. However, there is a special group option, REPEAT, that allows you to display and retrieve values from all the elements of a matrix. If the REPEAT option is specified, IML determines the maximum number of elements of any field-operand matrix, and then it repeats the group that number of times. If any field operand has fewer elements, the last element is repeated the required number of times (the last one becomes the data entered). Be sure to write your specifications so that the fields do not overlap. If the fields overlap, an error message results. Although the fields must be matrices, the positional operands are never treated as matrices.

The repeat feature can come in very handy in situations where you want to menu a list of items. For example, suppose you want to build a restaurant billing system and you have stored the menu items and prices in the matrices ITEM and PRICE. You want to obtain the quantity ordered in a matrix called AMOUNT. Enter the following code:

```
item={ "Hamburger", "Hot Dog", "Salad Bar", "Milk" };
price={1.10 .90 1.95 .45};
amount= repeat(0,nrow(item),1);
window menu
group=top
#1 a2 "Item" a44 "Price" a54 "Amount"
group=list
/ a2 item $10. a44 price 6.2 a54 amount 4.
;
display menu.top, menu.list repeat;
```

This defines the window

```
+-----Menu---------------------------------------+
+  Command --->                                  +
+                                                +
+  Item                 Price   Amount           +
+                                                +
+  Hamburger            1.10    0                +
+  Hot Dog              0.90    0                +
+  Salad Bar            1.95    0                +
+  Milk                 0.45    0                +
+                                                +
+------------------------------------------------ +
```

Example

The example below illustrates the following features:

□   multiple windows

□   the repeat feature

□   command- and message-line usage

□· a large display surface needing scrolling

□   windows linked to data set transactions.

This example uses two windows, FIND and ED. The FIND window instructs you to enter a name. Then a data set is searched for all the names starting with the entered value. If no observations are found, you receive the following message:

```
Not found, enter request
```

If any observations are found, they are displayed in the ED window. You can then edit all the fields. If several observations are found, you need to use the scrolling commands to view the entire display surface. If you enter the SUBMIT command, the data are updated in place in the data set. Otherwise, you receive the following message:

```
Not replaced, enter request
```

If you enter a blank field for the request, you are advised that EXIT is the keyword needed to exit the system.

```
start findedit;
    window ed rows=10 columns=40 icolumn=40 cmndline=c;
    window find rows=5 columns=35 icolumn=1 msgline=msg;
    edit user.class;
    display ed ( "Enter a name in the FIND window, and this"
    / "window will display the observations "
    / "starting with that name. Then you can"
    / "edit them and enter the submit command"
    / "to replace them in the data set. Enter cancel"
    / "to not replace the values in the data set."
    /
    / "Enter exit as a name to exit the program." );
    do while(1);
       msg=' ';
       again:
       name=" ";
       display find ("Search for name: " name);
       if name=" " then
          do;
             msg='Enter exit to end';
             goto again;
          end;
```

```
            if name="exit" then goto x;
            if name="PAUSE" then
               do;
                  pause;
                  msg='Enter again';
                  goto again;
               end;
            find all where(name=:name) into p;
            if nrow(p)=0 then
               do;
                  msg='Not found, enter request';
                  goto again;
               end;
            read point p;
            display ed (//" name: " name
               " sex: " sex
               " age: " age
               /" height: " height
               " weight: " weight ) repeat;
            if c='submit' then
               do;
                  msg="replaced, enter request";
                  replace point p;
               end;
            else
               do;
                  msg='Not replaced, enter request';
               end;
         end;
         x:
         display find ("Closing Data Set and Exiting");
         close user.class;
         window close=ed;
         window close=find;
      finish findedit;
      run findedit;
```

Chapter 12 Storage Features

Introduction

SAS/IML software can store user-defined modules and the values of matrices in special library storage on disk for later retrieval. The library storage feature allows you to perform the following tasks:

□   store and reload IML modules and matrices

□   save work for a later session

□   keep records of work

□   conserve space by saving large, intermediate results for later use

□   communicate data to other applications through the library

□   store and retrieve data generally.

SAS/IML Storage Catalogs

SAS/IML storage catalogs are specially structured SAS files that are located in a SAS data library. A SAS/IML catalog contains *entries* that are either matrices or modules. Like other SAS files, SAS/IML catalogs have two-level names in the form *libref.catalog*. The first-level name, *libref*, is a name assigned to the SAS data library to which the catalog belongs. The second-level name, *catalog*, is the name of the catalog file.

The default libref is initially SASUSER, and the default catalog is IMLSTOR. Thus, the default storage catalog is called SASUSER.IMLSTOR. You can change the storage catalog with the RESET STORAGE command (see the discussion of the RESET statement in Chapter 15, "SAS/IML Language Reference"). Using this command, you can change either the catalog or the libref.

When you store a matrix, IML automatically stores the matrix name, its type, its dimension, and its current values. Modules are stored in the form of their compiled code. Once modules are loaded, they do not need to be parsed again, making their use very efficient.

Catalog Management

IML provides you with all the commands necessary to reference a particular
storage catalog, to list the modules and matrices in that catalog, to store and
remove modules and matrices, and to load modules and matrices back to IML.
The following commands allow you to perform all necessary catalog management
functions:

LOAD	recalls entries from storage.
REMOVE	removes entries from storage.
RESET STORAGE	specifies the library name.
SHOW STORAGE	lists all entries currently in storage.
STORE	saves modules or matrices to storage.

Restoring Matrices and Modules

You can restore matrices and modules from storage back into the IML active
workspace using the LOAD command. The LOAD command has the general form

LOAD;
LOAD *matrices;*
LOAD MODULE=*module*;
LOAD MODULE=*(modules)*;
LOAD MODULE=*(modules) matrices*;

Some examples of valid LOAD commands are shown below:

```
load a b c;                     /* load matrices A,B, and C  */
load module=mymod1;             /* load module MYMOD1        */
load module=(mymod1 mymod2) a b; /* load modules and matrices */
```

The special operand _ALL_ can be used to load all matrices or modules or both.
For example, if you want to load all modules, use the statement

```
load module=_all_;
```

If you want to load all matrices and modules in storage, use the LOAD command
by itself:

```
load;                /* loads all matrices and modules  */
```

The LOAD command can be used with the STORE statement to save and
restore an IML environment between sessions.

Removing Matrices and Modules

You can remove modules or matrices from the catalog using the REMOVE command. The REMOVE command has the same form as the LOAD command. Some examples of valid REMOVE statements are shown below:

```
remove a b c;                     /* remove matrices A,B, and C  */
remove module=mymod1;             /* remove module MYMOD1        */
remove module=(mymod1 mymod2) a;  /* remove modules and matrices */
```

The special operand _ALL_ can be used to remove all matrices or modules or both. For example, if you want to remove all matrices, use the statement

```
remove _all_;
```

If you want to remove everything from storage, use the REMOVE command by itself:

```
remove;
```

Specifying the Storage Catalog

To specify the name of the storage catalog, use one of the following general forms of the STORAGE= option in the RESET statement:

RESET STORAGE=*catalog*;
RESET STORAGE=*libref.catalog*;

Each time you specify the STORAGE= option, the previously opened catalog is closed before the new one is opened.

You can have any number of catalogs, but you can have only one open at a time. A SAS data library can contain many IML storage catalogs, and an IML storage catalog can contain many entries (that is, many matrices and modules).

For example, you can change the name of the storage catalog without changing the libref with the statement

```
reset storage=mystor;
```

To change the libref as well, use the statement

```
reset storage=mylib.mystor;
```

Listing Storage Entries

You can list all modules and matrices in the current storage catalog using the SHOW STORAGE command, which has the general form

SHOW STORAGE;

Storing Matrices and Modules

You can save modules or matrices in the storage catalog using the STORE command. The STORE command has the same general form as the LOAD command. Several examples of valid STORE statements are shown below:

```
store a b c;                      /* store matrices A,B, and C    */
store module=mymod1;              /* store module MYMOD1          */
store module=(mymod1 mymod2) a;   /* storing modules and matrices */
```

The special operand _ALL_ can be used to store all matrices or modules. For example, if you want to store everything, use the statement

```
store _all_ module=_all_;
```

Alternatively, to store everything, you can also enter the STORE command by itself:

```
store;
```

This can help you to save your complete IML environment before exiting an IML session. Then you can use the LOAD statement in a subsequent session to restore the environment and resume your work.

Chapter **13** Using SAS/IML® Software to Generate IML Statements

Introduction

This chapter describes ways for using SAS/IML software to generate and execute statements from within IML. You can execute statements generated at run time, execute global SAS commands under program control, or create statements dynamically to get more flexibility.

Generating and Executing Statements

You can push generated statements into the input command stream (queue) with the PUSH, QUEUE, and EXECUTE subroutines. This can be very useful in situations that require added flexibility, such as menu-driven applications or interrupt handling.

The PUSH command inserts program statements at the front of the input command stream, whereas the QUEUE command inserts program statements at the back. In either case, if they are not input to an interactive application, the statements remain in the queue until IML enters a pause state, at which point they are executed. The pause state is usually induced by a program error or an interrupt control sequence. Any subsequent RESUME statement resumes execution of the module from the point where the PAUSE command was issued. For this reason, the last statement put into the command stream for PUSH or QUEUE is usually a RESUME command.

The EXECUTE statement also pushes program statements like PUSH and QUEUE, but it executes them immediately and returns. It is not necessary to push a RESUME statement when you use the CALL EXECUTE command.

Executing a String Immediately

The PUSH, QUEUE, and EXECUTE commands are especially useful when used in conjunction with the pause and resume features because they allow you to generate a pause-interrupt command to execute the code you push and return from it via a pushed RESUME statement. In fact, this is precisely how the EXECUTE subroutine is implemented generally.

▶ *Caution . . . . Note that the push and resume features work this way only in the context of being inside modules. You cannot resume an interrupted sequence of statements in immediate mode, that is, not inside a module.* ▲

For example, suppose you collect program statements in a matrix called CODE. You push the code to the command input stream along with a RESUME statement and then execute a PAUSE statement. The PAUSE statement interrupts the execution, parses and executes the pushed code, and returns back to the original execution via the RESUME statement.

```
proc iml;
start testpush;
   print '*** ENTERING MODULE TESTPUSH ***';
   print '*** I should be 1,2,3: ';
   /* constructed code * /
   code = ' do i = 1 to 3; print i; end;    ';
   /* push code+resume */
   call push (code, 'resume;');
   /* pause interrupt  */
   pause;
   print '*** EXITING MODULE TESTPUSH ***';
finish;
```

When the PAUSE statement interrupts the program, IML then parses and executes the line:

```
do i=1 to 3; print i; end; resume;
```

The RESUME command then causes IML to resume executing the module that issued the PAUSE.

Note: The EXECUTE routine is equivalent to a PUSH command, but it also adds the push of a RESUME command, then issues a pause automatically.

A CALL EXECUTE command should only be used from inside a module because pause and resume features do not support returning to a sequence of statements in immediate mode.

Feeding an Interactive Program

Suppose that an interactive program gets responses from the statement INFILE CARDS. If you want to feed it under program control, you can push lines to the command stream that is read.

For example, suppose a subroutine prompts a user to respond **YES** before performing some action. If you want to run the subroutine and feed the **YES** response without the user being bothered, you push the response as follows:

```
      /* the function that prompts the user */
   start delall;
      file log;
      put 'Do you really want to delete all records? (yes/no)';
      infile cards;
      input answer $;
      if upcase(answer)='YES' then
         do;
            delete all;
            purge;
            print "*** FROM DELALL:
            should see End of File (no records to list)";
            list all;
         end;
   finish;
```

The latter DO group is necessary so that the pushed **YES** is not read before the RUN statement. The following example illustrates the use of the module DELALL given above:

```
      /* Create a dummy data set for delall to delete records  */
   xnum = {1 2 3, 4 5 6, 7 8 0};
   create dsnum1 from xnum;
   append from xnum;
      do;
         call push ('yes');
         run delall;
      end;
```

Calling the Operating System

Suppose that you want to construct and execute an operating system command. Just push it to the token stream in the form of an X statement and have it executed under a pause interrupt.

The following module executes any system command given as an argument:

```
start system(command);
   call push(" x '",command,"'; resume;");
   pause;
finish;
run system('listc');
```

The call generates and executes a LISTC command under MVS:

```
x 'listc'; resume;
```

Calling Display Manager

The same strategy used for calling the operating system works for SAS global statements as well, including calling display manager by generating DM statements.

The following subroutine executes a display manager command:

```
start dm(command);
   call push(" dm '",command,"'; resume;");
   pause;
finish;
run dm('log; color source red');
```

The call generates and executes the statements

```
dm 'log; color source red'; resume;
```

which take you to the LOG window, where all source code is written in red.

Executing Any Command in an EXECUTE Call

The EXECUTE command executes the statements contained in the arguments using the same facilities as a sequence of CALL PUSH, PAUSE, and RESUME statements. The statements use the same symbol environment as that of the subroutine that calls them. For example, consider the following subroutine:

```
proc iml;
start exectest;
/*  IML STATEMENTS */
  call execute ("xnum = {1 2 3, 4 5 6, 7 8 0};");
  call execute ("create dsnum1 from xnum;");
  call execute ("append from xnum;");
  call execute ("print 'DSNUM should have 3 obs and 3 var:';");
  call execute ("list all;");
/*  global (options) statement */
  call execute ("options linesize=68;");
  call execute ("print 'Linesize should be 68';");
finish;
run exectest;
```

The output generated from EXECTEST is exactly the same as if you had entered the statements one at a time:

```
DSNUM should have 3 obs and 3 var:

OBS    COL1       COL2       COL3
------ ---------- ---------- ----------
    1   1.0000     2.0000     3.0000
    2   4.0000     5.0000     6.0000
    3   7.0000     8.0000          0

Linesize should be 68
```

CALL EXECUTE could almost be programmed in IML as shown here; the difference between this and the built-in command is that the following subroutine would not necessarily have access to the same symbols as the calling environment:

```
start execute(command1,...);
   call push(command1,...," resume;");
   pause;
finish;
```

Making Operands More Flexible

Suppose that you want to write a program that prompts a user for the name of a data set. Unfortunately the USE, EDIT, and CREATE commands expect the data set name as a hardcoded operand rather than an indirect one. However, you can construct and execute a function that prompts the user for the data set name for a USE statement.

```
/* prompt the user to give dsname for use statement */
start flexible;
   file log;
   put 'What data set shall I use?';
   infile cards;
   input dsname $;
   call execute('use', dsname, ';');
finish;
run flexible;
```

If you enter USER.A, the program generates and executes the line

```
use user.a;
```

Interrupt Control

Whenever a program error or interrupt occurs, IML automatically issues a pause, which places the module in a paused state. At this time, any statements pushed to the input command queue get executed. Any subsequent RESUME statement (including pushed RESUME statements) will resume executing the module from the point where the error or interrupt occurred.

If you have a long application such as reading a large data set and you want to be able to find out where the data processing is just by entering a break-interrupt (sometimes called an attention signal), you push the interrupt text. The pushed text can, in turn, push its own text on each interrupt, followed by a RESUME statement to continue execution.

For example, suppose you have a data set called TESTDATA that has 4096 observations. You want to print the current observation number if an attention signal is given. The following code does this:

```
start obsnum;
   use testdata;
   brkcode={"print 'now on observation number',i;"
              "if (i<4096) then do;"
              "call push(brkcode);"
              "resume;"
              "end;"
              };
   call push(brkcode);
   do i=1 to 4096;
      read point i;
   end;
finish;
run obsnum;
```

After the module has been run, enter the interrupt control sequence for your operating system. Type S to suspend execution. IML prints a message telling which observation is being processed. Because the pushed code is executed at the completion of the module, the message is also printed when OBSNUM ends.

Each time the attention signal is given, OBSNUM executes the code contained in the variable BRKCODE. This code prints the current iteration number and pushes commands for the next interrupt. Note that the PUSH and RESUME commands are inside a DO group, making them conditional and ensuring that they are parsed before the effect of the PUSH command is realized.

Specific Error Control

A PAUSE command is automatically issued whenever an execution error occurs, putting the module in a holding state. If you have some way of checking for specific errors, you can write an interrupt routine to correct them during the pause state.

In this example, if a singular matrix is passed to the INV function, IML pauses and executes the pushed code to make the result for the inverse be set to missing values. The code uses the variable SINGULAR to detect if the interrupt occurred during the INV operation. This is particularly necessary because the pushed code is executed on completion of the routine, as well as on interrupts.

```
proc iml;
a = {3 3, 3 3};                              /* singular matrix    */
/* If a singular matrix is sent to the INV function,               */
/* IML normally sets the resulting matrix to be empty              */
/* and prints an error message.                                    */
b = inv(a);
print "*** A should be non-singular", a;
start singtest;
   msg="    Matrix is singular - result set to missing ";
   onerror=
      "if singular then do; b=a#.; print msg; print b;
       call push(onerror); resume; end;";
   call push(onerror);
   singular = 1;
   b = inv(a);
   singular = 0;
finish ;
call singtest;
```

The resulting output is shown below:

```
ERROR: (execution) Matrix should be non-singular.

        Error occured in module SINGTEST at line      67 column   9
        operation : INV                 at line      67 column  16
        operands  : A

A               2 rows     2 cols     (numeric)

           3        3
           3        3

   stmt: ASSIGN                         at line      67 column   9

Paused in module SINGTEST.


           MSG
                Matrix is singular - result set to missing


                                          B
                                          .           .
                                          .           .



Resuming execution in module SINGTEST.
```

General Error Control

Sometimes, you may want to process or step over errors. To do this, put all the code into modules and push a code to abort if the error count goes above some maximum. Often, you may submit a batch job and get a trivial mistake that causes an error, but you do not want to cause the whole run to fail because of it. On the other hand, if you have many errors, you do not want to let the routine run.

In the following example, up to three errors are tolerated. A singular matrix **A** is passed to the INV function, which would, by itself, generate an error message and issue a pause in the module. This module pushes three RESUME statements, so that the first three errors are tolerated. Messages are printed and execution is resumed. The DO loop in the module OOPS is executed four times, and on the fourth iteration, an ABORT statement is issued and you exit IML.

```
proc iml;
a={3 3, 3 3};                            /*  singular matrix */
   /*                                                        */
   /*  GENERAL ERROR CONTROL -- exit iml for 3 or more errors   */
   /*                                                        */
start;                               /* module will be named MAIN */
   errcode = {" if errors >= 0 then do;",
              "    errors = errors + 1;",
              "    if errors > 2 then abort;",
              "    else do; call push(errcode); resume; end;",
              " end;" } ;
   call push (errcode);
   errors = 0;
   start oops;                            /* start module OOPS */
      do i = 1 to 4;
          b = inv(a);
      end;
   finish;                                /* finish OOPS */
   run oops;
finish;                                   /* finish MAIN */
errors=-1;                                /* disable   */
run;
```

The output generated from this example is shown below:

```
ERROR: (execution) Matrix should be non-singular.

Error occured in module OOPS      at line    41 column  17
called   from    module MAIN      at line    44 column  10
operation : INV                   at line    41 column  24
operands  : A

A              2 rows     2 cols    (numeric)

          3          3
          3          3


stmt: ASSIGN                      at line    41 column  17

Paused in module OOPS.


Resuming execution in module OOPS.
ERROR: (execution) Matrix should be non-singular.

Error occured in module OOPS      at line    41 column  17
called   from    module MAIN      at line    44 column  10
operation : INV                   at line    41 column  24
operands  : A

A              2 rows     2 cols    (numeric)

          3          3
          3          3


stmt: ASSIGN                      at line    41 column  17

Paused in module OOPS.


Resuming execution in module OOPS.
ERROR: (execution) Matrix should be non-singular.

Error occured in module OOPS      at line    41 column  17
called   from    module MAIN      at line    44 column  10
operation : INV                   at line    41 column  24
operands  : A

A              2 rows     2 cols    (numeric)

          3          3
          3          3


stmt: ASSIGN                      at line    41 column  17

Paused in module OOPS.

Exiting IML.
```

Actually, in this particular case it would probably be simpler to put three RESUME statements after the RUN statement to resume execution after each of the first three errors.

Macro Interface

The pushed text is scanned by the macro processor; therefore, the text can contain macro instructions. For example, here is an all-purpose routine that shows what the expansion of any macro is, assuming that it does not have embedded double quotes:

```
    /* function: y = macxpand(x);                    */
    /* will macro-process the text in x,             */
    /* and return the expanded text in the result.   */
    /* Do not use double quotes in the argument       */
    /*                                                */
start macxpand(x);
   call execute('Y="',x,'";');
   return(y);
finish;
```

Consider the following statements:

```
%macro verify(index);
    data _null_;
        infile junk&index;
        file print;
        input;
        put _infile_;
    run;
%mend;
y = macxpand('%verify(1)');
print y;
```

The output produced is shown below:

```
    Y

    DATA _NULL_;      INFILE JUNK1;      FILE PRINT;      INPUT;
    PUT _INFILE_;     RUN;
```

IML Line Pushing Contrasted with Using the Macro Facility

The SAS macro language is a language embedded in and running on top of another language; it generates text to feed the other language. Sometimes it is more convenient to generate the text using the primary language directly rather than embedding the text generation in macros. The examples above show that this can even be done at execution time, whereas pure macro processing is done only at parse time. The advantage of the macro language is its embedded, yet independent nature: it needs little quoting, and it works for all parts of the SAS language, not just IML. The disadvantage is that it is a separate language that has its own learning burden, and it uses extra reserved characters to mark its programming constructs and variables. Consider the quoting of IML versus the embedding characters of the macro facility: IML makes you quote every text constant, whereas the macro facility makes you use the special characters percent sign (%) and ampersand (&) on every macro item. There are some languages, such as REXX, that give you the benefits of both: no macro characters and no required quotes, but the cost is that the language forces you to discipline your naming so that names are not expanded inadvertently.

Example: Full-Screen Editing

The ability to form and submit statements dynamically provides a very powerful mechanism for making systems flexible. For example, consider the building of a data entry system for a file. It is straightforward to write a system using WINDOW and DISPLAY statements for the data entry and data processing statements for the I/O, but once you get the system built, it is good only for that one file. With the ability to push statements dynamically, however, it is possible to make a system that dynamically generates the components that are customized for each file. For example, you can change your systems from static systems to dynamic systems.

To illustrate this point, consider an IML system to edit an arbitrary file, a system like the FSEDIT procedure in SAS/FSP software but programmed in IML. You cannot just write it with open code because the I/O statements hardcode the filenames and the WINDOW and DISPLAY statements must hardcode the fields. However, if you generate just these components dynamically, the problem is solved for any file, not just one.

```
proc iml;
   /*              FSEDIT                                */
   /* This program defines and stores the modules FSEINIT,  */
   /* FSEDT, FSEDIT, and FSETERM in a storage catalog called */
   /* FSED.  To use it, load the modules and issue the command */
   /* RUN FSEDIT;  The system prompts or menus the files and */
   /* variables to edit, then runs a full screen editing    */
   /* routine that behaves similar to PROC FSEDIT           */
   /*                                                   */
   /* These commands are currently supported:           */
   /*                                                   */
   /* END       gets out of the system.  The user is prompted */
   /*           as to whether or not to close the files and */
   /*           window.                                 */
   /* SUBMIT    forces current values to be written out, */
   /*           either to append a new record or replace */
   /*           existing ones                           */
   /* ADD       displays a screen variable with blank values */
   /*           for appending to the end of a file      */
   /* DUP       takes the current values and appends them to */
   /*           the end of the file                     */
   /* number    goes to that line number                */
   /* DELETE    deletes the current record after confirmation */
   /*           by a Y response                         */
   /* FORWARD1  moves to the next record, unless at eof */
   /* BACKWARD1 moves to the previous record, unless at eof */
   /* EXEC      executes any IML statement              */
   /* FIND      finds records and displays them         */
   /*                                                   */
```

```
/* Use: proc iml;                                          */
/*       reset storage='fsed';                             */
/*       load module=_all_;                                */
/*       run fsedit;                                       */
/*                                                         */
/*---routine to set up display values for new problem---   */
start fseinit;
    window fsed0 rows=15 columns=60 icolumn=18 color='GRAY'
    cmndline=cmnd group=title +30 'Editing a data set' color='BLUE';
    /*---get file name---                                  */
    _file="                    ";
    msg =
       'Please Enter Data Set Name or Nothing For Selection List';
    display  fsed0.title,
       fsed0 ( / @5  'Enter Data Set:'
                 +1 _file
                 +4 '(or nothing to get selection list)' );
    if _file=' ' then
       do;
           loop:
           _f=datasets(); _nf=nrow(_f); _sel=repeat("_",_nf,1);
           display fsed0.title,
              fsed0 (/ "Select? File Name"/) ,
              fsed0 (/ @5  _sel +1 _f protect=yes ) repeat ;
           _l = loc(_sel¬='_');
           if nrow(_l)¬=1 then
              do;
                  msg='Enter one S somewhere';
                  goto loop;
              end;
           _file = _f[_l];
       end;
    /*---open file, get number of records---               */
    call queue(" edit ",_file,";
                setin ",_file," NOBS _nobs; resume;"); pause *;
    /*---get variables---                                  */
    _var = contents();
    _nv = nrow(_var);
    _sel = repeat("_",_nv,1);
    display fsed0.title,
            fsed0 (/ "File:" _file) noinput,
            fsed0 (/ @10 'Enter S to select each var, or select none
                      to get all.'
                      // @3 'select? Variable  ' ),
                      fsed0 ( / @5  _sel +5 _var protect=yes ) repeat;
    /*---reopen if subset of variables---                  */
    if any(_sel¬='_') then
       do;
           _var = _var[loc(_sel¬='_')];
           _nv = nrow(_var);
           call push('close ',_file,'; edit ',_file,' var
           _var;resume;');pause *;
       end;
```

```
      /*---close old window---                                      */
      window close=fsed0;
      /*---make the window---*/
      call queue('window fsed columns=55 icolumn=25 cmndline=cmnd
               msgline=msg ', 'group=var/@20 "Record " _obs
               protect=yes');
      call queue( concat('/"',_var,': " color="YELLOW" ',
               _var,' color="WHITE"'));
      call queue(';');
      /*---make a missing routine---*/
      call queue('start vmiss; ');
      do i=1 to _nv;
         val = value(_var[i]);
         if type(val)='N' then call queue(_var[i],'=.;');
         else call queue(_var[i],'="',
                         cshape(' ',1,1,nleng(val)),'";');
      end;
      call queue('finish; resume;');
      pause *;
      /*---initialize current observation---*/
      _obs = 1;
      msg = Concat('Now Editing File ',_file);
   finish;
    /*                                                              */
    /*---The Editor Runtime Controller---                          */
   start fsedt;
      _old = 0; go=1;
      do while(go);
      /*--get any needed data--*/
         if any(_obs¬=_old) then do; read point _obs; _old = _obs;
      end;
      /*---display the record---*/
      display fsed.var repeat;
      cmnd = upcase(left(cmnd));
      msg=' ';
      if cmnd='END' then go=0;
      else if cmnd='SUBMIT' then
         do;
            if _obs<=_nobs then
               do;
                  replace point _obs; msg='replaced';
               end;
            else do;
                  append;
                  _nobs=_nobs+nrow(_obs);
                  msg='appended';
               end;
         end;
```

```
        else if cmnd="ADD" then
           do;
              run vmiss;
              _obs = _nobs+1;
              msg='New Record';
           end;
        else if cmnd='DUP' then
           do;
              append;
              _nobs=_nobs+1;
              _obs=_nobs;
              msg='As Duplicated';
           end;
        else if cmnd>'0' & cmnd<'999999' then
           do;
              _obs = num(cmnd);
              msg=concat('record number ',cmnd);
           end;
        else if cmnd='FORWARD1' then _obs=min(_obs+1,_nobs);
        else if cmnd='BACKWARD1' then _obs=max(_obs-1,1);
        else if cmnd='DELETE' then
           do;
              records=cshape(char(_obs,5),1,1);
              msg=concat('Enter command Y to Confirm delete of'
                         ,records);
              display fsed.var repeat;
              if (upcase(cmnd)='Y') then
                 do;
                    delete point _obs;
                    _obs=1;
                    msg=concat('Deleted Records',records);
                 end;
              else msg='Not Confirmed, Not Deleted';
           end;
        else if substr(cmnd,1,4)='FIND' then
           do;
              call execute("find all where(",
                           substr(cmnd,5),
                           ") into _obs;" );
              _nfound=nrow(_obs);
              if _nfound=0 then
                 do;
                    _obs=1;
                    msg='Not Found';
                 end;
              else
                 do;
                    msg=concat("Found ",char(_nfound,5)," records");
                 end;
           end;
```

```
    else if substr(cmnd,1,4)='EXEC' then
       do;
          msg=substr(cmnd,5);
          call execute(msg);
       end;
    else msg='Unrecognized Command; Use END to exit.';
    end;
finish;
  /*---routine to close files and windows, clean up---*/
start fseterm;
    window close=fsed;
    call execute('close ',_file,';');
    free _q;
finish;
    /*---main routine for FSEDIT---*/
start fsedit;
  if (nrow(_q)=0) then
     do;
        run fseinit;
     end;
  else msg = concat('Returning to Edit File ',_file);
  run fsedt;
  _q='_';
  display fsed ( "Enter 'q' if you want to close files and windows"
                 _q " (anything else if you want to return later"
                 pause 'paused before termination';
  run fseterm;
finish;
reset storage='fsed';
store module=_all_;
```

In Conclusion

In this chapter you learned how to use SAS/IML software to generate IML statements. You learned how to use the PUSH, QUEUE, EXECUTE, and RESUME commands to interact with the operating system or with the SAS System in display manager. You also saw how to add flexibility to programs by adding interrupt control features and by modifying error control. Finally you learned how IML compares to the SAS macro language.

Chapter **14** Further Notes

Memory and Workspace

You do not need to be concerned with the details of memory usage because memory allocation is done automatically. However, if you are interested, the following sections explain how it works.

There are two logical areas of memory, *symbolspace* and *workspace*. *Symbolspace* contains symbol table information and compiled statements. *Workspace* contains matrix data values. *Workspace* itself is divided into one or more extents.

At the start of a session, the symbolspace and the first extent of workspace are allocated automatically. More workspace is allocated as the need to store data values grows. The SYMSIZE= and WORKSIZE= options in the PROC IML statement give you control over the size of symbolspace and the size of each extent of workspace. If you do not specify these options, PROC IML uses host dependent defaults. For example, you can begin an IML session and set the SYMSIZE= and WORKSIZE= options with the statement

```
proc iml symsize=n1 worksize=n2;
```

where *n1* and *n2* are specified in kilobytes.

If the symbolspace memory becomes exhausted, more memory is automatically acquired. The symbolspace is stable memory and is not compressible like workspace. Symbolspace is recycled whenever possible for reuse as the same type of object. For example, temporary symbols may be deleted after they are used in evaluating an expression. The symbolspace formerly used by these temporaries is added to a list of free symbol-table nodes. When allocating temporary variables to evaluate another expression, IML looks for symbol-table nodes in this list first before consuming unused symbolspace.

Workspace is compressible memory. Workspace extents fill up as more matrices are defined by operations. Holes in extents appear as you free matrices or as IML frees temporary intermediate results. When an extent fills up, compression reclaims the holes that have appeared in the extent. If compression does not reclaim enough memory for the current allocation, IML allocates a new extent. This procedure results in the existence of a list of extents, each of which contains a mixture of active memory and holes of unused memory. There is always a current extent, the one in which the last allocation was made.

For a new allocation, the search for free space begins in the current extent and proceeds around the extent list until finding enough memory or returning to the current extent. If the search returns to the current extent, IML begins a second transversal of the extent list, compressing each extent until either finding sufficient memory or returning to the current extent. If the second search returns to the current extent, IML opens a new extent and makes it the current one.

If the SAS System cannot provide enough memory to open a new extent with the full extent size, IML repeatedly reduces its request by 2K. In this case, the successfully opened extent will be smaller than the standard size.

If a single allocation is larger than the standard extent size, IML requests an allocation large enough to hold the matrix.

The WORKSIZE= and SYMSIZE= options offer tools for tuning memory usage. For data intensive applications involving a few large matrices, use a high WORKSIZE= value and a low SYMSIZE= value. For symbol intensive applications involving many matrices, perhaps through the use of many IML modules, use a high SYMSIZE= value.

You can use the SHOW SPACE command to display the current status of IML memory usage. This command also lists the total number of compressions done on all extents.

Setting the DETAILS option in the RESET command prints messages in the output file when IML compresses an extent, opens a new extent, allocates a large object, or acquires more symbolspace. These messages can be useful because these actions normally occur without the user's knowledge. The information can be used to tune WORKSIZE= and SYMSIZE= values for an application. However, the default WORKSIZE= and SYMSIZE= values should be appropriate in most applications.

Do not specify a very large value in the WORKSIZE= and SYMSIZE= options unless absolutely necessary. Many of the native functions and all of the DATA step functions used are dynamically loaded at execution time. If you use a large amount of the memory for symbolspace and workspace, there may not be enough remaining to load these functions, resulting in the error message

```
Unable to load module module-name.
```

Should you run into this problem, issue a SHOW SPACE command to examine current usage. You may be able to adjust the SYMSIZE= or WORKSIZE= values.

The amount of memory your system can provide depends on the capacity of your computer and on the products installed. The following techniques for efficient memory use are recommended when memory is at a premium:

□ Free matrices as they are no longer needed using the FREE command.

□ Store matrices you will need later in external library storage using the STORE command, and then FREE their values. You can restore the matrices later using the LOAD command. See Chapter 12, "Storage Features."

□ Plan your work to use smaller matrices

Accuracy

All numbers are stored and all arithmetic is done in double-precision. The algorithms used are generally very accurate numerically. However, when many operations are performed or when the matrices are ill-conditioned, matrix operations should be used in a numerically responsible way because numerical errors add up.

Error Diagnostics

When an error occurs, several lines of messages are printed. The error description, the operation being performed, and the line and column of the source for that operation are printed. The names of the operation's arguments are also printed. Matrix names beginning with a pound sign (#) or an asterisk (*) may appear; these are temporary names assigned by the IML procedure.

If an error occurs while you are in immediate mode, the operation is not completed and nothing is assigned to the result. If an error occurs while executing statements inside a module, a PAUSE command is automatically issued. You can correct the error and resume execution of module statements with a RESUME statement.

The most common errors are described below:

□ referencing a matrix that has not been set to a value, that is, referencing a matrix that has no value associated with the matrix name

□ making a subscripting error, that is, trying to refer to a row or column not present in the matrix

□ performing an operation with nonconformable matrix arguments, for example, multiplying two matrices together that do not conform, or using a function that requires a special scalar or vector argument

□ referencing a matrix that is not square for operations that require a square matrix (for example, INV, DET, or SOLVE)

□ referencing a matrix that is not symmetric for operations that require a symmetric matrix (for example, EIGEN)

□ referencing a matrix that is singular for operations that require a nonsingular matrix (for example, INV and SOLVE)

□ referencing a matrix that is not positive definite or positive semidefinite for operations that require such matrices (for example, ROOT and SWEEP)

□ not enough memory (see "Memory and Workspace" earlier in this chapter) to perform the computations and produce the result matrices.

These errors result from the actual dimensions or values of matrices and are caught only after a statement has begun to execute. Other errors, such as incorrect number of arguments or unbalanced parentheses, are syntax errors and resolution errors and are detected before the statement is executed.

Efficiency

The IML language is an interpretive language executor that can be characterized
as follows:

□ efficient and inexpensive to compile

□ inefficient and expensive for the number of operations executed

□ efficient and inexpensive within each operation.

Therefore, you should try to substitute matrix operations for iterative loops.
There is a high overhead involved in executing each instruction; however,
within the instruction IML runs very efficiently.

Consider four methods of summing the elements of a matrix:

```
s=0;                      /* method 1 */
do i=1 to m;
   do j=1 to n;
      s=s+x(i,j);
   end;
end;
s=j(1,m)*x*j(n,1);        /* method 2 */
s=x(+,+);                 /* method 3 */
s=sum(x);                 /* method 4 */
```

Method 1 is the least efficient, method 2 is more efficient, method 3 is more
efficient yet, and method 4 is the most efficient. The greatest advantage of using
IML is reducing human programming labor.

Missing Values

An IML numeric element can have a special value called a *missing value* that
indicates that the value is unknown or unspecified. (A matrix with missing
values should not be confused with an empty or unvalued matrix, that is, a
matrix with 0 rows and 0 columns.) A numeric matrix can have any mixture of
missing and nonmissing values.

SAS/IML software supports missing values in a limited way. The operators
listed below recognize missing values and propagate them. Most matrix
operators and functions do not support missing values. For example, matrix
multiplication or exponentiation involving a matrix with missing values is not
meaningful. Also, the inverse of a matrix with missing values has no meaning.

Missing values are coded in the bit pattern of very large negative numbers,
as an I.E.E.E. "NAN" code, or as a special string, depending on the host system.

In literals, a numeric missing value is specified as a single period. In data
processing operations, you can add or delete missing values. All operations that

move values around move missing values properly. The following arithmetic operators propagate missing values:

addition (+)	subtraction (−)
multiplication (#)	division (/)
maximum (<>)	minimum (><)
modulo (MOD)	exponentiation (##)

The comparison operators treat missing values as large negative numbers. The logical operators treat missing values as zeros. The operators SUM, SSQ, MAX, and MIN check for and exclude missing values.

The subscript reduction operators exclude missing values from calculations. If all of a row or column that is being reduced is missing, then the operator returns the result indicated in the table below.

Operator	Result If All Missing
addition (+)	0
multiplication (#)	1
maximum (<>)	large negative value
minimum (><)	large positive value
sum squares (##)	0
index maximum (<:>)	1
index minimum (>:<)	1
mean (:)	missing value

Also note that, unlike the SAS DATA step, IML does not distinguish between special and generic missing values; it treats all missing values alike.

Principles of Operation

This section presents various technical details on the operation of SAS/IML software. Statements in IML go through three phases:

1. The parsing phase includes text acquisition, word scanning, recognition, syntactical analysis, and enqueuing on the statement queue. This is performed immediately as IML reads the statements.

2. The resolution phase includes symbol resolution, label and transfer resolution, and function and call resolution. Symbol resolution connects the symbolic names in the statement with their descriptors in the symbol table. New symbols can be added or old ones recognized. Label and transfer resolution connects statements and references affecting the flow of control. This connects LINK and GOTO statements with labels; it connects IF with THEN and ELSE clauses; it connects DO with END. Function-call resolution identifies functions and call routines and loads them if necessary. Each reference is checked with respect to the number of arguments allowed. Resolution phase begins after a module definition is finished or a DO group

is ended. For all other statements outside of any module or DO group, resolution begins immediately after parsing.

3. The execution phase occurs when the statements are interpreted and executed. There are two levels of execution: statement and operation. Operation-level execution involves the evaluation of expressions within a statement.

Operation-Level Execution

Operations are executed from a chain of operation elements created at parse-time and resolved later. For each operation, the interpreter performs the following:

1. prints a record of the operation if the FLOW option is on.

2. looks at the operands to make sure they have values. Only certain special operators are allowed to tolerate operands that have not been set to a value. The interpreter checks whether any argument has character values.

3. inspects the operator and gives control to the appropriate execution routine. A separate set of routines is invoked for character values.

4. checks the operands to make sure they are valid for the operation. Then the routine allocates the result matrix and any extra workspace needed for intermediate calculations. Then the work is performed. Extra workspace is freed. A return code notifies IML if the operation was successful. If unsuccessful, it identifies the problem. Control is passed back to the interpreter.

5. checks the return code. If the return code is nonzero, diagnostic routines are called to explain the problem to the user.

6. associates the results with the result arguments in the symbol table. By keeping results out of the symbol table until this time, the operation does not destroy the previous value of the symbol if an error has occurred.

7. prints the result if RESET PRINT or RESET PRINTALL is specified. The PRINTALL option prints intermediate results as well as end results.

8. moves to the next operation.

Chapter **15** SAS/IML® Language Reference

Introduction

This chapter describes all operators, statements, functions, and subroutines that can be used in SAS/IML software. All necessary details, such as arguments and operands, are included. For information about the typographical conventions and syntax conventions used to give the general form for each entry, see "Using This Book" in the front of this book.

This chapter is divided into two sections. The first section contains operator descriptions. They are in alphabetic order according to the name of the operator. The second section contains descriptions of statements, functions, and subroutines. These also are arranged alphabetically by name.

Operators

All operators available in SAS/IML software are described in this section.

Addition Operator: +

adds corresponding matrix elements

Syntax

matrix1 + matrix2
matrix + scalar

Description

The addition infix operator (+) produces a new matrix whose elements are the sums of the corresponding elements of *matrix1* and *matrix2*. The element in the first row, first column of the first matrix is added to the element in the first row, first column of the second matrix, with the sum becoming the element in the first row, first column of the new matrix, and so on.

Addition Operator *continued*

For example, the statements

```
a={1 2,
   3 4};
b={1 1,
   1 1};
c=a+b;
```

produce the matrix **C**

```
        C        2 rows     2 cols    (numeric)

                        2          3
                        4          5
```

In addition to adding conformable matrices, you can also use the addition operator to add a matrix and a scalar or two scalars. When you use the *matrix+scalar* (or *scalar+matrix*) form, the scalar value is added to each element of the matrix to produce a new matrix.

For example, you can obtain the same result as you did in the previous example with the statement

```
c=a+1;
```

When a missing value occurs in an operand, IML assigns a missing value for the corresponding element in the result.

You can also use the addition operator on character operands. In this case, the operator does elementwise concatenation exactly as the CONCAT function.

Comparison Operators: $<$ $>$ $=$ $<=$ $>=$ $^=$

compare matrix elements

Syntax

matrix1<matrix2
matrix1<=matrix2
matrix1>matrix2
matrix1>=matrix2
matrix1=matrix2
matrix1^=matrix2

Description

The comparison operators compare two matrices element by element and produce a new matrix that contains only zeros and ones. If an element comparison is true, the corresponding element of the new matrix is 1. If the comparison is not true, the corresponding element is 0. Unlike in base SAS software or the MATRIX procedure, you cannot use the English equivalents GT and LT for the greater than and less than signs. Scalar values can be used instead of matrices in any of the forms shown above.

For example, let

```
a={1 7 3,
   6 2 4};
```

and

```
b={0 8 2,
   4 1 3};
```

Evaluation of the expression

```
c=a>b;
```

results in the matrix of values

```
       C          2 rows      3 cols      (numeric)


           1          0          1
           1          1          1
```

In addition to comparing conformable matrices, you can apply the comparison operators to a matrix and a scalar. If either argument is a scalar, the comparison is between each element of the matrix and the scalar.

For example the expression

```
d=(a>=2);
```

yields the result

```
       D          2 rows      3 cols      (numeric)


           0          1          1
           1          1          1
```

If the element lengths of two character operands are different, the shorter elements are padded on the right with blanks for the comparison.

If a numeric missing value occurs in an operand, IML treats it as lower than any valid number for the comparison.

When you are making conditional comparisons, all values of the result must be nonzero for the condition to be evaluated as true.

Consider the following statement:

```
if x>=y then goto loop1;
```

The GOTO statement is executed only if every element of **x** is greater than or equal to the corresponding element in **y**. See also the descriptions of the ALL and ANY functions.

Concatenation Operator, Horizontal: ||

concatenates matrices horizontally

Syntax

matrix1 || *matrix2*

Description

The horizontal concatenation operator (||) produces a new matrix by horizontally joining *matrix1* and *matrix2*. *Matrix1* and *matrix2* must have the same number of rows, which is also the number of rows in the new matrix. The number of columns in the new matrix is the number of columns in *matrix1* plus the number of columns in *matrix2*.

For example, the statements

```
a={1 1 1,
   7 7 7};
b={0 0 0,
   8 8 8};
c=a||b;
```

result in

```
         C           2 rows      6 cols     (numeric)

      1        1        1        0        0        0
      7        7        7        8        8        8
```

Also, if

```
b={A B C,
   D E F};
```

and

```
c={"GH" "IJ",
   "KL" "MN"};
```

then

```
a=b||c;
```

results in

```
      A           2 rows      5 cols    (character, size 2)

      A   B   C   GH IJ
      D   E   F   KL MN
```

For character operands, the element size in the result matrix is the larger of the two operands. In the above example, **A** has element size 2.

You can use the horizontal concatenation operator when one of the arguments has no value. For example, if **A** has not been defined and **B** is a matrix, **A** || **B** results in a new matrix equal to **B**.

Quotation marks (") are needed around matrix elements only if you want to embed blanks or maintain uppercase and lowercase distinctions.

Concatenation Operator, Vertical: //

concatenates matrices vertically

Syntax

matrix1//matrix2

Description

The vertical concatenation operator (//) produces a new matrix by vertically joining *matrix1* and *matrix2*. *Matrix1* and *matrix2* must have the same number of columns, which is also the number of columns in the new matrix. For example, if **A** has three rows and two columns and **B** has four rows and two columns, then **A**//**B** produces a matrix with seven rows and two columns. Rows 1 through 3 of the new matrix correspond to **A**; rows 4 through 7 correspond to **B**.

For example, the statements

```
a={1 1 1,
   7 7 7};
b={0 0 0,
   8 8 8};
c=a//b;
```

result in

```
C           4 rows      3 cols      (numeric)

            1           1           1
            7           7           7
            0           0           0
            8           8           8
```

Concatenation Operator, Vertical *continued*

Also let

```
b={"AB" "CD",
   "EF" "GH"};
```

and

```
c={"I" "J",
   "K" "L",
   "M" "N"};
```

Then the statement

```
a=b//c;
```

produces the new matrix

```
        A           5 rows     2 cols    (character, size 2)

                              AB CD
                              EF GH
                              I  J
                              K  L
                              M  N
```

For character matrices, the element size of the result matrix is the larger of the element sizes of the two operands.

You can use the vertical concatenation operator when one of the arguments has not been assigned a value. For example, if **A** has not been defined and **B** is a matrix, **A//B** results in a new matrix equal to **B**.

Quotation marks (") are needed around matrix elements only if you want to embed blanks or maintain uppercase and lowercase distinctions.

Direct Product Operator: @

takes the direct product of two matrices

Syntax

matrix1@matrix2

Description

The direct product operator (@) produces a new matrix that is the direct product (also called the *Kronecker product*) of *matrix1* and *matrix2*, usually denoted by A⊗B. The number of rows in the new matrix equals the product of the number of rows in *matrix1* and the number of rows in *matrix2*; the number of columns in the new matrix equals the product of the number of columns in *matrix1* and the number of columns in *matrix2*.

For example, the statements

```
a={1 2,
   3 4};
b={0 2};
c=a@b;
```

result in

```
        C          2 rows      4 cols      (numeric)

             0          2          0          4
             0          6          0          8
```

The statement

```
d=b@a;
```

results in

```
        D          2 rows      4 cols      (numeric)

             0          0          2          4
             0          0          6          8
```

Division Operator: /

performs elementwise division

Syntax

matrix1/matrix2
matrix/scalar

Description

The division operator (/) divides each element of *matrix1* by the corresponding element of *matrix2*, producing a matrix of quotients.

In addition to dividing elements in conformable matrices, you can also use the division operator to divide a matrix by a scalar. If either operand is a scalar, the operation does the division for each element and the scalar value.

When a missing value occurs in an operand, IML assigns a missing value for the corresponding element in the result.

If a divisor is zero, the procedure prints a warning and assigns a missing value for the corresponding element in the result. An example of a valid statement using this operater follows:

```
c=a/b;
```

Element Maximum Operator: < >

selects the larger of two elements

Syntax

matrix1<>matrix2

Description

The element maximum operator (<>) compares each element of *matrix1* to the corresponding element of *matrix2*. The larger of the two values becomes the corresponding element of the new matrix that is produced.

When either argument is a scalar, the comparison is between each matrix element and the scalar.

The element maximum operator can take as operands two character matrices of the same dimensions or a character matrix and a character string. If the element lengths of the operands are different, the shorter elements are padded on the right with blanks. The element length of the result is the longer of the two operand element lengths.

When a missing value occurs in an operand, IML treats it as smaller than any valid number.

For example, the statements

```
a={2 4 6, 10 11 12 };
b={1 9 2, 20 10 40};
c=a<>b;
```

produce the result

```
            C           2 rows     3 cols     (numeric)

                          2          9          6
                         20         11         40
```

Element Minimum Operator: > <

selects the smaller of two elements

Syntax

matrix1><matrix2

Description

The element minimum operator (><) compares each element of *matrix1* with the corresponding element of *matrix2*. The smaller of the values becomes the corresponding element of the new matrix that is produced.

When either argument is a scalar, the comparison is between the scalar and each element of the matrix.

The element minimum operator can take as operands two character matrices of the same dimensions or a character matrix and a character string. If the element lengths of the operands are different, the shorter elements are padded on the right with blanks. The element length of the result is the longer of the two operand element lengths.

When a missing value occurs in an operand, IML treats it as smaller than any valid numeric value.

For example, the statements

```
a={2 4 6, 10 11 12 };
b={1 9 2, 20 10 40};
c=a><b;
```

produce the result

```
C              2 rows     3 cols     (numeric)

               1          4          2
               10         10         12
```

Index Creation Operator: :

creates an index vector

Syntax

value1:value2

Description

The index creation operator (:) creates a row vector whose first element is *value1*. The second element is *value1*+1, and so on, as long as the elements are less than or equal to *value2*. For example, the statement

```
I=7:10;
```

results in

```
I              1 row      4 cols     (numeric)

               7          8          9          10
```

Index Creation Operator *continued*

If *value1* is greater than *value2*, a reverse order index is created. For example, the statement

    r=10:6;

results in the row vector

            R   1 row   5 cols   (numeric)

            10   9   8   7   6

The index creation operator also works on character arguments with a numeric suffix. For example, the statement

    varlist='var1':'var5';

results in

        VARLIST   1 row   5 cols   (character, size 4)

                    var1   var2   var3   var4   var5

Use the DO function if you want an increment other than 1 or −1. See the description of the DO function later in this chapter.

Logical Operators: & | ^

perform elementwise logical comparisons

Syntax

matrix1&matrix2
matrix&scalar
matrix1 | matrix2
matrix | scalar
^matrix

Description

The AND logical operator (&) compares two matrices, element by element, to produce a new matrix. An element of the new matrix is 1 if the corresponding elements of *matrix1* and *matrix2* are both nonzero; otherwise, it is a zero.

An element of the new matrix produced by the OR operator (|) is 1 if either of the corresponding elements of *matrix1* and *matrix2* is nonzero. If both are zero, the element is zero.

The NOT prefix operator (^) examines each element of a matrix and produces a new matrix whose elements are ones and zeros. If an element of *matrix* equals 0, the corresponding element in the new matrix is 1. If an element of *matrix* is nonzero, the corresponding element in the new matrix is 0.

The statements below illustrate the use of these logical operators:

```
z=x&r;
if a|b then print c;
if ^m then link x1;
```

Note: The ¬ symbol is the same as the ^ symbol for ASCII and the "not" symbol for EBCDIC keyboards.

Multiplication Operator, Elementwise:

performs elementwise multiplication

Syntax

matrix1#matrix2
matrix#scalar
matrix#vector

Description

The elementwise multiplication operator (#) produces a new matrix whose elements are the products of the corresponding elements of *matrix1* and *matrix2*.

For example, the statements

```
a={1 2,
   3 4};
b={4 8,
   0 5};
c=a#b;
```

result in the matrix

```
        C        2 rows      2 cols    (numeric)

                    4          16
                    0          20
```

In addition to multiplying conformable matrices, you can use the elementwise multiplication operator to multiply a matrix and a scalar. When either argument is a scalar, the scalar value is multiplied by each element in *matrix* to form the new matrix.

Multiplication Operator, Elementwise: # *continued*

You can also multiply vectors by matrices. You can multiply matrices as long as they either conform in each dimension or one operand has dimension value 1. For example, a 2×3 matrix can be multiplied on either side by a 2×3, a 1×3, a 2×1, or a 1×1 matrix. Multiplying the 2×2 matrix **A** by the column vector **D**, as in

```
d={10,100};
ad=a#d;
```

produces the matrix

```
        AD          2 rows      2 cols      (numeric)

                        10          20
                       300         400
```

whereas the statements

```
d={10 100};
ad=a#d;
```

produce the matrix

```
        AD          2 rows      2 cols      (numeric)

                        10         200
                        30         400
```

The result of elementwise multiplication is also known as the Schur or Hadamard product. Element multiplication (using the # operator) should not be confused with matrix multiplication (using the * operator).

When a missing value occurs in an operand, IML assigns a missing value in the result.

Multiplication Operator, Matrix: *

performs matrix multiplication

Syntax

*matrix1*matrix2*

Description

The matrix multiplication infix operator (*) produces a new matrix by performing matrix multiplication. The first matrix must have the same number of columns as the second matrix has rows. The new matrix has the same number of rows as the first matrix and the same number of columns as the second matrix. The matrix multiplication operator does not consistently propagate missing values.

For example, the statements

```
a={1 2,
   3 4};
b={1 2};
c=b*a;
```

result in

```
            C           1 row       2 cols      (numeric)

                            7          10
```

and the statement

```
d=a*b`;
```

results in

```
            D           2 rows      1 col       (numeric)

                            5
                           11
```

Power Operator, Elementwise:

raises each element to a power

Syntax

matrix1##matrix2
matrix##scalar

Description

The elementwise power operator (##) creates a new matrix whose elements are the elements of *matrix1* raised to the power of the corresponding element of *matrix2*. If any value in *matrix1* is negative, the corresponding element in *matrix2* must be an integer.

In addition to handling conformable matrices, the elementwise power operator allows either operand to be a scalar. In this case, the operation takes the power for each element and the scalar value. Missing values are propagated if they occur.

Power Operator, Elementwise: ## *continued*

For example, the statements

```
a={1 2 3};
b=a##3;
```

result in

```
             B           1 row      2 cols      (numeric)

                             1          8          27
```

The statement

```
b=a##.5;
```

results in

```
             B           1 row      3 cols      (numeric)

                  1 1.4142136 1.7320508
```

Power Operator, Matrix: **

raises a matrix to a power

Syntax

matrix**scalar

Description

The matrix power operator (**) creates a new matrix that is *matrix* multiplied by itself *scalar* times. *Matrix* must be square; *scalar* must be an integer greater than or equal to −1. Large scalar values cause numerical precision problems. If the scalar is not an integer, it is truncated to an integer.

For example, the statements

```
a={1 2,
   1 1};
c=a**2;
```

result in

```
C          2 rows     2 cols    (numeric)

           3          4
           2          3
```

If the matrix is symmetric, it is preferable to power its eigenvalues rather than using the matrix power operator directly on the matrix (see the description of the EIGEN call). Note that the expression

A**(−1)

is permitted and is equivalent to INV(A).

The matrix power operater does not support missing values.

Sign Reverse Operator: −

reverses the signs of elements

Syntax

−*matrix*

Description

The sign reverse prefix operator (−) produces a new matrix whose elements are formed by reversing the sign of each element in *matrix*. A missing value is assigned if the element is missing.

For example, the statements

```
a={-1  7   6,
    2  0  -8};
b=-a;
```

result in the matrix

```
B          2 rows     3 cols    (numeric)

           1         -7        -6
          -2          0         8
```

Subscripts: [ ]

select submatrices

Syntax

matrix[*rows,columns*]
matrix[*elements*]

Description

Subscripts are used with matrices to select submatrices, where *rows* and *columns* are expressions that evaluate to scalars or numeric vectors. These expressions contain valid subscript values of rows and columns in the argument matrix. A subscripted matrix can appear on the left side of the equal sign. The dimensions of the target submatrix must conform to the dimensions of the source matrix. See Chapter 4, "Working with Matrices," for further information.

For example, the statements

```
x={1 2 3,
   4 5 6,
   7 8 9};
a=3;
m=x[2,a];
```

select the element in the second row and third column of **X** and produce the matrix **M**:

```
M          1 row     1 col     (numeric)

                        6
```

The statements

```
a=1:3;
m=x[2,a];
```

select row 2, and columns 1 through 3 of **X**, producing the matrix **M**:

```
M          1 row     3 cols    (numeric)

             4          5          6
```

Subtraction Operator: −

subtracts corresponding matrix elements

Syntax

matrix1 − matrix2
matrix − scalar

Description

The subtraction infix operator (−) produces a new matrix whose elements are formed by subtracting the corresponding elements of *matrix2* from those of *matrix1*.

In addition to subtracting conformable matrices, you can also use the subtraction operator to subtract a matrix and a scalar. When either argument is a scalar, the operation is performed by using the scalar against each element of the matrix argument.

When a missing value occurs in an operand, IML assigns a missing value for the corresponding element in the result.

An example of a valid statement follows:

```
c=a-b;
```

Transpose Operator: `

transposes a matrix

Syntax

matrix`

Description

The transpose operator (`) (denoted by the backquote character) exchanges the rows and columns of *matrix*, producing the transpose of *matrix*. For example, if an element in *matrix* is in the first row and second column, it is in the second row and first column of the transpose; an element in the first row and third column of *matrix* is in the third row and first column of the transpose, and so on. If *matrix* contains three rows and two columns, its transpose has two rows and three columns.

Transpose Operator *continued*

For example, the statements

```
a={1 2,
   3 4,
   5 6};
b=a`;
```

result in

```
            B          2 rows     3 cols    (numeric)

                         1          3          5
                         2          4          6
```

If your keyboard does not have a backquote character, you can transpose a matrix with the T (transpose) function, documented later in this chapter.

Statements, Functions, and Subroutines

This section presents descriptions of all statements, functions, and subroutines available in IML.

ABORT Statement

stops execution and exits IML

Syntax

ABORT;

Description

The ABORT statement instructs IML to stop executing statements. It also stops IML from parsing any further statements, causing IML to close its files and exit. See also the description of the STOP statement.

ABS Function

takes the absolute value

returns the absolute value of every element in the input matrix

Syntax

ABS(*matrix*)

where *matrix* is a numeric matrix or literal.

Description

The ABS function is a scalar function that returns the absolute value of every element of the argument matrix. An example of how to use the ABS function is shown below.

```
c=abs(a);
```

ALL Function

checks for all elements nonzero

returns a value of 0 or 1

Syntax

ALL(*matrix*)

where *matrix* is a numeric matrix or literal.

Description

The ALL function returns a value of 1 if all elements in *matrix* are nonzero. If any element of *matrix* is zero, the ALL function returns a value of 0. Missing values in *matrix* are treated as zeros.

You can use the ALL function to express the results of a comparison operator as a single 1 or 0. For example, the comparison operation **A**>**B** yields a matrix whose elements can be either ones or zeros. All the elements of the new matrix are ones only if each element of **A** is greater than the corresponding element of **B**.

ALL Function *continued*

For example, consider the statement

```
if all(a>b) then goto loop;
```

IML executes the GOTO statement only if every element of **A** is greater than the corresponding element of **B**. The ALL function is implicitly applied to the evaluation of all conditional expressions. The statements

```
if (a>b) then goto loop;
```

and

```
if all(a>b) then goto loop;
```

have the same effect.

ANY Function

checks for any nonzero element

returns a value of 0 or 1

Syntax

ANY(*matrix*)

where *matrix* is a numeric matrix or literal.

Description

The ANY function returns a value of 1 if any of the elements in *matrix* are nonzero. If all the elements of *matrix* are zeros, the ANY function returns a value of 0. Missing values in *matrix* are treated as zeros.

For example, consider the statement

```
if any(a=b) then print a b;
```

The matrices **A** and **B** are printed if at least one value in **A** is the same as the corresponding value in **B**. In the next example, the statements below do not print the message:

```
a={-99 99};
b={-99 98};
if a^=b then print 'a^=b';
```

However, the following statement prints the message:

```
if any(a^=b) then print 'a^=b';
```

APPEND Statement

adds observations to the end of a SAS data set

Syntax

APPEND <VAR *operand*>;
APPEND <FROM *from-name* <[ROWNAME=*row-name*]>>;

where

operand   can be specified as one of the following:

- □ a literal containing variable names

- □ a character matrix containing variable names

- □ an expression in parentheses yielding variable names

- □ one of the keywords described below:

ALL	for all variables
CHAR	for all character variables
NUM	for all numeric variables.

from-name   is the name of a matrix containing data to append.

row-name   is a character matrix or quoted literal containing descriptive row names.

Description

Use the APPEND statement to add data to the end of the current output data set. The appended observations are from either the variables specified in the VAR clause or variables created from the columns of the FROM matrix. The FROM clause and the VAR clause should not be specified together.

You can specify a set of variables to use with the VAR clause. Below are examples showing each possible way you can use the VAR clause.

```
var {time1 time5 time9};   /* a literal giving the variables  */
var time;                  /* a matrix containing the names   */
var('time1':'time9');      /* an expression                   */
var _all_;                 /* a keyword                       */
```

If the VAR clause includes a matrix with more than one row and column, the APPEND statement adds one observation for each element in the matrix with the greatest number of elements. Elements are appended in row-major order. Variables in the VAR clause with fewer than the maximum number of elements contribute missing values to observations after all of their elements have been used.

The default variables for the APPEND statement are all matrices that match variables in the current data set with respect to name and type.

The ROWNAME= operand to the FROM clause specifies the name of a character matrix to contain row titles. The first *nrow* values of this matrix become values of a variable with the same name in the output data set; *nrow* is

APPEND Statement *continued*

the number of rows in the FROM matrix. The procedure uses the first *nrow* elements in row-major order.

Examples using the APPEND statement follow. The first example shows use of the FROM clause when creating a new data set. See also the description of the CREATE statement.

```
x={1 2 3, 4 5 6};
create mydata from x[colname={x1 x2 x3}];
append from x;
show contents;
   /* shows 3 variables (x1 x2 x3) and 2 observations */
```

The next example shows use of the VAR clause for selecting variables from which to append data.

```
names={'Jimmy' 'Sue' 'Ted'};
sex={m f m};
create folks var{names sex};
append;
show contents;
   /* shows 2 variables (names,sex) and 3 observations in FOLKS */
```

You could achieve the same result with the statements

```
dsvar={names sex};
create folks var dsvar;
append;
```

APPLY Function

applies an IML module to its arguments

returns a matrix of results

Syntax

APPLY(*modname, argument1* <,*argument2,...,argument15*>)

where

modname   is the name of an existing module, supplied in quotes, as a matrix containing the module name, or an expression rendering the module name.

argument   is an argument passed to the module. You must have at least one argument. You can specify up to 15 arguments.

Description

The APPLY function applies a user-defined IML module to each element of the argument matrix or matrices. The first argument to APPLY is the name of the module. The module must already be defined before the APPLY function is executed. The module must be a function module, capable of returning a result.

The subsequent arguments to the APPLY function are the arguments passed to the module. They all must have the same dimension. If the module takes n arguments, *argument1* through *argumentn* should be passed to APPLY, where $1 \leq n \leq 15$. The APPLY function effectively calls the module. The result has the same dimension as the input arguments, and each element of the result corresponds to the module applied to the corresponding elements of the argument matrices. The APPLY function can work on numeric as well as character arguments. For example, the statements below define module ABC and then call the APPLY function, with matrix **A** as an argument:

```
start abc(x);
   r=x+100;
   return (r);
finish abc;

a={6 7 8,
   9 10 11};
r=apply("ABC",a);
```

The result is

```
             R        2 rows      3 cols     (numeric)

                       106         107        108
                       109         110        111
```

In the next example, the statements define the module SWAP and call the APPLY function:

```
start swap(a,b,c);
   r=a*b*c;
   a=b;
   if r<0 then return(0);
   return(r);
finish swap;

a={2  3, 4 5};
b={4  3, 5 6};
c={9 -1, 3 7};
mod={swap};
r=apply(mod,a,b,c);
print a r;
```

The results are

```
             A                      R
             4          3          72          0
             5          6          60         210
```

ARMACOV Call

computes an autocovariance sequence for an ARMA model

returns autocovariance sequences

Syntax

CALL ARMACOV(*auto,cross,convol,phi,theta,num*);

where

phi	refers to a $1 \times (p+1)$ matrix containing the autoregressive parameters. The first element is assumed to have the value 1.
theta	refers to a $1 \times (q+1)$ matrix containing the moving-average parameters. The first element is assumed to have the value 1.
num	refers to a scalar containing n, the number of autocovariances to be computed, which must be a positive number.

The ARMACOV subroutine returns the following values:

auto	specifies a variable to contain the returned $1 \times n$ matrix containing the autocovariances of the specified ARMA model, assuming unit variance for the innovation sequence.
cross	specifies a variable to contain the returned $1 \times (q+1)$ matrix containing the covariances of the moving-average term with lagged values of the process.
convol	specifies a variable to contain the returned $1 \times (q+1)$ matrix containing the autocovariance sequence of the moving-average term.

Description

The ARMACOV subroutine computes the autocovariance sequence that corresponds to a given autoregressive moving-average (ARMA) time-series model. An arbitrary number of terms in the sequence can be requested. Two related covariance sequences are also returned.

The model notation for the ARMACOV and ARMALIK subroutines is the same. The ARMA(p,q) model is denoted

$$\Sigma_{j=0}^{p} \varphi_j y_{t-j} = \Sigma_{i=0}^{q} \theta_i \varepsilon_{t-i}$$

with $\theta_0 = \varphi_0 = 1$. The notation is the same as that of Box and Jenkins (1976) except the model parameters are opposite in sign. The innovations $\{\varepsilon_t\}$ satisfy

$E(\varepsilon_t)=0$ and $E(\varepsilon_t\varepsilon_{t-k})=1$ if $k=0$, and are zero otherwise. The formula for the kth element of the *convol* argument is

$$\Sigma_{i=k-1}^{q} \theta_i\, \theta_{i-k+1}$$

for $k=1, 2, \ldots, q+1$. The formula for the kth element of the *cross* argument is

$$\Sigma_{i=k-1}^{q} \theta_i\, \psi_{i-k+1}$$

for $k=1, 2, \ldots, q+1$, where ψ_i is the ith impulse response value. The ψ_i sequence, if desired, can be computed with the RATIO function. It can be shown that ψ_k is the same as $E(Y_{t-k}\varepsilon_t^2)/\sigma$, which is used by Box and Jenkins (1976, p. 75) in their formulation of the autocovariances. The kth autocovariance, denoted γ_k and returned as the $k+1$ element of the *auto* argument ($k=0, 1, \ldots, n-1$), is defined implicitly for $k>0$ by

$$\Sigma_{i=0}^{p} \gamma_{k-i}\varphi_i = \eta_k$$

where η_k is the kth element of the *cross* argument. See Box and Jenkins (1976) or McLeod (1975) for more information.

To compute the autocovariance function at lags zero through four for the model

$$y_t = .5y_{t-1} + e_t + .8e_{t-1}$$

use the following statements:

```
proc iml;
     /* an arma(1,1) model */
   phi  ={1 -.5};
   theta={1 .8};
   call armacov(auto,cross,convol,phi,theta,5);
   print auto,,cross convol;
```

The result is

```
          AUTO
3.2533333 2.4266667 1.2133333 0.6066667 0.3033333


          CROSS          CONVOL
          2.04      0.8   1.64      0.8
```

ARMALIK Call

computes the log likelihood and residuals for an ARMA model

returns the concentrated log likelihood function for an ARMA model

Syntax

CALL ARMALIK(*lnl,resid,std,x,phi,theta*);

where

x	is an $n \times 1$ or $1 \times n$ matrix containing values of the time series (assuming mean zero).
phi	is a $1 \times (p+1)$ matrix containing the autoregressive parameter values. The first element is assumed to have the value 1.
theta	is a $1 \times (q+1)$ matrix containing the moving-average parameter values. The first element is assumed to have the value 1.

The ARMALIK subroutine returns the following values:

lnl	specifies a 3×1 matrix containing the log likelihood concentrated with respect to the innovation variance; the estimate of the innovation variance (the unconditional sum of squares divided by n); and the log of the determinant of the variance matrix, which is standardized to unit variance for the innovations.
resid	specifies an $n \times 1$ matrix containing the standardized residuals. These values are uncorrelated with a constant variance if the specified ARMA model is the correct one.
std	specifies an $n \times 1$ matrix containing the scale factors used to standardize the residuals (see "Description" below). The actual residuals from the one-step-ahead predictions using the past values can be computed as *std#resid*.

Description

ARMALIK subroutine computes the concentrated log likelihood function for an ARMA model. The unconditional sum of squares is readily available, as are the one-step-ahead prediction residuals. Factors that can be used to generate confidence limits associated with prediction from a finite past sample are also returned.

The notational conventions for the ARMALIK subroutine are the same as those used by the ARMACOV subroutine. See the description of the ARMACOV call for the model employed. In addition, the condition $\Sigma_{i=0}^{q}\theta_{iz}^{i} \neq 0$ for $|z| < 1$ should be satisfied to guard against floating-point overflow.

If the column vector $\mathbf{x}$ contains n values of a time series and the variance matrix is denoted $\Sigma = \sigma^2\mathbf{V}$, where σ^2 is the variance of the innovations, then, up to additive constants, the log likelihood, concentrated with respect to σ^2, is

$$(-n/2)\log(\mathbf{x'}\mathbf{V}^{-1}\mathbf{x}) - (1/2)\log|\mathbf{V}| \quad .$$

The matrix **V** is a function of the specified ARMA model parameters. If **L** is the lower Cholesky root of **V** (that is, $\mathbf{V}=\mathbf{L}^*\mathbf{L}'$), then the standardized residuals are computed as $resid=\mathbf{L}^{-1}\mathbf{x}$. The elements of *std* are the diagonal elements of **L**. The variance estimate is $\mathbf{x}'\mathbf{V}^{-1}\mathbf{x}/n$, and the log determinant is $\log|\mathbf{V}|$. See Ansley (1979) for further details. To compute the log-likelihood for the model

$$y_t - y_{t-1} + 0.25y_{t-2} = e_t + .5e_{t-1}$$

use the following IML code:

```
proc iml;
   phi={ 1 -1 .25} ;
   theta={ 1 .5} ;
   x={ 1 2 3 4 5} ;
   call armalik(lnl,resid,std,x,phi,theta);
   print lnl resid std;
```

The printed output is

```
        LNL      RESID       STD
  -0.822608  0.4057513  2.4645637
   0.8721154 0.9198158  1.2330147
   2.3293833 0.8417343  1.0419028
              1.0854175  1.0098042
              1.2096421  1.0024125
```

ARMASIM Function

simulates a univariate ARMA series

returns the generated time series

Syntax

ARMASIM(*phi,theta,mu,sigma,n,<seed>*)

where

phi   is a $1\times(p+1)$ matrix containing the autoregressive parameters. The first element is assumed to have the value 1.

theta   is a $1\times(q+1)$ matrix containing the moving-average parameters. The first element is assumed to have the value 1.

mu   is a scalar containing the overall mean of the series.

sigma   is a scalar containing the standard deviation of the innovation series.

n   is a scalar containing *n*, the length of the series. The value of *n* must be greater than 0.

ARMASIM Function *continued*

seed   is a scalar containing the random number seed. If not supplied, the system clock is used to generate the seed. If negative, then the absolute value is used as the starting seed; otherwise, subsequent calls ignore the value of *seed* and use the last seed generated internally.

The ARMASIM function returns the following value:

y   is an $n \times 1$ matrix containing the generated time series.

Description

The ARMASIM function generates a series of length n from a given autoregressive moving-average (ARMA) time series model. The notational conventions for the ARMASIM function are the same as those used by the ARMACOV subroutine. See the description of the ARMACOV call for the model employed. The ARMASIM function uses an exact simulation algorithm as described in Woodfield (1988). A sequence $Y_0, Y_1, \ldots, Y_{p+q-1}$ of starting values is produced using an expanded covariance matrix, and then the remaining values are generated using the recursion form of the model, namely

$$Y_t = -\Sigma_{i=1}^{p} \varphi_i Y_{t-i} + \varepsilon_t + \Sigma_{i=1}^{q} \theta_i \varepsilon_{t-i} \quad t = p+q, p+q+1, \ldots, n-1 \quad .$$

The random number generator RANNOR is used to generate the noise component of the model. Note that the statement

```
armasim(1,1,0,1,n,seed);
```

returns n standard normal pseudo-random deviates.

For example, to generate a time series of length 10 from the model

$$y_t = .5y_{t-1} + e_t + .8e_{t-1}$$

use the following code to produce the result shown:

```
proc iml;
    phi={1 -.8};
    theta={1 0.5};
    y=armasim(phi, theta, 0, 1, 10, -1234321);
    print y;
```

```
                Y        10 rows      1 col      (numeric)

                              3.0764594
                              1.8931735
                              0.9527984
                              0.0892395
                             -1.811471
                              -2.8063
                             -2.52739
                             -2.865251
                             -1.332334
                              0.1049046
```

BLOCK Function

forms block-diagonal matrices

returns a block-diagonal matrix

Syntax

BLOCK(*matrix1* <,*matrix2*,...,*matrix15*>)

where *matrix* is a numeric matrix or literal.

Description

The BLOCK function creates a new block-diagonal matrix from all the matrices specified in the argument matrices. Up to 15 matrices can be specified. The matrices are combined diagonally to form a new matrix. For example, the statement

```
block(a,b,c);
```

produces a matrix of the form

$$
\begin{bmatrix}
A & 0 & 0 \\
0 & B & 0 \\
0 & 0 & C
\end{bmatrix}
$$

The statements

```
a={2 2,
   4 4} ;
b={6 6,
   8 8} ;
c=block(a,b);
```

result in the matrix

```
         C          4 rows      4 cols    (numeric)

              2          2          0          0
              4          4          0          0
              0          0          6          6
              0          0          8          8
```

BRANKS Function

computes bivariate ranks

returns a matrix with tied and bivariate ranks

Syntax

BRANKS(*matrix*)

where *matrix* is an $n\times2$ numeric matrix.

Description

The BRANKS function calculates the tied ranks and the bivariate ranks for an $n\times2$ matrix. The tied ranks of the first column of *matrix* are contained in the first column of the result matrix; the tied ranks of the second column of *matrix* are contained in the second column of the result matrix; and the bivariate ranks of *matrix* are contained in the third column of the result matrix.

The tied rank of an element x_i of a vector is defined as

$$\mathbf{R}_i = 1/2 + \Sigma_j\, u(x_i - x_j)$$

where

$$u(t) = 1 \quad \text{if } t>0$$
$$ = 1/2 \quad \text{if } t = 0$$
$$ = 0 \quad \text{if } t<0.$$

The bivariate rank of a pair (x_j,y_j) is defined as

$$\mathbf{Q}_i = 3/4 + \Sigma_j\, u(x_i - x_j)\, u(y_i - y_j) \quad .$$

For example, the following statements produce the result shown below:

```
x=(1 0,
   4 2,
   3 4,
   5 3,
   6 3);
f=branks(x);
```

F	5 rows	3 cols	(numeric)
1	1	1
3	2	2
2	5	2
4	3.5	3
5	3.5	3.5

BTRAN Function

computes the block transpose

returns a block transposed matrix

Syntax

BTRAN(*x,n,m*)

where

x   is an $(i^*nx)\times(j^*mx)$ numeric matrix.

n   is a scalar whose value specifies the row dimension of the submatrix blocks.

m   is a scalar whose value specifies the column dimension of the submatrix blocks.

Description

The BTRAN function computes the block transpose of a partitioned matrix. The argument *x* is a partitioned matrix formed from submatrices of dimension $n\times n$. If the *i*th, *j*th submatrix of the argument *x* is denoted $\mathbf{A}_{ij}$, then the *i*th, *j*th submatrix of the result is $\mathbf{A}_{ji}$.

The value returned by the BTRAN function is a $(j^*n)\times(i^*m)$ matrix, the block tranpose of *x*, where the blocks are $n\times m$. For example, the statements

```
proc iml;
   z=btran({1 2 3 4,
           5 6 7 8},2,2);
   print z;
```

produce the result

```
            Z            4 rows     2 cols     (numeric)


                           1          2
                           5          6
                           3          4
                           7          8
```

BYTE Function

translates numbers to ordinal characters

returns a matrix containing translated values

Syntax

BYTE(*matrix*)

where *matrix* is a numeric matrix or literal.

Description

The BYTE function returns a character matrix with the same shape as the numeric argument. Each element of the result is a single character whose ordinal position in the computer's character set is specified by the corresponding numeric element in the argument. These numeric elements should generally be in the range 0 to 255.

For example, in the ASCII character set,

```
a=byte(47);
```

sets

```
a="/";        /* the slash character */
```

while the lowercase alphabet can be generated with

```
y=byte(97:122);
```

which produces

```
        Y              1 row       26 cols    (character, size 1)

        a b c d e f g h i j k l m n o p q r s t u v w x y z
```

This function simplifies the use of special characters and control sequences that cannot be entered directly using the keyboard into IML source code. Consult the character set tables for the computer you are using to determine the printable and control characters that are available and their ordinal positions.

CALL Statement

calls a subroutine or function

Syntax

CALL *name* <(*arguments*)>;

where

name   is the name of a user-defined module or an IML subroutine or function.

arguments   are arguments to the module or subroutine.

Description

The CALL statement executes a subroutine. The order of resolution for the CALL statement is

1. An IML built-in subroutine

2. A user-defined module.

This resolution order needs to be considered only if you have defined a module with the same name as an IML built-in subroutine.
     See also the description of the RUN statement.

CHANGE Call

search and replace text in an array

Syntax

CALL CHANGE(*matrix,old,new*<,*numchange*>);

where

matrix   is a character matrix or quoted literal.

old   is the string to be changed.

new   is the string to replace the *old* string.

numchange   is the number times to make the change.

CHANGE Call *continued*

Description

The CHANGE subroutine changes the first *numchange* occurrences of the substring *old* in each element of the character array *matrix* to the form *new*. If *numchange* is not specified, it defaults to 1. If *numchange* is 0, the routine changes all occurrences of *old*. If no occurrences are found, the matrix is not changed. For example, the statements

```
a="It was a dark and stormy night.";
call change(a, "night","day");
```

produce

```
          A="It was a dark and stormy day."
```

In the *old* operand, the following characters are reserved:

> % $ [ ] { } < > — ? * # @ ' `(backquote) ^

CHAR Function

produces a character representation of a numeric matrix

returns a matrix containing character representations of numeric values

Syntax

CHAR(*matrix* <,*w*<,*d*>>)

where

matrix   is a numeric matrix or literal.

w   is the field width.

d   is the number of decimal positions.

Description

The CHAR function takes a numeric matrix as an argument and, optionally, a field width *w* and a number of decimal positions *d*. The CHAR function produces a character matrix whose dimensions are the same as the dimensions of the argument matrix and whose elements are character representations of the corresponding numeric elements.

The CHAR function can take one, two, or three arguments. The first argument is the name of a numeric matrix and must always be supplied. The second argument is the field width of the result. If the second argument is not supplied, the system default field width is used. The third argument is the number of decimal positions in the result. If no third argument is supplied, the

BEST. representation is used. See also the description of the NUM function, which does the reverse conversion.

For example, the statements

```
a={1 2 3 4};
f=char(a,4,1);
```

produce the result

```
F               1 row       4 cols     (character, size 4)

                   1.0   2.0   3.0   4.0
```

CHOOSE Function

conditionally chooses and changes elements

returns a new matrix

Syntax

CHOOSE(*condition,result-for-true,result-for-false*)

where

condition	is checked for being true or false for each element.
result-for-true	is returned when *condition* is true.
result-for-false	is returned when *condition* is false.

Description

The CHOOSE function examines each element of the first argument for being true (nonzero and not missing) or false (zero or missing). For each true element, it returns the corresponding element in the second argument. For each false element, it returns the corresponding element in the third argument. Each argument must be conformable with the others or be a single element to be propagated.

CHOOSE Function *continued*

For example, suppose that you want to choose between *x* and *y* according to whether *x*#*y* is odd or even, respectively. The statements

```
x={1, 2, 3, 4, 5};
y={101, 205, 133, 806, 500};
r=choose(mod(x#y,2)=1,x,y);
print x y r;
```

result in

X	Y	R
1	101	1
2	205	205
3	133	3
4	806	806
5	500	500

Suppose you want all missing values in *x* to be changed to zeros. Submit the following statements to produce the result shown below:

```
x={1 2 ., 100 . -90, . 5 8};
print x;
```

X	3 rows	3 cols	(numeric)
1	2	.
100	.	-90
.	5	8

The following statement replaces the missing values in **X** with zeros:

```
x=choose(x=.,0,x);
print x;
```

X	3 rows	3 cols	(numeric)
1	2	0
100	0	-90
0	5	8

CLOSE Statement

closes a SAS data set

Syntax

CLOSE <*SAS-data-set*>;

where *SAS-data-set* can be specified with a one-word name (for example, A) or a two-word name (for example, SASUSER.A). For more information on specifying SAS data sets, see Chapter 6, "SAS Files," in *SAS Language: Reference*. More than one SAS data set can be listed in a CLOSE statement.

Description

The CLOSE statement is used to close one or more SAS data sets opened with the USE, EDIT, or CREATE statements. To find out which data sets are open, use the statement SHOW DATASETS. See also the description of the SAVE statement. IML automatically closes all open data sets when a QUIT statement is executed. See Chapter 6, "Working with SAS Data Sets," for more information. Examples of the CLOSE statement are shown below.

```
close mydata;
close mylib.mydata;
close;                    /* closes the current data set */
```

CLOSEFILE Statement

closes an input or output file

Syntax

CLOSEFILE *files*;

where *files* can be names (for defined filenames), literals, or expressions in parentheses (for filepaths).

Description

The CLOSEFILE statement is used to close files opened by the INFILE or FILE statements. The file specification should be the same as when the file was opened. File specifications are either a name (for a defined filename), a literal, or an expression in parentheses (for a filepath). To find out what files are open, use the statement SHOW FILES. For further information, consult Chapter 7, "File Access." See also the description of the SAVE statement. IML automatically closes all files when a QUIT statement is executed.

CLOSEFILE Statement *continued*

Examples of the CLOSEFILE statement are shown below.

```
filename in1 'mylib.mydata';
closefile in1;
```

or

```
closefile 'mylib.mydata';
```

or

```
in='mylib/mydata';
closefile(in);
```

CONCAT Function

performs elementwise string concatenation

returns a concatenated matrix

Syntax

CONCAT(*argument1,argument2* <*,...,argument15*>)

where *arguments* are character matrices or quoted literals.

Description

The CONCAT function produces a character matrix whose elements are the concatenations of corresponding element strings from each argument. The CONCAT function accepts up to 15 arguments, where each argument is a character matrix or a scalar. All nonscalar arguments must conform. Any scalar arguments are used repeatedly to concatenate to all elements of the other arguments. The element length of the result equals the sum of the element lengths of the arguments. Trailing blanks of one matrix argument appear before elements of the next matrix argument in the result matrix. For example, if you specify

```
b={"AB" "C ",
   "DE" "FG"};
```

and

```
c={"H " "IJ",
   " K" "LM"};
```

then the statement

```
a=concat(b,c);
```

produces the new 2×2 matrix

```
         A          2 rows     2 cols    (character, size 4)

                    ABH  C IJ
                    DE K FGLM
```

Quotation marks (") are needed only if you want to embed blanks or maintain uppercase and lowercase distinctions. You can also use the addition infix operator to concatenate character operands. See the description of the addition operator.

CONTENTS Function

obtains the variables in a SAS data set

returns a character matrix containing the variable names

Syntax

CONTENTS(<*libref*>, <*SAS-data-set*>)

where *SAS-data-set* can be specified with a one-word name or with a libref and a SAS-data-set name. For more information on specifying SAS data sets, see Chapter 6, "SAS Files," in *SAS Language: Reference*.

Description

The CONTENTS function returns a character matrix containing the variable names for *SAS-data-set*. The result is a character matrix with n rows, one column, and 8 characters per element, where n is the number of variables in the data set. The variable list is returned in the order in which the variables occur in the data set. If a one word name is provided, IML uses the default SAS data library (as specified in the DEFLIB= option). If no arguments are specified, the current open input data set is used. Some examples follow.

```
x=contents();             /* current open input data set */

x=contents('work','a');   /* contents of data set A in   */
                          /* WORK library                */
```

See also the description of the SHOW CONTENTS statement.

CONVMOD Function

converts modules to character matrices

returns a character matrix containing module statements

Syntax

CONVMOD(*module-name*)

where *module-name* is a character matrix or quoted literal containing the name of an IML module.

Description

The CONVMOD function returns a character matrix with *n* rows and 1 column, where *n* is the number of statements in the module. The element length is determined from the longest statement in the module. The CONVMOD function is supported to maintain compatibility with Version 5 SAS/IML software. It should be used only in conjunction with the STORE and PARSE statements. For example, consider the statements

```
start abc;
   statements
finish;
r=convmod('abc');      /* convert module ABC to matrix R   */
store r;               /* store module as character matrix */
```

Note that this can also be done in just one step with the statement

```
store module='abc';
```

This statement stores module ABC in the storage library. You should use the STORE MODULE= command instead to store modules. See Chapter 12, "Storage Features," for details concerning storage of modules.

COVLAG Function

computes autocovariance estimates for a vector time series

returns a sequence of lagged crossproduct matrices

Syntax

COVLAG(*x*,*k*)

where

x   is an $n \times nv$ matrix of time series values; n is the number of observations, and nv is the dimension of the random vector.

k   is a scalar, the absolute value of which specifies the number of lags desired. If *k* is positive, a mean correction is made. If *k* is negative, no mean correction is made.

Description

The COVLAG function computes a sequence of lagged crossproduct matrices. This function is useful for computing sample autocovariance sequences for scalar or vector time series.

The value returned by the COVLAG function is an $nv \times (k^*nv)$ matrix. The *i*th $nv \times nv$ block of the matrix is the sum

$$(1/n) \sum_{j=i}^{n} x_j' x_{j-i+1} \quad \text{if } k < 0$$

where x_j is the *j*th row of *x*. If k>0, then the *i*th $nv \times nv$ block of the matrix is

$$(1/n) \sum_{j=i}^{n} (x_j - \overline{x})'(x_{j-i+1} - \overline{x})$$

where $\overline{x}$ is a row vector of the column means of *x*. For example, the statements

```
x={-9,-7,-5,-3,-1,1,3,5,7,9};
cov=covlag(x,4);
```

produce the matrix

```
        COV           1 row      4 cols     (numeric)

              33      23.1      13.6       4.9
```

CREATE Statement

creates a new SAS data set

Syntax

CREATE *SAS-data-set* <VAR *operand*>;
CREATE *SAS-data-set* FROM *matrix-name*
        <[COLNAME=*column-name* ROWNAME=*row-name*]>;

where

SAS-data-set   can be specified with a one-word name (for example, A) or a two-word name (for example, SASUSER.A). For more information on specifying SAS data sets, see Chapter 6, "SAS Files," in *SAS Language: Reference*.

operand   gives a set of existing IML variables to become data set variables.

CREATE Statement *continued*

matrix-name	names a matrix containing the data.
column-name	is a character matrix or quoted literal containing descriptive names to associate with data set variables.
row-name	is a character matrix or quoted literal containing descriptive names to associate with observations on the data set.

Description

The CREATE statement creates a new SAS data set and makes it both the current input and output data sets. The variables in the new SAS data set are either the variables listed in the VAR clause or variables created from the columns of the FROM matrix. The FROM clause and the VAR clause should not be specified together.

You can specify a set of variables to use with the VAR clause, where *operand* can be specified as one of the following:

□ a literal containing variable names

□ the name of a matrix containing variable names

□ an expression in parentheses yielding variable names

□ one of the keywords described below:

ALL	for all variables
CHAR	for all character variables
NUM	for all numeric variables.

Below are examples showing each possible way you can use the VAR clause.

```
var {time1 time5 time9};    /* a literal giving the variables  */
var time;                   /* a matrix containing the names   */
var('time1':'time9');       /* an expression                   */
var _all_;                  /* a keyword                       */
```

You can specify a COLNAME= and a ROWNAME= matrix in the FROM clause. The COLNAME= matrix gives names to variables in the SAS data set being created. The COLNAME= operand specifies the name of a character matrix. The first *ncol* values from this matrix provide the variable names in the data set being created, where *ncol* is the number of columns in the FROM matrix. The CREATE statement uses the first *ncol* elements of the COLNAME= matrix in row-major order.

The ROWNAME= operand adds a variable to the data set to contain row titles. The operand must be a character matrix that exists and has values. The length of the data set variable added is the length of a matrix element of the operand. The same ROWNAME= matrix should be used on any subsequent APPEND statements for this data set.

The variable types and lengths are the current attributes of the matrices specified in the VAR clause or the matrix in the FROM clause. The default type is numeric when the name is undefined and unvalued. The default, when no variables are specified, is all active variables. To add observations to your data set, you must use the APPEND statement.

For example, the following statements create a new SAS data set CLASS having variables NAME, SEX, AGE, HEIGHT, and WEIGHT. The data come from IML matrices with the same names. You must initialize the character variables (NAME and SEX) and set the length prior to invoking the CREATE statement. NAME and SEX are character variables of lengths 12 and 1, respectively. AGE, HEIGHT, and WEIGHT are, by default, numeric.

```
name="123456789012";
sex="M";
create class var {name sex age height weight};
append;
```

In the next example, you use the FROM clause with the COLNAME= operand to create a SAS data set named MYDATA. The new data set has variables named with the COLNAME= operand. The data are in the FROM matrix **X**, and there are two observations because **X** has two rows of data. The COLNAME= operand gives descriptive names to the data set variables.

```
x={1 2 3, 4 5 6};
varnames='x1':'x3';
    /* creates data set MYDATA with variables X1, X2, X3  */
create mydata from x [colname=varnames];
append;
```

CSHAPE Function

reshapes and repeats character values

returns a reshaped matrix

Syntax

CSHAPE(*matrix,nrow,ncol,size*<,*padchar*>)

where

matrix	is a character matrix or quoted literal.
nrow	is the number of rows.
ncol	is the number of columns.
size	is the element length.
padchar	is a padding character.

CSHAPE Function *continued*

Description

The CSHAPE function shapes character matrices. See also the description of the SHAPE function, which is used with numeric data. The dimension of the matrix created by the CSHAPE function is specified by *nrow* (the number of rows), *ncol* (the number of columns), and *size* (the element length). A padding character is specified by *padchar*.

The CSHAPE function works by looking at the source matrix as if the characters of the source elements had been concatenated in row-major order. The source characters are then regrouped into elements of length *size*. These elements are assigned to the result matrix, once again in row-major order. If there are not enough characters for the result matrix, the source of the remaining characters depends on whether padding was specified with *padchar*. If no padding was specified, the source matrix's characters are cycled through again. If a padding character was specified, the remaining characters are all the padding character.

If one of the dimension arguments (*nrow, ncol,* or *size*) is zero, the function computes the dimension of the output matrix by dividing the number of elements of the input matrix by the product of the nonzero arguments.

Some examples follow. The statement

```
r=cshape('abcd',2,2,1);
```

results in

```
          R            2 rows      2 cols    (character, size 1)

                                   a b
                                   c d
```

The statement

```
r=cshape('a',1,2,3);
```

results in

```
          R            1 row       2 cols    (character, size 3)


                                   aaa aaa
```

The statement

```
r=cshape({'ab' 'cd',
          'ef' 'gh',
          'ij' 'kl'}, 2, 2, 3);
```

results in

```
R           2 rows     2 cols    (character, size 3)

                        abc def
                        ghi jkl
```

The statement

```
r=cshape('XO',3,3,1);
```

results in

```
R           3 rows     3 cols    (character, size 1)

                        X O X
                        O X O
                        X O X
```

And finally, the statement

```
r=cshape('abcd',2,2,3,'*');
```

results in

```
R           2 rows     2 cols    (character, size 3)

                        abc d**
                        *** ***
```

CUSUM Function

calculates cumulative sums

returns a matrix of cumulative sums

Syntax

CUSUM(*matrix*)

where *matrix* is a numeric matrix or literal.

CUSUM Function *continued*

Description

The CUSUM function returns a matrix of the same dimension as the argument matrix. The result contains the cumulative sums obtained by scanning the argument and summing in row-major order.

For example, the statements

```
a=cusum({1 2 4 5});
b=cusum({5 6, 3 4});
```

produce the result

A	1 row	4 cols	(numeric)
1	3	7	12

B	2 rows	2 cols	(numeric)
5	11		
14	18		

CVEXHULL Function

finds a convex hull of a set of planar points

returns a matrix of indicies

Syntax

CVEXHULL(*matrix*)

where *matrix* is an $n \times 2$ matrix of (x,y) points.

Description

The argument for the CVEXHULL function is an $n \times 2$ matrix of (x,y) points. The result matrix is an $n \times 1$ matrix of indicies. The indicies of points in the convex hull in counter-clockwise order are returned as the first part of the result matrix, and the negative of the indicies of the internal points are returned as the remaining elements of the result matrix. Any points that lie on the convex hull but lie on a line segment joining two other points on the convex hull are not included as part of the convex hull. The result matrix can be split into positive and negative parts using the LOC function. For example, the statements

```
z=cvexhull(x);
c=z[loc(z>0),];
```

yield the index vector for the convex hull.

DATASETS Function

obtains the names of SAS data sets in a SAS data library

returns a matrix containing data set names

Syntax

DATASETS(<*libref*>)

where *libref* is the name of a SAS data library. For more information on specifying a SAS data library, see Chapter 6, "SAS Files," in *SAS Language: Reference*.

Description

The DATASETS function returns a character matrix containing the names of the SAS data sets in the specified SAS data library. The result is a character matrix with *n* rows and one column, where *n* is the number of data sets in the library. If no argument is specified, IML uses the default libname. (See the DEFLIB = option in the description of the RESET statement.)

For example, suppose you have several data sets in the SAS data library SASUSER. You can list the names of the data sets in SASUSER by using the DATASETS function as follows:

```
lib={sasuser};
a=datasets(lib);
```

```
          A            6 rows      1 col      (character, size 8)

                                    CLASS
                                    FITNESS
                                    GROWTH
                                    HOUSES
                                    SASPARM
                                    TOBACCO
```

DELETE Statement

marks records for deletion

Syntax

DELETE <*range*> <WHERE(*expression*)>;

where

range   specifies a range of observations.

expression   is an expression that is evaluated for being true or false.

Description

Use the DELETE statement to mark records for deletion in the current output data set. To delete records and renumber the remaining observations, use the PURGE statement.

You can specify *range* by using a keyword or by record number using the POINT operand. The following keywords are valid values for *range*:

ALL   specifies all observations.

CURRENT   specifies the current observation.

NEXT <*number*> specifies the next observation or the next *number* of observations.

AFTER   specifies all observations after the current one.

POINT *operand* specifies observations by number, where *operand* is one of the following:

Operand	Example
a single record number	`point 5`
a literal giving several record numbers	`point {2 5 10}`
the name of a matrix containing record numbers	`point p`
an expression in parentheses	`point (p+1)`

CURRENT is the default value for *range*. If the current data set has an index in use, the POINT option is invalid.

The WHERE clause conditionally selects observations that are contained within the *range* specification. The general form of the WHERE clause is

WHERE(*variable comparison-op operand*)

where

variable   is a variable in the SAS data set.

comparison-op   is one of the following comparison operators:

<	less than
<=	less than or equal to
=	equal to
>	greater than
>=	greater than or equal to
^=	not equal to
?	contains a given string
^?	does not contain a given string
= :	begins with a given string
= *	sounds like or is spelled similar to a given string.

operand   is a literal value, a matrix name, or an expression in
                    parentheses.

WHERE comparison arguments can be matrices. For the following operators, the WHERE clause succeeds if *all* the elements in the matrix satisfy the conditon:

    ^=   ^?   <   <=   >   >=

For the following operators, the WHERE clause succeeds if *any* of the elements in the matrix satisfy the condition:

    =   ?   = :   = *

Logical expressions can be specified within the WHERE clause using the AND (&) and OR (|) operators. The general form is

clause&clause   (for an AND clause)
clause | clause   (for an OR clause)

where *clause* can be a comparison, a parenthesized clause, or a logical expression clause that is evaluated using operator precedence.

DELETE Statement *continued*

Note: The expression on the left-hand side refers to values of the data set variables and the expression on the right-hand side refers to matrix values.

Here are several examples of DELETE statements:

```
delete;                    /* deletes the current obs  */
delete point 34;           /* deletes obs 34           */
delete all where(age<21);  /* deletes obs where age<21 */
```

You can use the SETOUT statement with the DELETE statement as follows:

```
setout class point 34;     /* makes CLASS current output */
delete;                    /* deletes ob 34              */
```

Observations deleted using the DELETE statement are not physically removed from the data set until a PURGE statement is issued.

DELETE Call

deletes a SAS data set

Syntax

CALL DELETE(<*libname,*> *member-name*);

where

libname   is a character matrix or quoted literal containing the name of a SAS data library.

member-name   is a character matrix or quoted literal containing the name of a data set.

Description

The DELETE subroutine deletes a SAS data set in the specified library. If a one word name is specified, the default SAS data library is used. (See the DEFLIB= option in the description of the RESET statement.) Some examples follow.

```
call delete(work,a);    /* deletes WORK.A */

reset deflib=work;      /* sets default libname to WORK */
call delete(a);         /* also deletes WORK.A */

d=datasets('work');     /* returns all data sets in WORK */
call delete(work,d[1]);     /* deletes data set whose name is */
                        /* first element of matrix D */
```

DESIGN Function

creates a design matrix

returns a matrix whose elements have a value of 0 or 1

Syntax

DESIGN(*column-vector*)

where *column-vector* is a numeric column vector or literal.

Description

The DESIGN function creates a design matrix from *column-vector*. Each unique value of the vector generates a column of the design matrix. This column contains ones in elements whose corresponding elements in the vector are the current value; it contains zeros elsewhere. The columns are arranged in the sort order of the original values.

For example, the statements

```
a={1,1,2,2,3,1};
a=design(a);
```

produce the design matrix

```
           A          6 rows       3 cols      (numeric)

                         1            0           0
                         1            0           0
                         0            1           0
                         0            1           0
                         0            0           1
                         1            0           0
```

DESIGNF Function

creates a full-rank design matrix

returns a matrix whose elements have a value of 0 or 1

Syntax

DESIGNF(*column-vector*)

where *column-vector* is a numeric column vector or literal.

DESIGNF Function *continued*

Description

The DESIGNF function works similar to the DESIGN function; however, the
result matrix is one column smaller and can be used to produce full-rank design
matrices. The result of the DESIGNF function is the same as if you took the last
column off the DESIGN function result and subtracted it from the other columns
of the result.

For example, the statements

```
a={1,1,2,2,3,3};
b=designf(a);
```

produce the design matrix

```
            B           6 rows      2 cols      (numeric)

                              1           0
                              1           0
                              0           1
                              0           1
                             -1          -1
                             -1          -1
```

DET Function

computes the determinant of a square matrix

returns a scalar

Syntax

DET(*square-matrix*)

where *square-matrix* is a numeric matrix or literal.

Description

The DET function computes the determinant of *square-matrix*, which must be
square. The determinant, the product of the eigenvalues, is a single numeric
value. If the determinant of a matrix is zero, then that matrix is singular; that is,
it does not have an inverse.

The method performs an LU decomposition and collects the product of the diagonals (Forsythe, Malcolm, and Moler 1967). For example, the statements

```
a={1 1 1,1 2 4,1 3 9};
c=det(a);
```

produce the matrix **C** containing the determinant:

```
C          1 row      1 col     (numeric)

                         2
```

DIAG Function

creates a diagonal matrix

returns a diagonal matrix

Syntax

DIAG(*argument*)

where *argument* can be either a numeric square matrix or a vector.

Description

If *argument* is a square matrix, the DIAG function creates a matrix with diagonal elements equal to the corresponding diagonal elements. All off-diagonal elements in the new matrix are zeros.

If *argument* is a vector, the DIAG function creates a matrix whose diagonal elements are the values in the vector. All off-diagonal elements are zeros.

For example, the statements

```
a={4 3,
   2 1};
c=diag(a);
```

result in

```
C          2 rows     2 cols    (numeric)

                         4        0
                         0        1
```

DIAG Function *continued*

The statements

```
b={1 2 3};
d=diag(b);
```

result in

```
            D          3 rows      3 cols     (numeric)

                          1           0           0
                          0           2           0
                          0           0           3
```

DISPLAY Statement

displays fields in display windows

Syntax

DISPLAY <*group-spec group-options*, <...,*group-spec group-options*>>;

where

group-spec	specifies a group. It can be specified as either a compound name of the form *windowname.groupname* or a window name followed by a group of the form *window-name (field-specs)*, where *field-specs* is as defined for the WINDOW statement.
group-options	can be any of the following:

	NOINPUT	displays the group with all fields protected so that no data can be entered in the fields.
	REPEAT	repeats the group for each element of the matrices specified as field operands.
	BELL	rings the bell, sounds the alarm, or beeps the speaker on your workstation when the window is displayed.

Description

The DISPLAY statement directs IML to gather data into fields defined on the screen for purposes of display, data entry, or menu selection. The DISPLAY statement always refers to a window that has been previously opened by a WINDOW statement. The statement is described completely in Chapter 11, "Windows and Display Features."

Below are several examples of using the DISPLAY statement:

```
display;
display w(i);
display w ("BELL") bell;
display w.g1 noinput;
display w (i protect=yes
           color="blue"
           j color="yellow");
```

DO Function

produces an arithmetic series

returns a row vector

Syntax

DO(*start,stop,increment*)

where

start	is the starting value for the series.
stop	is the stopping value for the series.
increment	is an increment value.

Description

The DO function creates a row vector containing a sequence of numbers starting with *start* and incrementing by *increment* as long as the elements are less than or equal to *stop* (greater than or equal to *stop* for a negative increment). This function is a generalization of the index creation operator (:).

For example, the statement

```
i=do(3,18,3);
```

yields the result

```
        I          1 row      6 cols     (numeric)

        3       6       9       12       15       18
```

DO Function *continued*

The statement

```
j=do(3,-1,-1);
```

yields the result

```
        J           1 row     5 cols    (numeric)

                 3        2        1        0       -1
```

DO and END Statements

groups statements as a unit

Syntax

DO;
  statements
END;

Description

The DO statement specifies that the statements following the DO statement are executed as a group until a matching END statement appears. DO statements often appear in IF-THEN/ELSE statements, where they designate groups of statements to be performed when the IF condition is true or false.

For example, consider the following statements:

```
if x=y then
   do;
      i=i+1;
      print x;
   end;
print y;
```

The statements between the DO and END statements (called the DO group) are performed only if **X=Y**; that is, only if all elements of **X** are equal to the corresponding elements of **Y**. If any element of **X** is not equal to the corresponding element of **Y**, the statements in the DO group are skipped and the next statement is executed, in this case

```
print y;
```

DO groups can be nested. Any number of nested DO groups is allowed. Here is an example of nested DO groups:

```
if y>z then
    do;
        if z=0 then
            do;
                z=b*c;
                x=2#y;
            end;
    end;
```

It is good practice to indent the statements in a DO group as shown above so that their positions indicate their levels of nesting.

DO Statement, Iterative

iteratively executes a DO group

Syntax

DO *variable*=*start* TO *stop*<BY *increment*>;

where

variable	is the name of a variable indexing the loop.
start	is the starting value for the looping variable.
stop	is the stopping value for the looping variable.
increment	is an increment value.

Description

When the DO group has this form, the statements between the DO and END statements are executed repetitively. The number of times the statements are executed depends on the evaluation of the expressions given in the DO statement.

The *start, stop,* and *increment* values should be scalars or expressions whose evaluation yields scalars. The *variable* is given a new value for each repetition of the group. The index variable starts with the *start* value, then is incremented by the *increment* value each time. The iterations continue as long as the index variable is less than or equal to the *stop* value. If a negative increment is used, then the rules reverse so that the index variable decrements to a lower bound. Note that the *start, stop,* and *increment* expressions are evaluated only once before the looping starts.

DO Statement, Iterative *continued*

For example, the statements

```
do i=1 to 5 by 2;
   print 'THE VALUE OF I IS:' i;
end;
```

produce the output

```
                                          I
                  THE VALUE OF I IS:       1


                                          I
                  THE VALUE OF I IS:       3


                                          I
                  THE VALUE OF I IS:       5
```

DO DATA Statement

repeats a loop until an end of file occurs

Syntax

DO DATA <*variable=start* TO *stop*>;

where

variable   is the name of a variable indexing the loop.

start   is the starting value for the looping variable.

stop   is the stopping value for the looping variable.

Description

The DO DATA statement is used for repetitive DO loops that need to be exited upon the occurrence of an end of file for an INPUT, READ, or other I/O statement. This form is common for loops that read data from either a sequential file or a SAS data set. When an end of file is reached inside the DO DATA group, IML immediately jumps from the group and starts executing the statement following the END statement. DO DATA groups can be nested, where each end of file causes a jump from the most local DO DATA group. The DO DATA loop simulates the end-of-file behavior of the SAS DATA step. You should avoid using GOTO and LINK statements to jump out of a DO DATA group.

Examples of valid statements follow. The first example inputs the variable NAME from an external file for the first 100 lines or until the end of file, whichever occurs first.

```
do data i=1 to 100;
   input name $8.;
end;
```

Or, if reading from a SAS data set, then the code can be

```
do data;                /* read next obs until eof is reached */
   read next var{x};    /* read only variable X                */
end;
```

DO Statement with an UNTIL Clause

conditionally executes statements iteratively

Syntax

DO UNTIL(*expression*);
DO *variable*=*start* TO *stop* <BY *increment*> **UNTIL**(*expression*);

where

expression	is an expression that is evaluated at the bottom of the loop for being true or false.
variable	is the name of a variable indexing the loop.
start	is the starting value for the looping variable.
stop	is the stopping value for the looping variable.
increment	is an increment value.

Description

Using an UNTIL expression makes possible the conditional execution of a set of statements iteratively. The UNTIL expression is evaluated at the bottom of the loop, and the statements inside the loop are executed repeatedly as long as the expression yields a zero or missing value. In the example that follows, the body of the loop executes until the value of X exceeds 100:

```
x=1;
do until (x>100);
   x+1;
end;
print x;                 /* x=101 */
```

DO Statement with a WHILE Clause

conditionally executes statements iteratively

Syntax

DO WHILE(*expression*);
DO *variable=start* TO *stop* <BY *increment*> **WHILE**(*expression*);

where

expression	is an expression that is evaluated at the top of the loop for being true or false.
variable	is the name of a variable indexing the loop.
start	is the starting value for the looping variable.
stop	is the stopping value for the looping variable.
increment	is an increment value.

Description

Using a WHILE expression makes possible the conditional execution of a set of statements iteratively. The WHILE expression is evaluated at the top of the loop, and the statements inside the loop are executed repeatedly as long as the expression yields a nonzero or nonmissing value.

Note that the incrementing is done before the WHILE expression is tested. The following example demonstrates the incremeting:

```
x=1;
do while(x<100);
   x+1;
end;
print x;          /* x=100                    */
```

The next example increments the starting value by 2:

```
y=1;
do x=1 to 100 by 2 while(y<200);
   y=y#x;
end;               /* at end of loop, x=11 and y=945  */
```

ECHELON Function

reduces a matrix to row-echelon normal form

returns a row-echlon normal matrix

Syntax

ECHELON(*matrix*)

where *matrix* is a numeric matrix or literal.

Description

The ECHELON function uses elementary row operations to reduce a matrix to row-echelon normal form as in the following example (Graybill 1969, p. 286):

```
a={3  6  9,
   1  2  5,
   2  4 10};
e=echelon(a);
```

The resulting matrix is

```
        E        3 rows      3 cols    (numeric)

                   1           2          0
                   0           0          1
                   0           0          0
```

If the argument is a square matrix, then the row-echelon normal form can be obtained from the Hermite normal form by rearranging rows that are all zeros.

EDIT Statement

opens a SAS data set for editing

Syntax

EDIT *SAS-data-set* <VAR *operand*> <WHERE(*expression*)> <NOBS *name*>;

where

SAS-data-set	can be specified with a one-word name (for example, A) or a two-word name (for example, SASUSER.A). For more information on specifying SAS data sets, see Chapter 6, "SAS Files," in *SAS Language: Reference*.
operand	selects a set of variables.

EDIT Statement *continued*

expression	selects observations conditionally.
name	names a variable to contain the number of observations.

Description

The EDIT statement opens a SAS data set for reading and updating. If the data set has already been opened, the EDIT statement makes it the current input and output data sets.

You can specify a set of variables to use with the VAR clause, where *operand* can be specified as one of the following:

□   a literal containing variable names

□   the name of a matrix containing variable names

□   an expression in parentheses yielding variable names

□   one of the keywords described below:

ALL	for all variables
CHAR	for all character variables
NUM	for all numeric variables.

Below are examples showing each possible way you can use the VAR clause.

```
var {time1 time5 time9};   /* a literal giving the variables  */
var time;                  /* a matrix containing the names   */
var('time1':'time9');      /* an expression                   */
var _all_;                 /* a keyword                       */
```

The WHERE clause conditionally selects observations, within the range specification, according to conditions given in the clause. The general form of the WHERE clause is

WHERE (*variable comparison-op operand*)

where

variable	is a variable in the SAS data set.
comparison-op	is any one of the following comparison operators:

<	less than
<=	less than or equal to
=	equal to
>	greater than

$>=$	greater than or equal to
$^=$	not equal to
?	contains a given string
$^?$	does not contain a given string
$=:$	begins with a given string
$=*$	sounds like or is spelled similar to a given string.

operand   is a literal value, a matrix name, or an expression in parentheses.

WHERE comparison arguments can be matrices. For the following operators, the WHERE clause succeeds if *all* the elements in the matrix satisfy the condition:

$^=$ $^?$ $<$ $<=$ $>$ $>=$

For the following operators, the WHERE clause succeeds if *any* of the elements in the matrix satisfy the condition:

$=$? $=:$ $=*$

Logical expressions can be specified within the WHERE clause using the AND (&) and OR (|) operators. The general form is

clause&clause   (for an AND clause)
clause | clause   (for an OR clause)

where *clause* can be a comparison, a parenthesized clause, or a logical expression clause that is evaluated using operator precedence.

Note: The expression on the left-hand side refers to values of the data set variables and the expression on the right-hand side refers to matrix values.

The EDIT statement can define a set of variables and the selection criteria that are used to control access to data set observations. The NOBS clause returns the total number of observations in the data set in the variable *name*.

The VAR and WHERE clauses are optional and can be specified in any order. The NOBS clause is also optional.

See Chapter 6, "Working with SAS Data Sets," for more information on editing SAS data sets.

To edit the data set DAT, or WORK.DAT, use the statements

```
edit dat;
edit work.dat;
```

To control the variables you want to edit and conditionally select observations for editing, use the VAR and WHERE clauses. For example, to read and update observations for variable I where I is greater than 9, use the statement

```
edit work.dat var{i} where (i>9);
```

EDIT Statement *continued*

Below is an example using the NOBS option.

```
/* if MYDATA has 10 observations,          */
/* then ct is a numeric matrix with value 10     */
edit mydata nobs ct;
```

EIGEN Call

computes eigenvalues and eigenvectors

returns eigenvalues and eigenvectors of a matrix

Syntax

CALL EIGEN(*eigenvalues,eigenvectors,symmetric-matrix*);

where

eigenvalues	names a matrix to contain the eigenvalues of the input matrix.
eigenvectors	names a matrix to contain the eigenvectors of the input matrix.
symmetric-matrix	is a symmetric numeric matrix.

Description

The EIGEN subroutine computes *eigenvalues*, a column vector containing the eigenvalues of *symmetric-matrix* arranged in descending order. The EIGEN subroutine also computes *eigenvectors*, a matrix containing the orthonormal column eigenvectors of *symmetric-matrix* arranged so that the matrices correspond; that is, the first column of *eigenvectors* is the eigenvector corresponding to the largest eigenvalue, and so forth.

The result of the statement

```
call eigen(m,e,a);
```

has the properties

$$A*E = E*DIAG(M)$$
$$E'*E = I(N)$$

that is,

$$E' = INV(E)$$

so that

$$A = E*DIAG(M)*E' \quad .$$

The QL method is used to compute the eigenvalues (Wilkinson and Reinsch 1971).

IML cannot directly compute the eigenvalues of a general nonsymmetric matrix because some of the eigenvalues may be imaginary. In statistical applications, nonsymmetric matrices for which eigenvalues are desired are usually of the form $\mathbf{E}^{-1}\mathbf{H}$, where $\mathbf{E}$ and $\mathbf{H}$ are symmetric. The eigenvalues $\mathbf{L}$ and eigenvectors $\mathbf{V}$ of $\mathbf{E}^{-1}\mathbf{H}$ can be obtained by using the GENEIG subroutine or as follows:

```
f=root(einv);
a=f*h*f';
call eigen(l,w,a);
v=f'*w;
```

The computation can be checked by forming the residuals:

```
r=einv*h*v-v*diag(l);
```

The values in **R** should be of the order of round-off error.

EIGVAL Function

computes eigenvalues

returns a column vector containing eigenvalues

Syntax

EIGVAL(*symmetric-matrix*)

where *symmetric-matrix* is a symmetric numeric matrix.

Description

The EIGVAL function returns a column vector of the eigenvalues of *symmetric-matrix*. The eigenvalues are arranged in descending order. See the description of the EIGEN subroutine for more details. An example of a valid statement follows:

```
x={1 1,1 2,1 3,1 4};
xpx=t(x)*x;
a=eigval(xpx);          /* xpx is a symmetric matrix */
```

The matrix produced containing the eigenvalues is

```
                A         2 rows     1 col     (numeric)

                        33.401219
                        0.5987805
```

EIGVEC Function

computes eigenvectors

returns a matrix containing eigenvectors of a matrix

Syntax

EIGVEC(*symmetric-matrix*)

where *symmetric-matrix* is a symmetric numeric matrix.

Description

The EIGVEC function creates a matrix containing the orthonormal eigenvectors of *symmetric-matrix*. The columns of the new matrix are the eigenvectors. See the description of the EIGEN subroutine for more details. An example of a valid statement follows:

```
x={1 1,1 2,1 3,1 4};
xpx=t(x)*x;
a=eigvec(xpx);        /* xpx is a symmetric matrix   */
```

The matrix produced containing the eigenvectors is

```
         A          2 rows     2 cols    (numeric)

                  0.3220062 0.9467376
                  0.9467376 -0.322006
```

END Statement

ends a DO loop or DO statement

Syntax

END;

Description

See the description of the DO and END statements.

EXECUTE Call

executes SAS statements immediately

Syntax

CALL EXECUTE(*operands*);

where *operands* are character matrices or quoted literals containing valid SAS statements.

Description

The EXECUTE subroutine pushes character arguments to the input command stream, executes them, and then returns to the calling module. You can specify up to 15 arguments. The subroutine should be called from a module rather than from the immediate environment (because it uses the *resume* mechanism that works only from modules). The strings you push do not appear on the log.

Below are examples of valid EXECUTE subroutines:

```
call execute("x={1 2 3, 4 5 6};");
call execute(" x 'ls';");
call execute(" dm 'log; color source red';");
call execute(concat(" title '",string,"';"));
```

For more details on the EXECUTE subroutine, see Chapter 13, "Using SAS/IML Software to Generate IML Statements."

EXP Function

calculates the exponential

returns a matrix

Syntax

EXP(*matrix*)

where *matrix* is a numeric matrix or literal.

EXP Function *continued*

Description

The EXP function is a scalar function that takes the exponential function of every element of the argument matrix. The exponential is the natural number *e* raised to the indicated power. An example of a valid statement follows:

```
b={2  3  4};
a=exp(b);
```

```
          A              1 row      3 cols      (numeric)

              7.3890561 20.085537  54.59815
```

FFT Function

performs the finite Fourier transform

returns sine and cosine coefficients

Syntax

FFT(*x*)

where *x* is a $1 \times n$ or $n \times 1$ numeric vector.

Description

The FFT function returns the cosine and sine coefficients for the expansion of a vector into a sum of cosine and sine functions.

The argument of the FFT function, *x*, is a $1 \times n$ or $n \times 1$ vector. The value returned is the resulting transform, an $np \times 2$ matrix, where $np = \text{floor}(n/2+1)$.

The elements of the first column of the returned matrix are the cosine coefficients; that is, the *i*th element of the first column is

$$\Sigma_{j=1}^{n} x_j \cos((2\pi/n)(i-1)(j-1))$$

for $i=1, \ldots, np$, where the elements of *x* are denoted as x_j. The elements of the second column of the returned matrix are the sine coefficients; that is, the *i*th element of the second column is

$$\Sigma_{j=1}^{n} x_j \sin((2\pi/n)(i-1)(j-1))$$

for $i=1, \ldots, np$.

Note: For most efficient use of the FFT function, *n* should be a power of 2. If *n* is a power of 2, a fast Fourier transform is used (Singleton 1969); otherwise, a Chirp-Z algorithm is used (Monro and Branch 1976). The FFT function can be used to compute the periodogram of a time series. In conjunction with the inverse finite Fourier transform routine IFFT, the FFT function can be used to efficiently compute convolutions of large vectors (Gentleman and Sande 1966; Nussbaumer 1982). An example of a valid statement follows:

```
a=fft(c);
```

FILE Statement

opens or points to an external file

Syntax

FILE *file-name* <RECFM=N> <LRECL=*operand*>;

where

file-name	is a name (for defined filenames), a quoted literal, or an expression in parentheses (for filepaths).
RECFM=N	specifies that the file is to be written as a pure binary file without record-separator characters.
LRECL=*operand*	specifies the record length of the output file. The default record length is 512.

Description

You can use the FILE statement to open a file for output, or if the file is already open, to make it the current output file so that subsequent PUT statements write to it. The FILE statement is similar in syntax and operation to the INFILE statement. The FILE statement is described in detail in Chapter 7, "File Access."

The *file-name* is either a predefined filename or a quoted string or character expression in parentheses referring to the filepath. There are two ways to refer to an input or output file: by a filepath and by a filename. The filepath is the name as known to the operating system. The filename is a SAS reference to the file established directly through a connection made with the FILENAME statement. You can specify a file in either way in the FILE and INFILE statements. To specify a filename as the operand, just give the name. The name must be one already connected to a filepath by a previously issued FILENAME statement. There are, however, two special filenames that are recognized by IML: LOG and PRINT. These refer to the standard output streams for all SAS sessions. To specify a filepath, put it in quotes or specify an expression yielding the filepath in parentheses.

When the filepath is specified, there is a limit of 64 characters to the operand.

FILE Statement *continued*

Below are several valid uses of FILE statement.

```
file "student.dat";              /* by literal filepath  */

filename out "student.dat";      /* specify filename OUT  */
file out;                        /* refer to by filename  */

file print;                      /* standard print output */
file log;                        /* output to log         */

file "student.dat" recfm=n;      /* for a binary file     */
```

FIND Statement

finds observations

returns a numeric matrix containing the observation numbers

Syntax

FIND <*range*> <WHERE(*expression*)> INTO *matrix-name*;

where

range	specifies a range of observations.
expression	is an expression that is evaluated for being true or false.
matrix-name	names a matrix to contain the observation numbers.

Description

The FIND statement finds the observation numbers of records in *range* that satisfy the conditions of the WHERE clause. The FIND statement places these observation numbers in the numeric matrix whose name follows the INTO keyword.

You can specify a *range* of observations with a keyword or by record number using the POINT option. You can use any of the following keywords to specify *range:*

ALL	all observations
CURRENT	the current observation
NEXT <*number*>	the next observation or the next *number* of observations
AFTER	all observations after the current one

POINT *operand*   observations specified by number, where *operand* is one of the following:

Operand	Example
a single record number	`point 5`
a literal giving several record numbers	`point {2 5 10}`
the name of a matrix containing record numbers	`point p`
an expression in parentheses	`point (p+1)`

If the current data set has an index in use, the POINT option is invalid.

The WHERE clause conditionally selects observations, within the range specification, according to conditions given in the clause. The general form of the WHERE clause is

WHERE(*variable comparison-op operand*)

where

variable   is a variable in the SAS data set.

comparison-op   is one of the following comparison operators:

    $<$ less than

    $<=$ less than or equal to

    $=$ equal to

    $>$ greater than

    $>=$ greater than or equal to

    $\verb|^|=$ not equal to

    ?   contains a given string

    $\verb|^|?$ does not contain a given string

    $=:$ begins with a given string

    $=*$ sounds like or is spelled similar to a given string.

operand   is a literal value, a matrix name, or an expression in parentheses.

WHERE comparison arguments can be matrices. For the following operators, the WHERE clause succeeds if *all* the elements in the matrix satisfy the conditon:

    $\verb|^|=$ $\verb|^|?$ $<$ $<=$ $>$ $>=$

For the following operators, the WHERE clause succeeds if *any* of the elements in the matrix satisfy the condition:

    $=$? $=:$ $=*$

FIND Statement *continued*

Logical expressions can be specified within the WHERE clause using the AND (&) and OR (|) operators. The general form is

clause&clause (for an AND clause)
clause | clause (for an OR clause)

where *clause* can be a comparison, a parenthesized clause, or a logical expression clause that is evaluated using operator precedence.

Note: The expression on the left-hand side refers to values of the data set variables, and the expression on the right-hand side refers to matrix values.

Below are some valid examples of the FIND statement:

```
find all  where(name=:"Smith") into p;
find next where(age>30) into p2;
```

P and **P2** are column vectors containing the observation numbers that satisfy the WHERE clause in the given range. The default range is all observations.

FINISH Statement

denotes the end of a module

Syntax

FINISH <*module-name*>;

where *module-name* is the name of a user-defined module.

Description

The FINISH statement signals the end of a module and the end of module definition mode. Optionally, the FINISH statement can take the module name as its argument. See the description of the START statement and consult Chapter 5, "Programming," for further information on defining modules. Some examples follow.

```
finish;
finish mod1;
```

FORCE Statement

see the description of the SAVE statement

FREE Statement

frees matrix storage space

Syntax

FREE *matrices*
FREE / <*matrices*>;

where *matrices* are names of matrices.

Description

The FREE statement causes the specified matrices to lose their values; the memory is then freed for other uses. After execution of the FREE statement, the matrix does not have a value, and it returns 0 for the NROW and NCOL functions. Any printing attributes (assigned by the MATTRIB statement) are not released.

The FREE statement is used mostly in large applications or under tight memory constraints to make room for more data (matrices) in the workspace.

For example, to free the matrices **a, b,** and **c,** use the statement

```
free a b c;
```

If you want to free all matrices, specify a slash (/) after the keyword FREE. If you want to free all matrices except a few, then list the ones you do not want to free after the slash. For example, to free all matrices except **d** and **e,** use the statement

```
free / d e;
```

For more information, see the discussion of workspace storage in Chapter 14, "Further Notes."

GBLKVP Call

defines a blanking viewport

Syntax

CALL GBLKVP(*viewport* <,*inside*>);

where

viewport   is a numeric matrix or literal defining a viewport. This rectangular area's boundary is specified in normalized coordinates, where you specify the coordinates of the lower left corner and the upper right corner of the rectangular area in the form

       {*minimum-x minimum-y maximum-x maximum-y*}

inside   is a numeric argument that specifies whether graphics output is to be clipped inside or outside the blanking area. The default is to clip outside the blanking area.

Description

The GBLKVP subroutine defines an area, called the blanking area, in which nothing is drawn until the area is released. This routine is useful for clipping areas outside the graph or for blanking out inner portions of the graph. If *inside* is set to 0 (the default), no graphics output appears outside the blanking area. Setting *inside* to 1 clips inside the blanking areas.

Note that the blanking area (as specified by the viewport argument) is defined on the current viewport, and it is released when the viewport is changed or popped. At most one blanking area is in effect at any time. The blanking area can also be released by the GBLKVPD subroutine or another GBLKVP call. The coordinates in use for this graphics command are given in normalized coordinates because it is defined relative to the current viewport.

For example, to blank out a rectangular area with corners at the coordinates (20,20) and (80,80), relative to the currently defined viewport, use the statement

```
call gblkvp({20 20, 80 80});
```

No graphics or text can be written outside this area until the blanking viewport is ended.

Alternatively, if you want to clip inside of the rectangular area as above, use the *inside* parameter:

```
call gblkvp({20 20, 80 80},1);
```

See also the description of the CLIP option to the RESET statement.

GBLKVPD Call

deletes the blanking viewport

Syntax

CALL GBLKVPD;

Description

The GBLKVPD subroutine releases the current blanking area. It allows graphics output to be drawn in the area previously blanked out by a call to the GBLKVP subroutine.

To release an area previously blanked out, as in the example for the GBLKVP subroutine, use the statements shown below:

```
    /* define blanking viewport                      */
call gblkvp({20 20,80 80});
    more graphics statements
    /*  now release the blanked out area             */
call gblkvpd;
    /*  graphics or text can now be written to the area */
    continue graphics statements
```

See also the description of the CLIP option to the RESET statement.

GCLOSE Call

closes the graphics segment

Syntax

CALL GCLOSE;

Description

The GCLOSE subroutine closes the current graphics segment. Once a segment is closed, no other primitives can be added to it. The next call to a graph-generating function begins building a new graphics segment. However, the GCLOSE subroutine does not have to be called explicitly to terminate a segment; the GOPEN subroutine causes GCLOSE to be called.

GDELETE Call

deletes a graphics segment

Syntax

CALL GDELETE(*segment-name*);

where *segment-name* is a character matrix or quoted literal containing the name of the segment.

Description

The GDELETE subroutine searches the current catalog and deletes the first segment found with the name *segment-name*. An example of a valid statement follows:

```
    /* SEG_A is defined as a character matrix      */
    /* that contains the name of the segment to delete */
call gdelete(seg_a);
```

The segment can also be specified as a quoted literal:

```
call delete("plot_13");
```

GDRAW Call

draws a polyline

Syntax

CALL GDRAW(*x,y*,<*style*>,<*color*>,<*window*>,<*viewport*>);

where

x	is a vector containing the *x* coordinates of points used to draw a sequence of lines.
y	is a vector containing the *y* coordinates of points used to draw a sequence of lines.
style	is a numeric matrix or literal that specifies an index corresponding to a valid line style.
color	is a valid SAS color, where *color* can be specified as a quoted text string (such as 'RED'), the name of a character matrix containing a valid color as an element, or a color number (such as 1). A color number *n* refers to the *n*th color in the color list.

window is a numeric matrix or literal specifying a window. This is given in world coordinates and has the form

    {*minimum-x minimum-y maximum-x maximum-y*}

viewport is a numeric matrix or literal specifying a viewport. This is given in normalized coordinates and has the form

    {*minimum-x minimum-y maximum-x maximum-y*}

Description

The GDRAW subroutine draws a sequence of connected lines from points represented by values in *x* and *y*, which must be vectors of the same length. If *x* and *y* have *n* points, there will be $n-1$ lines. The first line will be from the point $(x(1),y(1))$ to $(x(2),y(2))$. The lines are drawn in the same color and line style. The coordinates in use for this graphics command are world coordinates. An example using the GDRAW subroutine follows:

```
    /* line from (50,50) to (75,75) - x and y take */
    /* default window range of 0 to 100            */
call gdraw({50 75},{50 75});
call gshow;
```

GDRAWL Call

draws individual lines

Syntax

CALL GDRAWL(*xy1,xy2*,<*style*>,<*color*>,<*window*>,<*viewport*>);

where

xy1 is a matrix of points used to draw a sequence of lines.

xy2 is a matrix of points used to draw a sequence of lines.

style is a numeric matrix or literal that specifies an index corresponding to a valid line style.

color is a valid SAS color, where *color* can be specified as a quoted text string (such as 'RED'), the name of a character matrix containing a valid color as an element, or a color number (such as 1). A color number *n* refers to the *n*th color in the color list.

window is a numeric matrix or literal specifying a window. This is given in world coordinates and has the form

    {*minimum-x minimum-y maximum-x maximum-y*}

GDRAWL Call *continued*

viewport   is a numeric matrix or literal specifying a viewport. This is given in normalized coordinates and has the form

```
{minimum-x minimum-y maximum-x maximum-y}
```

Description

The GDRAWL subroutine draws a sequence of lines specified by their beginning and ending points. The matrices *xy1* and *xy2* must have the same number of rows and columns. The first two columns (other columns are ignored) of *xy1* give the *x,y* coordinates of the beginning points of the line segment, and the first two columns of *xy2* have *x,y* coordinates of the corresponding end points. If *xy1* and *xy2* have *n* rows, *n* lines are drawn. The first line is from $(xy1(1,1),xy1(1,2))$ to $(xy2(1,1),xy2(1,2))$. The lines are drawn in the same color and line style. The coordinates in use for this graphics command are world coordinates. An example using the GDRAWL call follows:

```
    /* line from (25,25) to (50,50) - x and y take */
    /* default window range of 0 to 100      */
call gdrawl({25 25},{50 50});
call gshow;
```

GENEIG Call

computes eigenvalues and eigenvectors of a generalized eigenproblem

returns eigenvalues and eigenvectors

Syntax

CALL GENEIG(*eigenvalues,eigenvectors,symmetric-matrix1,symmetric-matrix2*);

where

eigenvalues   is a returned vector containing the eigenvalues.

eigenvectors   is a returned matrix containing the corresponding eigenvectors.

symmetric-matrix1 is a symmetric numeric matrix.

symmetric-matrix2 is a positive definite symmetric matrix.

Description

The GENEIG subroutine computes eigenvalues and eigenvectors of the generalized eigenproblem. The statement

```
call geneig (m,e,a,b);
```

computes eigenvalues **M** and eigenvectors **E** of the generalized eigenproblem **A** ***E**=**B*****E***diag(**M**), where **A** and **B** are symmetric and **B** is positive definite. The vector **M** contains the eigenvalues arranged in descending order, and the matrix **E** contains the corresponding eigenvectors in the columns.

The following example is from Wilkinson and Reinsch (1971, p. 311).

```
a={10   2   3   1   1,
    2  12   1   2   1,
    3   1  11   1  -1,
    1   2   1   9   1,
    1   1  -1   1  15};

b={12   1  -1   2   1,
    1  14   1  -1   1,
   -1   1  16  -1   1,
    2  -1  -1  12  -1,
    1   1   1  -1  11};

call geneig(m,e,a,b);
```

The matrices produced are shown below.

```
M
1.49235
1.10928
0.94385
0.66366
0.43278

E
-0.07638    0.14201    0.19171   -0.08292   -0.13459
 0.01709    0.14242   -0.15899   -0.15314    0.06129
-0.06666    0.12099    0.07483    0.11860    0.15790
 0.08604    0.12553   -0.13746    0.18281   -0.10946
 0.28943    0.00769    0.08897   -0.00356    0.04147
```

GGRID Call

draws a grid

Syntax

CALL GGRID(*x,y*,<*style*>,<*color*>,<*window*>,<*viewport*>);

where

x and *y*	are vectors of points used to draw sequences of lines.
style	is a numeric matrix or literal that specifies an index corresponding to a valid line style.
color	is a valid SAS color, where *color* can be specified as a quoted text string (such as 'RED'), the name of a character matrix containing a valid color as an element, or a color number (such as 1). A color number *n* refers to the *n*th color in the color list.
window	is a numeric matrix or literal specifying a window. This is given in world coordinates and has the form

```
{minimum-x minimum-y maximum-x maximum-y}
```

viewport	is a numeric matrix or literal specifying a viewport. This is given in normalized coordinates and has the form

```
{minimum-x minimum-y maximum-x maximum-y}
```

Description

The GGRID subroutine draws a sequence of vertical and horizontal lines specified by the *x* and *y* vectors, respectively. The start and end of the vertical lines are implicitly defined by the minimum and maximum of the *y* vector. Likewise, the start and end of the horizontal lines are defined by the minimum and maximum of the *x* vector. The grid lines are drawn in the same color and line style. The coordinates in use for this graphics command are world coordinates.

For example, use the following statements to place a grid in the lower left corner of the screen:

```
x={10,20,30,40,50};
y=x;

   /* The following GGRID command will place a GRID   */
   /* in lower left   corner of the screen            */
   /* assuming the default window and viewport        */
call ggrid(x,y);
call gshow;
```

GINCLUDE Call

includes a graphics segment

Syntax

CALL GINCLUDE(*segment-name*);

where *segment-name* is a character matrix or quoted literal specifying a graphics segment.

Description

The GINCLUDE subroutine includes into the current graph a previously defined graph named *segment-name* from the same catalog. The included segment is defined in the current viewport but not the current window.

The implementation of the GINCLUDE subroutine makes it possible to include other segments to the current segment and reposition them in different viewports. Furthermore, a segment can be included by different graphs, thus effectively reducing storage space. Examples of valid statements follow:

```
    /* segment1 is a character variable      */
    /*containing the segment name            */
segment1={myplot};
call ginclude(segment1);

    /* specify the segment with quoted literal */
call ginclude("myseg");
```

GINV Function

computes the generalized inverse

returns the generalized inverse

Syntax

GINV(*matrix*)

where *matrix* is a numeric matrix or literal.

GINV Function *continued*

Description

The GINV function creates the Moore-Penrose generalized inverse of *matrix*. This inverse, known as the four-condition inverse, has these properties:
   If

$$\mathbf{G} = \text{GINV}(\mathbf{A})$$

then

$$\mathbf{AGA} = \mathbf{A} \qquad \mathbf{GAG} = \mathbf{G} \qquad (\mathbf{AG})' = \mathbf{AG} \qquad (\mathbf{GA})' = \mathbf{GA} \ .$$

The generalized inverse is also known as the *pseudoinverse*, usually denoted by $\mathbf{A}^-$. It is computed using the singular value decomposition (Wilkinson and Reinsch 1971).
   Least-squares regression for the model

$$\mathbf{Y} = \mathbf{X}\boldsymbol{\beta} + \boldsymbol{\varepsilon}$$

can be performed by using

```
b=ginv(x)*y;
```

as the estimate of $\boldsymbol{\beta}$. This solution has minimum $\mathbf{b}'\mathbf{b}$ among all solutions minimizing $\boldsymbol{\varepsilon}'\boldsymbol{\varepsilon}$, where $\boldsymbol{\varepsilon} = \mathbf{Y} - \mathbf{Xb}$.
   Projection matrices can be formed by specifying GINV($\mathbf{X}$)*$\mathbf{X}$ (*row space*) or $\mathbf{X}$*GINV($\mathbf{X}$) (*column space*).
   See Rao and Mitra (1971) for a discussion of properties of this function.

GOPEN Call

opens a graphics segment

Syntax

CALL GOPEN(<*segment-name*>,<*replace*>,<*description*>);

where

segment-name	is a character matrix or quoted literal specifying the name of a graphics segment.
replace	is a numeric argument.
description	is a character matrix or quoted text string with a maximum length of 40 characters.

Description

The GOPEN subroutine starts a new graphics segment. The window and viewport are reset to the default values ({0 0 100 100}) in both cases. Any attribute modified using a GSET call is reset to its default value, which is set by the attribute's corresponding GOPTIONS value.

A nonzero value for *replace* indicates that the new segment should replace the first found segment with the same name, and zero indicates otherwise. If you do not specify the *replace* flag, the flag set by a previous GSTART call is used. By default, the GSTART subroutine sets the flag to NOREPLACE.

The *description* is a text string of up to 40 characters that you want to store with the segment to describe the graph.

Two graphs cannot have the same name. If you try to create a segment, say PLOT_A, twice, the second segment is named using a name generated by IML.

To open a new segment named COSINE, set *replace* to replace a like-named segment, and attach a description to the segment, use the statement

```
call gopen('cosine',1,'Graph of Cosine Curve');
```

GOTO Statement

jumps to a new statement

Syntax

GOTO *label*;

where *label* is a labeled statement. Execution jumps to this statement. A label is a name followed by a colon (:).

Description

The GOTO (or GO TO) statement directs IML to jump immediately to the statement with the given *label* and begin executing statements from that point. Any IML statement can have a label, which is a name followed by a colon preceding any executable statement.

GOTO statements are usually clauses of IF statements, for example,

```
if x>y then goto skip;
y=log(y-x);
yy=y-20;
skip: if y<0 then
    do;
        more statements
    end;
```

GOTO Statement *continued*

The function of GOTO statements is usually better performed by DO groups. For example, the statements above could be better written

```
if x<=y then
    do;
        y=log(y-x);
        yy=y-20;
    end;
more statements
```

▶ *Caution . . . . You can only use the GOTO statement inside a module or a DO group.*
As good programming practice, you should avoid GOTO statements when they refer to a label above the GOTO statement; otherwise, an infinite loop is possible. ▲

GPIE Call

draws pie slices

Syntax

CALL GPIE(*x,y,r*,<*angle1*>,<*angle2*>,<*color*>,<*outline*>,<*pattern*>,
        <*window*>,<*viewport*>);

where

x and *y*	are numeric scalars (or possibly vectors) defining the center (or centers) of the pie (or pies).
r	is a scalar or vector giving the radii of the pie slices.
angle1	is a scalar or vector giving the start angles. It defaults to 0.
angle2	is a scalar or vector giving the terminal angles. It defaults to 360.
color	is a valid SAS color, where *color* can be specified as a quoted text string (such as 'RED'), the name of a character matrix containing a valid color as an element, or a color number (such as 1). A color number *n* refers to the *n*th color in the color list.
outline	is an index indicating the side of the slice to draw. The default is 3.
pattern	is a character matrix or quoted literal that specifies the pattern with which to fill the interior of a closed curve.

window   is a numeric matrix or literal specifying a window. This is given in world coordinates and has the form

    `{minimum-x minimum-y maximum-x maximum-y}`

viewport   is a numeric matrix or literal specifying a viewport. This is given in normalized coordinates and has the form

    `{minimum-x minimum-y maximum-x maximum-y}`

Description

The GPIE subroutine draws one or more pie slices. The number of pie slices is the maximum dimension of the first five vectors. The angle arguments are specifed in degrees. The start angle (*angle1*) defaults to 0, and the terminal angle (*angle2*) defaults to 360. *Outline* is an index that indicates the side of the slice to draw. The *outline* specification can be one of the following:

<0   uses absolute value as the line style and draws no line segment from center to arc.

0   draws no line segment from center to arc.

1   draws an arc and line segment from the center to the starting angle point.

2   draws an arc and line segment from the center to the ending angle point.

3   draws all sides of the slice. This is the default.

Color, *outline*, and *pattern* can have more than one element. The coordinates in use for this graphics command are world coordinates. An example using the GPIE subroutine follows:

```
/* draws a pie with 4 slices of equal size */
call gpie(50,50,30,{0 90 180 270},{90 180 270 0});
```

GPIEXY Call

converts from polar to world coordinates

Syntax

CALL GPIEXY(*x,y,fract-radii,angles*,<*center*>,<*radius*>,<*window*>);

where

x and y   are vectors of coordinates returned by GPIEXY.

fract-radii   is a vector of fractions of the radius of the reference circle.

GPIEXY Call *continued*

angles	is the vector of angle coordinates in degrees.
center	defines the reference circle.
radius	defines the reference circle.
window	is a numeric matrix or literal specifying a window. This is given in world coordinates and has the form

$$\{minimum\text{-}x \quad minimum\text{-}y \quad maximum\text{-}x \quad maximum\text{-}y\}$$

Description

The GPIEXY subroutine computes the world coordinates of a sequence of points relative to a circle. The *x* and *y* arguments are vectors of new coordinates returned by the GPIEXY subroutine. Together, the vectors *fract-radii* and *angles* define the points in polar coordinates. Each pair from the *fract-radii* and *angles* vectors yields a corresponding pair in the *x* and *y* vectors. For example, suppose *fract-radii* has two elements, .5 and .3, and the corresponding two elements of *angles* are 90 and 30. The GPIEXY subroutine returns two elements in the *x* vector and two elements in the *y* vector. The first *x,y* pair locates a point half way from the center to the reference circle on the vertical line through the center, and the second *x,y* pair locates a point one-third of the way on the line segment from the center to the reference circle, where the line segment slants 30 degrees from the horizontal. The reference circle can be defined by an earlier GPIE call or another GPIEXY call, or it can be defined by specifying *center* and *radius*.

Graphics devices can have diverse aspect ratios; thus, a circle may appear distorted when drawn on some devices. The SAS graphics system adjusts computations to compensate for this distortion. Thus, for any given point, the transformation from polar coordinates to world coordinates may need an equivalent adjustment. The GPIEXY subroutine ensures that the same adjustment applied in the GPIE subroutine is applied to the conversion. An example using the GPIEXY call follows:

```
    /* add labels to a pie with 4 slices of equal size */
call gpie(50,50,30,{0 90 180 270},{90 180 270 0});
call gpiexy(x,y,1.2,{45 135 225 315},{50 50},30,{0 0 100 100});

    /* adjust for label size: */
x [4,]=x[4,]-3;
x [1,]=x[1,]-4;
x [2,]=x[2,]+1;
call gscript(x,y,{'QTR1' 'QTR2' 'QTR3' 'QTR4'});
call gshow;
```

GPOINT Call

plots points

Syntax

CALL GPOINT(*x,y,<symbol>,<color>,<height>,<window>,<viewport>*);

where

x   is a vector containing the *x* coordinates of points.

y   is a vector containing the *y* coordinates of points.

symbol   is a character vector or quoted literal that specifies a valid plotting symbol or symbols.

color   is a valid SAS color, where *color* can be specified as a quoted text string (such as 'RED'), the name of a character matrix containing a valid color as an element, or a color number (such as 1). A color number *n* refers to the *n*th color in the color list.

height   is a numeric matrix or literal specifying the character height.

window   is a numeric matrix or literal specifying a window. This is given in world coordinates and has the form

```
{minimum-x minimum-y maximum-x maximum-y}
```

viewport is a numeric matrix or literal specifying a viewport. This is given in normalized coordinates and has the form

```
{minimum-x minimum-y maximum-x maximum-y}
```

Description

The GPOINT subroutine marks one or more points with symbols. The *x* and *y* vectors define the points where the markers are to be placed. *The symbol* and *color* arguments can have from one up to as many elements as there are well-defined points. The coordinates in use for this graphics command are world coordinates.

In the example that follows, points on the line Y=X are generated for 30≤X≤80 and then plotted with the GPOINT call:

```
x=30:80;
y=x;
call gpoint(x,y);
call gshow;
```

As another example, you can plot symbols at specific locations on the screen using the GPOINT subroutine. To print **i** in the lower left corner and **j** in the upper right corner, use the statements

```
call gpoint({10 80},{5 95},{i j});
call gshow;
```

See Chapter 9, "Introduction to Graphics," for examples using the GPOINT subroutine.

GPOLY Call

draws and fills a polygon

Syntax

CALL GPOLY(*x,y,*<*style*>,<*ocolor*>,
            <*pattern*>,<*color*>,<*window*>,<*viewport*>);

where

x	is a vector defining the *x* coordinates of the corners of the polygon.
y	is a vector defining the *y* coordinates of the corners of the polygon.
style	is a numeric matrix or literal that specifies an index corresponding to a valid line style.
ocolor	is a matrix or literal specifying a valid outline color. The *ocolor* argument can be specified as a quoted text string (such as 'RED'), the name of a character matrix containing a valid color as an element, or a color number (such as 1). A color number *n* refers to the *n*th color in the color list.
pattern	is a character matrix or quoted literal that specifies the pattern to fill the interior of a closed curve.
color	is a valid SAS color used in filling the polygon. The *color* argument can be specified as a quoted text string (such as 'RED'), the name of a character matrix containing a valid color as an element, or a color number (such as 1). A color number *n* refers to the *n*th color in the color list.
window	is a numeric matrix or literal specifying a window. This is given in world coordinates and has the form

$$\{minimum\text{-}x \ minimum\text{-}y \ maximum\text{-}x \ maximum\text{-}y\}$$

viewport	is a numeric matrix or literal specifying a viewport. This is given in normalized coordinates and has the form

$$\{minimum\text{-}x \ minimum\text{-}y \ maximum\text{-}x \ maximum\text{-}y\}$$

Description

The GPOLY subroutine fills an area enclosed by a polygon. The polygon is defined by the set of points given in the vectors *x* and *y*. The *color* argument is the color used in shading the polygon, and *ocolor* is the outline color. By default, the shading color and the outline color are the same, and the interior pattern is empty. The coordinates in use for this graphics command are world coordinates. An example using the GPOLY subroutine follows:

```
xd={20 20 80 80};
yd={35 85 85 35};
call gpoly (xd,yd, , ,'X','red');
```

GPORT Call

defines a viewport

Syntax

CALL GPORT(*viewport*);

where *viewport* is a numeric matrix or literal defining the viewport. The rectangular area's boundary is specified in normalized coordinates, where you specify the coordinates of the lower left corner and the upper right corner of the rectangular area in the form

```
{minimum-x minimum-y maximum-x maximum-y}
```

Description

The GPORT subroutine changes the current viewport. The *viewport* argument defines the new viewport using device coordinates (always 0 to 100). Changing the viewport may affect the height of the character fonts; if so, you may want to modify the HEIGHT parameter. An example of a valid statement follows:

```
call gport({20 20 80 80});
```

The default values for viewport are {0 0 100 100}.

GPORTPOP Call

pops the viewport

Syntax

CALL GPORTPOP;

Description

The GPORTPOP subroutine deletes the top viewport from the stack.

GPORTSTK Call

stacks the viewport

Syntax

CALL GPORTSTK(*viewport*);

where *viewport* is a numeric matrix or literal defined in normalized coordinates
in the form

```
{minimum-x minimum-y maximum-x maximum-y}
```

Description

The GPORTSTK subroutine stacks the viewport defined by the matrix *viewport*
onto the current viewport; that is, the new viewport is defined relative to the
current viewport. The coordinates in use for this graphics command are world
coordinates. An example of a valid statement follows:

```
call gportstk({5 5 95 95});
```

GSCALE Call

calculates round numbers for labeling axes

Syntax

CALL GSCALE(*scale,x,nincr,<nicenum>,<fixed-end>*);

where

scale	is a returned vector containing the scaled minimum data value, the scaled maximum data value, and a grid increment.
x	is a numeric matrix or literal.
nincr	is the number of intervals desired.
nicenum	is numeric and provides up to ten numbers to use for scaling. By default, *nicenum* is (1,2,2.5,5).
fixed-end	is a character argument and specifies which end of the scale is held fixed. The default is **x**.

Description

The GSCALE subroutine obtains simple (round) numbers with uniform grid
interval sizes to use in scaling a linear axis. The GSCALE subroutine implements
algorithm 463 of Collected Algorithms from CACM. The scale values are integer
multiples of the interval size. They are returned in the first argument, a vector

with three elements. The first element is the scaled minimum data value. The second element is the scaled maximum data value. The third element is the grid increment.

The required input parameters are *x*, a matrix of data values, and *nincr*, the number of intervals desired. If *nincr* is positive, the scaled range includes approximately *nincr* intervals. If *nincr* is negative, the scaled range includes exactly ABS(*nincr*) intervals. The *nincr* parameter cannot be zero.

The *nicenum* and *fixed-end* arguments are optional. The *nicenum* argument provides up to ten numbers, all between 1 and 10 (inclusive of the end points), to be used for scaling. The default for *nicenum* is 1, 2, 2.5, and 5. The linear scale with this set of numbers is a scale whose interval size is the product of an integer power of 10 and 1, 2, 2.5, or 5. Changing these numbers alters the rounding of the scaled values.

For *fixed-end*, **U** fixes the upper end; **L** fixes the lower end; **X** allows both ends to vary from the data values. The default is **X**. An example using the GSCALE subroutine follows:

```
/* scalemat is set to {0,1000,100} */
call gscale(scalmat, {1 1000}, 10);
```

GSCRIPT Call

writes multiple text strings with special fonts

Syntax

CALL GSCRIPT(*x,y,text*,<*angle*>,<*rotate*>,<*height*>,<*font*>,<*color*>,
                <*window*>,<*viewport*>);

where

x	is a scalar or vector containing the *x* coordinates of the lower left starting position of the text string's first character.
y	is a scalar or vector containing the *y* coordinates of the lower left starting position of the text string's first character.
text	is a character vector of text strings.
angle	is the slant of each text string.
rotate	is the rotation of individual characters.
height	is a real number specifying the character height.
font	is a character matrix or quoted literal that specifies a valid font name.
color	is a valid SAS color. The *color* argument can be specified as a quoted text string (such as 'RED'), the name of a character matrix containing a valid color as an element, or a color number (such as 1). A color number *n* refers to the *n*th color in the color list.

GSCRIPT Call *continued*

window   is a numeric matrix or literal specifying a window. This is given in world coordinates and has the form

```
{minimum-x minimum-y maximum-x maximum-y}
```

viewport   is a numeric matrix or literal specifying a viewport. This is given in normalized coordinates and has the form

```
{minimum-x minimum-y maximum-x maximum-y}
```

Description

The GSCRIPT subroutine writes multiple text strings with special character fonts. The *x* and *y* vectors describe the coordinates of the lower left starting position of the text string's first character. The *color* argument can have more than one element.

Note: Hardware characters cannot always be obtained if you change the HEIGHT or ASPECT parameters or if you use a viewport.

The coordinates in use for this graphics command are world coordinates. Examples of valid statements follow:

```
call gscript(7,y,names);
call gscript(50,50,"plot of height vs weight");
call gscript(10,90,"yaxis",-90,90);
```

GSET Call

sets attributes for a graphics segment

Syntax

CALL GSET(*attribute* <,*value*>);

where

attribute   is a graphics attribute. The *attribute* argument can be a character matrix or quoted literal.

value   is the value to which the attribute is set. The *value* argument is specified as a matrix or quoted literal.

Description

The GSET subroutine allows you to change the following attributes for the current graphics segment:

aspect   a numeric matrix or literal that specifies the aspect ratio (width relative to height) for characters.

color   a valid SAS color. The *color* argument can be specified as a quoted text string (such as 'RED'), the name of a character matrix containing a valid color as an element, or a color number (such as 1). A color number *n* refers to the *n*th color in the color list.

font   a character matrix or quoted literal that specifies a valid font name.

height   a numeric matrix or literal that specifies the character height.

pattern   a character matrix or quoted literal that specifies the pattern to use to fill the interior of a closed curve.

style   a numeric matrix or literal that specifies an index corresponding to a valid line style.

thick   an integer specifying line thickness.

To reset the IML default value for any one of the attributes, omit the second argument. Attributes are reset back to the default with a call to the GOPEN or GSTART subroutines. Single or double quotes can be used around this argument. For more information on the attributes, see Chapter 9, "Introduction to Graphics." Examples of valid statements follow:

```
call gset('pattern','m1n45');
call gset('font','simplex');

f='font';
s='simplex';
call gset(f,s);
```

For example, the statement

```
call gset("color");
```

resets *color* to its default.

GSHOW Call

shows a graph

Syntax

CALL GSHOW<(*segment-name*)>;

where *segment-name* is a character matrix or literal specifying a graphics segment.

Description

If you do not specify *segment-name*, the GSHOW subroutine displays the current graph. If the current graph is active at the time that the GSHOW subroutine is

GSHOW Call *continued*

called, it remains active after the call; that is, graphics primitives can still be added to the segment. On the other hand, if you specify *segment-name*, the GSHOW subroutine closes any active graphics segment, searches the current catalog for a segment with the given name, and then displays that graph. Examples of valid statements follow:

```
call gshow;
call gshow("plot_a5");

seg={myplot};
call gshow(seg);
```

GSORTH Call

computes the Gram-Schmidt orthonormalization

Syntax

CALL GSORTH(*p,t,lindep,a*);

where

p	is an $m \times n$ column-orthonormal output matrix.
t	is an upper triangular $n \times n$ output matrix.
lindep	is a flag with a value of 0 if columns of *a* are independent and a value of 1 if they are dependent. The *lindep* argument is an output scalar.
a	is an input $m \times n$ matrix.

Description

The GSORTH subroutine computes the Gram-Schmidt orthonormal factorization of the $m \times n$ matrix **A**, where m is greater than or equal to n; that is, the GSORTH subroutine computes the column-orthonormal $m \times n$ matrix **P** and the upper triangular $n \times n$ matrix **T** such that

$$\mathbf{A} = \mathbf{P}^*\mathbf{T} \quad .$$

If the columns of **A** are linearly independent (that is, rank(**A**)=n), then **P** is full-rank column-orthonormal: $\mathbf{P}'^*\mathbf{P}=\mathbf{I}_w$, **T** is nonsingular, and the value of *lindep* (a scalar) is set to 0. If the columns of **A** are linearly dependent (say rank(**A**)=$k<n$) then $n-k$ columns of **P** are set to 0, the corresponding rows of **T** are set to 0 (**T** is singular), and *lindep* is set to 1. The pattern of zero columns in **P** corresponds to the pattern of linear dependencies of the columns of **A** when columns are considered in left-to-right order.

The GSORTH subroutine is not recommended for the construction of matrices of values of orthogonal polynomials; the ORPOL function should be used for that purpose.

If *lindep* is 1, you can rearrange the columns of **P** and rows of **T** so that the zero columns of **P** are right-most, that is, **P**=(P(,1), P(,k), 0, . . . , 0), where *k* is the column rank of **A** and **A**=**P*T** is preserved. The following statements make this rearrangement:

```
d=rank((ncol(t)-(1:ncol(t))`)#(vecdiag(t)=0));
temp=p;
p[,d]=temp;
temp=t;
t[,d]=temp;
```

An example of a valid GSORTH call follows:

```
x={1 1 1, 1 2 4, 1 3 9};
xpx=x`*x;
call gsorth(p, t, 1, xpx);
```

These statements produce the output matrices

```
P                 3 rows      3 cols     (numeric)

            0.193247 -0.753259 0.6286946
            0.386494 -0.530521 -0.754434
            0.9018193 0.3887787 0.1886084


T                 3 rows      3 cols     (numeric)

          15.524175 39.035892 104.99753
                  0 2.0491877 8.4559365
                  0         0 0.1257389


L                 1 row       1 col      (numeric)

                           0
```

See "Acknowledgments" in the front of this book for authorship of the GSORTH subroutine.

GSTART Call

initializes the graphics system

Syntax

CALL GSTART(<*catalog*> <,*replace*>);

where

catalog   is a character matrix or quoted literal specifying the SAS catalog for saving the graphics segments.

replace   is a numeric argument.

Description

The GSTART subroutine activates the graphics system the first time it is called. A catalog is opened to capture the graphics segments to be generated in the session. If you do not specify a catalog, IML uses the temporary catalog WORK.GSEG.

The *replace* argument is a flag; a nonzero value indicates that the new segment should replace the first found segment with the same name. The *replace* flag set by the GSTART subroutine is a global flag, as opposed to the *replace* flag set by the GOPEN subroutine. When set by GSTART, this flag is applied to all subsequent segments created for this catalog, whereas with GOPEN, the *replace* flag is applied only to the segment that is being created. The GSTART subroutine sets the *replace* flag to 0 when the *replace* argument is omitted. The *replace* option can be very inefficient for a catalog with many segments. In this case, it is better to create segments with different names (if necessary) than to use the *replace* option.

The GSTART subroutine must be called at least once to load the graphics subsystem. Any subsequent GSTART calls are generally to change graphics catalogs or reset the global *replace* flag.

The GSTART subroutine resets the defaults for all graphics attributes that can be changed by the GSET subroutine. It does not reset GOPTIONS back to their defaults unless the GOPTION corresponds to a GSET parameter. The GOPEN subroutine also resets GSET parameters. An example of a valid statement follows:

```
call gstart;
```

GSTOP Call

deactivates the graphics system

Syntax

CALL GSTOP;

Description

The GSTOP subroutine deactivates the graphics system. The graphics subsystem is disabled until the GSTART subroutine is called again.

GSTRLEN Call

finds the string length

returns the length of a string

Syntax

CALL GSTRLEN(*length,text,<height>,<font>,<window>***;**

where

length	is a matrix of lengths specified in world coordinates.
text	is a matrix of text strings.
height	is a numeric matrix or literal specifying the character height.
font	is a character matrix or quoted literal that specifies a valid font name.
window	is a numeric matrix or literal specifying a window. This is given in world coordinates and has the form

        `{minimum-x minimum-y maximum-x maximum-y}`

Description

The GSTRLEN subroutine returns in world coordinates the graphics text lengths in a given font and for a given character height. The *length* argument is the returned matrix. It has the same shape as the matrix *text*. Thus, if *text* is an

GSTRLEN Call *continued*

$n \times m$ matrix of text strings, then *length* will be an $n \times m$ matrix of lengths in world coordinates. If you do not specify *font*, the default font is assumed. If you do not specify *height*, the default height is assumed. An example using the GSTRLEN subroutine follows:

```
    /* centers text strings about coordinate   */
    /* points (50, 90) assume font=simplex      */
ht=2;
x=30;
y=90;
str='Nonparametric Cluster Analysis';
call gstrlen(len, str, ht, 'simplex');
call gscript(x-(len/2), y, str, ,,ht,'simplex');
```

GTEXT and GVTEXT Calls

place text horizontally or vertically on a graph

Syntax

CALL GTEXT(*x,y,text*,<*color*>,<*window*>,<*viewport*>);
CALL GVTEXT(*x,y,text*,<*color*>,<*window*>,<*viewport*>);

where

x	is a scalar or vector containing the *x* coordinates of the lower left starting position of the text string's first character.
y	is a scalar or vector containing the *y* coordinates of the lower left starting position of the text string's first character.
text	is a vector of text strings
color	is a valid SAS color. The *color* argument can be specified as a quoted text string (such as 'RED'), the name of a character matrix containing a valid color as an element, or a color number (such as 1). A color number *n* refers to the *n*th color in the color list.
window	is a numeric matrix or literal specifying a window. This is given in world coordinates and has the form

$$\{minimum\text{-}x \ minimum\text{-}y \ maximum\text{-}x \ maximum\text{-}y\}$$

viewport	is a numeric matrix or literal specifying a viewport. This is given in normalized coordinates and has the form

$$\{minimum\text{-}x \ minimum\text{-}y \ maximum\text{-}x \ maximum\text{-}y\}$$

Description

The GTEXT subroutine places text horizontally across a graph; the GVTEXT subroutine places text vertically on a graph. Both subroutines use hardware characters when possible. The number of text strings drawn is the maximum dimension of the first three vectors. The *color* argument can have more than one element. Hardware characters cannot always be obtained if you change the HEIGHT or ASPECT parameters (using GSET or GOPTIONS) or if you use a viewport. The coordinates in use for this graphics command are world coordinates. Examples of the GTEXT and GVTEXT subroutines follow:

```
call gopen;
call gport((0 0 50 50));
call gset('height',4);  /* shrink to a 4th of the screen */
call gtext(50,50,'Testing GTEXT: This will start in the
                  center of the viewportp ');
call gshow;
call gopen;
call gvtext(.35,4.6,'VERTICAL STRING BY GVTEXT',
           'white',(0.2 -1,1.5 6.5),(0 0,100 100));
call gshow;
```

GWINDOW Call

defines the data window

Syntax

CALL GWINDOW(*window*);

where *window* is a numeric matrix or literal specifying a window. The rectangular area's boundary is given in world coordinates, where you specify the lower left and upper right corners in the form

```
{minimum-x minimum-y maximum-x maximum-y}
```

Description

The GWINDOW subroutine sets up the window for scaling data values in subsequent graphics primitives. It is in effect until the next GWINDOW call or until the segment is closed. The coordinates in use for this graphics command are world coordinates. An example using the GWINDOW subroutine follows:

```
ydata={2.358,0.606,3.669,1.000,0.981,1.192,0.926,1.590,
       1.806,1.962,4.028,3.148,1.836,2.845,1.013,0.414};
xdata={1.215,0.930,1.152,1.138,0.061,0.696,0.686,1.072,
       1.074,0.934,0.808,1.071,1.009,1.142,1.229,0.595};

   /* WD shows the actual range of the data */
wd=(min(xdata)||min(ydata))//(max(xdata)||max(ydata));
call gwindow(wd);
```

GXAXIS and GYAXIS Calls

draw a horizontal or vertical axis

Syntax

CALL GXAXIS(*starting-point,length,nincr,*<*nminor*>,<*noticklab*>,<*format*>,
<*height*>,<*font*>,<*color*>,<*fixed-end*>,<*window*>,<*viewport*>);
CALL GYAXIS(*starting-point,length,nincr,*<*nminor*>,<*noticklab*>,<*format*>,
<*height*>,<*font*>,<*color*>,<*fixed-end*>,<*window*>,<*viewport*>);

where

starting-point	is the (*x,y*) starting point of the axis, specified in world coordinates.
length	is a numeric scalar giving the length of the axis.
nincr	is a numeric scalar giving the number of major tick marks on the axis.
nminor	is an integer specifying the number of minor tick marks between major tick marks.
noticklab	is a flag that is nonzero if the tick marks are not labeled. The default is to label tick marks.
format	is a character scalar that gives a valid SAS numeric format used in formatting the tick-mark labels. The default format is 8.2.
height	is a numeric matrix or literal that specifies the character height. This is used for the tick-mark labels.
font	is a character matrix or quoted literal that specifies a valid font name. This is used for the tick-mark labels.
color	is a valid color. The *color* argument can be specified as a quoted text string (such as 'RED'), the name of a character matrix containing a valid color as an element, or a color number (such as 1). A color number *n* refers to the *n*th color in the color list.
fixed-end	allows one end of the scale to be held fixed. **U** fixes the upper end; **L** fixes the lower end; **X** and allows both ends to vary from the data values. In addition, you may specify **N**, which cause the axis routines to bypass the scaling routine. The interval between tick marks is *length* divided by (*nincr*−1). The default is **X**.
window	is a numeric matrix or literal specifying a window. This is given in world coordinates and has the form

{*minimum-x minimum-y maximum-x maximum-y*}

viewport   is a numeric matrix or literal specifying a viewport. This is given in normalized coordinates and has the form

```
{minimum-x minimum-y maximum-x maximum-y}
```

Description

The GXAXIS subroutine draws a horizontal axis; the GYAXIS subroutine draws a vertical axis. The first three arguments are required.

The *starting-point* argument is a matrix of two numbers given in world coordinates. The matrix is the (x,y) starting point of the axis.

The *length* argument is a scalar value giving the length of the x axis or y axis in world coordinates along the x or y direction.

The *nincr* argument is a scalar value giving the number of major tick marks shown on the axis. The first tick mark will be on the starting point as specified.

The axis routines use the same scaling algorithm as the GSCALE subroutine. For example, if the x starting point is 10 and the length of the axis is 44, and if you call the GSCALE subroutine with the x vector containing the two elements, 10 and 44, the scale obtained should be the same as that obtained by the GXAXIS subroutine. Sometimes, it may be helpful to use the GSCALE subroutine in conjunction with the axis routines to get more precise scaling and labeling.

For example, suppose you want to draw the axis for $-2 \leq X \leq 2$ and $-2 \leq Y \leq 2$. The code below draws these axes. Each axis is 4 units long. Note that the x axis begins at the point $(-2,0)$ and the y axis begins at the point $(0,-2)$. The tick marks can be set at each integer value, with minor tick marks in between the major tick marks. The *noticklab* option is turned off, so that the tick marks are not labeled.

```
call gport({20 20 80 80});
call gwindow({-2 -2 2 2});
call gxaxis({-2,0},4,5,2,1);
call gyaxis({0,-2},4,5,2,1);
```

HALF Function

computes Cholesky decomposition

returns an upper triangular matrix

Syntax

HALF(*matrix*)

where *matrix* is a numeric matrix or literal.

Description

The HALF function is the same as the ROOT function. See the description of the ROOT function for Cholesky decomposition.

HANKEL Function

generates a Hankel matrix

returns a Hankel matrix

Syntax

HANKEL(*matrix*)

where *matrix* is a numeric matrix or literal.

Description

The HANKEL function generates a Hankel matrix from a vector, or a block Hankel matrix from a matrix. A block Hankel matrix has the property that all matrices on the reverse diagonals are the same. The argument matrix is an $(n^*p) \times p$ or $p \times (n^*p)$ matrix; the value returned is the $(n^*p) \times (n^*p)$ result.

The Hankel function uses the first $p \times p$ submatrix $\mathbf{A}_1$ of the argument matrix as the blocks of the first reverse diagonal. The second $p \times p$ submatrix $\mathbf{A}_2$ of the argument matrix forms the second reverse diagonal. The remaining reverse diagonals are formed accordingly. After the values in the argument matrix have all been placed, the rest of the matrix is filled in with 0. If $\mathbf{A}$ is $(n^*p) \times p$, then the first p columns of the returned matrix, $\mathbf{R}$, will be the same as $\mathbf{A}$. If $\mathbf{A}$ is $p \times (n^*p)$, then the first p rows of $\mathbf{R}$ will be the same as $\mathbf{A}$. The HANKEL function is especially useful in time-series applications, where the covariance matrix of a set of variables representing the present and past and a set of variables representing the present and future is often assumed to be a block Hankel matrix.

If

$$\mathbf{A} = \mathbf{A}_1 \,|\, \mathbf{A}_2 \,|\, \mathbf{A}_3 \,|\, \ldots \,|\, \mathbf{A}_n$$

and if $\mathbf{R}$ is the matrix formed by the HANKEL function, then

$$
\begin{aligned}
\mathbf{R} = \ &\mathbf{A}_1 \,|\, \mathbf{A}_2 \,|\, \mathbf{A}_3 \,|\, \ldots \,|\, \mathbf{A}_n \\
&\mathbf{A}_2 \,|\, \mathbf{A}_3 \,|\, \mathbf{A}_4 \,|\, \ldots \,|\, \mathbf{0} \\
&\mathbf{A}_3 \,|\, \mathbf{A}_4 \,|\, \mathbf{A}_5 \,|\, \ldots \,|\, \mathbf{0} \\
&\quad \cdot \\
&\quad \cdot \\
&\quad \cdot \\
&\mathbf{A}_n \,|\, \mathbf{0} \quad\;|\, \mathbf{0} \,|\, \ldots \,|\, \mathbf{0} \quad.
\end{aligned}
$$

If

$$
\begin{aligned}
\mathbf{A} = \ &\mathbf{A}_1 \\
&\mathbf{A}_2 \\
&\;\cdot \\
&\;\cdot \\
&\;\cdot \\
&\mathbf{A}_n
\end{aligned}
$$

and if **R** is the matrix formed by the HANKEL function, then

$$\mathbf{R} = \begin{array}{l} \mathbf{A}_1 \mathbf{A}_2 \,|\, \mathbf{A}_3 \,|\, \ldots \,|\, \mathbf{A}_n \\ \mathbf{A}_2 \,|\, \mathbf{A}_3 \,|\, \ldots \,|\, 0 \\ \quad \cdot \\ \quad \cdot \\ \quad \cdot \\ \mathbf{A}_n \,|\, 0 \,\,\,|\, 0 \,|\, \ldots \,|\, 0 \end{array} \quad .$$

For example, the IML code

```
r=hankel({1 2 3 4 5});
```

results in

```
             R            5 rows      5 cols    (numeric)

                 1            2           3         4           5
                 2            3           4         5           0
                 3            4           5         0           0
                 4            5           0         0           0
                 5            0           0         0           0
```

The statement

```
r=hankel({1 2 ,
          3 4 ,
          5 6 ,
          7 8});
```

returns the matrix

```
             R            4 rows      4 cols    (numeric)

                 1            2           5         6
                 3            4           7         8
                 5            6           0         0
                 7            8           0         0
```

And the statement

```
r=hankel({1 2 3 4 ,
          5 6 7 8});
```

returns the result

```
             R            4 rows      4 cols    (numeric)

                 1            2           3         4
                 5            6           7         8
                 3            4           0         0
                 7            8           0         0
```

HDIR Function

performs a horizontal direct product

returns the horizontal direct product

Syntax

HDIR(*matrix1,matrix2*)

where *matrix1* and *matrix2* are numeric matrices or literals.

Description

The HDIR function performs a direct product on all rows of *matrix1* and *matrix2* and creates a new matrix by stacking these row vectors into a matrix. This operation is useful in constructing design matrices of interaction effects. The *matrix1* and *matrix2* arguments must have the same number of rows. The result has the same number of rows as *matrix1* and *matrix2*. The number of columns is equal to the product of the number of columns in *matrix1* and *matrix2*.

For example, the statements

```
a={1 2,
   2 4,
   3 6};
b={0  2,
   1  1,
   0 -1};
c=hdir(a,b);
```

produce a matrix containing the values

```
        C        3 rows     4 cols    (numeric)

                    0          2         0         4
                    2          2         4         4
                    0         -3         0        -6
```

The HDIR function is useful for constructing crossed and nested effects from main effect design matrices in *ANOVA* models.

HERMITE Function

reduces a matrix to Hermite normal form

returns a reduced matrix

Syntax

HERMITE(*matrix*)

where *matrix* is a numeric matrix or literal.

Description

The HERMITE function uses elementary row operations to reduce a matrix to Hermite normal form. For square matrices this normal form is upper-triangular and idempotent.

If the argument is square and nonsingular, the result will be the identity matrix. In general the result satisfies the following four conditions (Graybill 1969, p. 120):

□ It is upper-triangular.

□ It has only values of 0 and 1 on the diagonal.

□ If a row has a 0 on the diagonal, then every element in that row is 0.

□ If a row has a 1 on the diagonal, then every off-diagonal element is 0 in the column in which the 1 appears.

Consider the following example (Graybill 1969, p. 288):

```
a={3  6  9,
   1  2  5,
   2  4 10};
h=hermite(a);
```

These statements produce

```
        H          3 rows    3 cols    (numeric)

                   1         2         0
                   0         0         0
                   0         0         1
```

If the argument is a square matrix, then the Hermite normal form can be transformed into the row echelon form by rearranging rows in which all values are 0.

HOMOGEN Function

solves homogeneous linear systems

returns a solution vector or vectors

Syntax

HOMOGEN(*matrix*)

where *matrix* is a numeric matrix or literal.

Description

The HOMOGEN function solves the homogeneous system of linear equations $A*X=0$ for X. For at least one solution vector X to exist, the $m \times n$ matrix A, $m \geq n$, has to be of rank $r < n$. The HOMOGEN function computes an $n \times (n-r)$ column orthonormal matrix X with the property $A*X=0$, $X'X=I$. If $A'A$ is ill-conditioned, rounding-error problems can occur in determining the correct rank of A and in determining the correct number of solutions X. Consider the following example (Wilkinson and Reinsch 1971, p. 149):

```
a={22  10   2   3   7,
   14   7  10   0   8,
   -1  13  -1 -11   3,
   -3  -2  13  -2   4,
    9   8   1  -2   4,
    9   1  -7   5  -1,
    2  -6   6   5   1,
    4   5   0  -2   2};
x=homogen(a);
```

These statements produce the solution

```
        X          5 rows      2 cols     (numeric)

                -0.419095           0
                0.4405091   0.4185481
                -0.052005   0.3487901
                0.6760591    0.244153
                0.4129773  -0.802217
```

In addition, this function could be used to determine the rank of an $m \times n$ matrix A, $m \geq n$.

I Function

creates an identity matrix

returns an identity matrix

Syntax

I(*dimension*)

where *dimension* specifies the size of the identity matrix.

Description

The I function creates an identity matrix with *dimension* rows and columns. The diagonal elements of an identity matrix are 1s; all other elements are 0s. The value of *dimension* must be an integer greater than or equal to 1. Noninteger operands are truncated to their integer part.

For example, the statement

```
a=I(3);
```

yields the result

```
A
1  0  0
0  1  0
0  0  1
```

IF-THEN/ELSE Statement

conditionally executes statements

Syntax

IF *expression* **THEN** *statement1*;
ELSE *statement2*;

where

expression	is an expression that is evaluated for being true or false.
statement1	is a statement executed when *expression* is true.
statement2	is a statement executed when *expression* is false.

Description

IF statements contain an expression to be evaluated, the keyword THEN, and an action to be taken when the result of the evaluation is true.

IF-THEN/ELSE Statement *continued*

The ELSE statement optionally follows the IF statement and gives an action to be taken when the IF expression is false. The expression to be evaluated is often a comparison, for example,

```
if max(a)<20 then p=0;
else p=1;
```

The IF statement results in the evaluation of the condition MAX(A)<20. If the largest value found in matrix **A** is less than 20, **P** is set to 0. Otherwise, **P** is set to 1. See the description of the MAX function for details.

When the condition to be evaluated is a matrix expression, the result of the evaluation is another matrix. If all values of the result matrix are nonzero and nonmissing, the condition is true; if any element in the result matrix is 0 or missing, the condition is false. This evaluation is equivalent to using the ALL function.

For example, writing

```
if x<y then
   do;
```

produces the same result as writing

```
if all(x<y) then
   do;
```

IF statements can be nested within the clauses of other IF or ELSE statements. Any number of nesting levels is allowed. Below is an example.

```
if x=y then if abs(y)=z then
   do;
```

▶ *Caution* . . . . *Execution of THEN clauses occurs as if you were using the ALL function.*
The statements

```
if a^=b then do;
```

and

```
if ^(a=b) then do;
```

are both valid, but the THEN clause in each case is only executed when all corresponding elements of **A** and **B** are unequal; that is, when none of the corresponding elements are equal.

Evaluation of the statement

```
if any(a^=b) then do;
```

requires only one element of **A** and **B** to be unequal for the expression to be true. ▲

IFFT Function

computes the inverse finite Fourier transform

returns a sequence

Syntax

IFFT(*f*)

where *f* is an $np \times 2$ numeric matrix.

Description

The IFFT function expands a set of sine and cosine coefficients into a sequence equal to the sum of the coefficients times the sine and cosine functions. The argument *f* is an $np \times 2$ matrix; the value returned is an $n \times 1$ vector.

Note: If the element in the last row and second column of *f* is exactly 0, *n* is $2{*}np - 2$; otherwise, *n* is $2{*}np - 1$.

The inverse finite Fourier transform of a two column matrix **F**, denoted by the vector **x** is

$$x_i = F_{1,1} + 2 \sum_{j=2}^{n} (F_{j,1}\cos((2\pi/n)(j-1)(i-1)) + F_{j,2}\sin((2\pi/n)(j-1)(i-1))) + q_i$$

for $i = 1, \ldots, n$, where $q_i = (-1)^i F_{np,1}$ if *n* is even, or $q = 0$ if *n* is odd.

Note: For most efficient use of the IFFT function, *n* should be a power of 2. If *n* is a power of 2, a fast Fourier transform is used (Singleton 1969); otherwise, a Chirp-Z algorithm is used (Monro and Branch 1976). IFFT(FFT(X)) returns *n* times **x**, where *n* is the dimension of **x.** If *f* is not the Fourier transform of a real sequence, then the vector generated by the IFFT function is not a true inverse Fourier transform. However, applications exist where the FFT and IFFT functions may be used for operations on multidimensional or complex data (Gentleman and Sande 1966; Nussbaumer 1982).

The convolution of two vectors **x** $(n \times 1)$ and **y** $(m \times 1)$ can be accomplished using the following statements:

```
a=fft(x//j(nice-nrow(x),1,0));
b=fft(y//j(nice-nrow(y),1,0));
z=(a#b)[,+];
b[,2]=-b[,2];
z=ifft(z||((a#(b[,2])))[,+]));
```

where NICE is a number chosen to allow efficient use of the FFT and IFFT functions and also is greater than $n+m$.

Windowed spectral estimates and inverse autocorrelation function estimates can also be readily obtained.

INDEX Statement

indexes a variable in a SAS data set

Syntax

INDEX *variables* | NONE

where *variables* are the names of variables for which indexes are to be built.

Description

You can use the INDEX statement to create an index for the named variables in the current input SAS data set. An index is created for each variable listed if it does not already have an index. Current retrieval is set to the last variable indexed. Subsequent I/O operations such as LIST, READ, FIND, and DELETE may use this index to retrieve observations from the data set if IML determines that indexed retrieval will be faster. The indices are automatically updated when a data set is edited with the APPEND, DELETE, or REPLACE statements. Only one index is in effect at any given time. The SHOW CONTENTS command indicates which index is in use.

For example, the following statement creates indexes for the SAS data set CLASS in the order of NAME and the order of SEX:

```
index name sex;
```

Current retrieval is set to use SEX. A LIST ALL statement would list females before males.

An INDEX NONE statement can be used to set retrieval back to physical order.

When a WHERE clause is being processed, IML automatically determines which index to use, if any. The decision is based on the variables and operators involved in the WHERE clause, and the decision criterion is based on the efficiency of retrieval.

INFILE Statement

opens a file for input

Syntax

INFILE *operand* <*options*>;

where

operand   is either a predefined filename or a quoted string containing the filename or character expression in parentheses referring to the filepath.

options   are explained below.

Description

You can use the INFILE statement to open an external file for input or, if the file is already open, to make it the current input file so that subsequent INPUT statements read from it.

The options available for the INFILE statement are described below.

LENGTH=variable

specifies a variable in which the length of a record will be stored as IML reads it in.

RECFM=N

specifies that the file is to be read in as a pure binary file rather than as a file with record separator characters. To do this, you must use the byte operand (<) on the INPUT statement to get new records rather than using separate input statements or the new line (/) operator.

The following options control how IML behaves when an INPUT statement tries to read past the end of a record. The default is STOPOVER.

FLOWOVER

allows the INPUT statement to go to the next record to obtain values for the variables.

MISSOVER

tolerates attempted reading past the end of the record by assigning missing values to variables read past the end of the record.

STOPOVER

treats going past the end of a record as an error condition, which triggers an end-of-file condition.

Several examples of INFILE statements are given below:

```
filename in1 'student.dat';     /* specify filename IN1   */
infile in1;                     /* infile filepath        */

infile 'student.dat';           /* path by quoted literal */

infile 'student.dat' missover;     /* using options  */
```

See Chapter 7, "File Access," for further information.

INPUT Statement

inputs data

Syntax

INPUT *<variables>* *<informats>* *<record-directives>* *<positionals>*;

where the clauses and options are explained below.

Description

You can use the INPUT statement to input records from the current input file, placing the values into IML variables. The INFILE statement sets up the current input file. See Chapter 7, "File Access," for details.

The INPUT statement contains a sequence of arguments that include the following:

variables

specify the variable or variables you want to read from the current position in the record. Each variable can be followed immediately by an input format specification.

informats

specify an input format. These are of the form *w.d* or *$w.* for standard numeric and character informats, respectively, where *w* is the width of the field and *d* is the decimal parameter, if any. You can also use a SAS format of the form *NAMEw.d*, where *NAME* is the name of the format. Also, you can use a single $ or & for list input applications. If the width is unspecified, the informat uses list-input rules to determine the length by searching for a blank (or comma) delimiter. The special format $RECORD. is used for reading the rest of the record into one variable. For more information on formats, see *SAS Language: Reference, Version 6, First Edition.*

Record holding is always implied for RECFM = N binary files, as if the INPUT statement has a trailing @ sign. For more information, see Chapter 7, "File Access."

Examples of valid INPUT statements are shown below:

```
input x y;
input ə1 name $ ə20 sex $ ə(20+2) age 3.;

eight=8;
input >9 <eight  number2 ib8.;
```

Below is an example using binary input:

```
proc iml;
   file 'out2.dat' recfm=n ;
   number=499; at=1;
   do i = 1 to 5;
      number=number+1;
      put >at number ib8.; at=at+8;
   end;
   closefile 'out2.dat';

   infile 'out2.dat' recfm=n;
   size=8;       /* 8 bytes */
   do pos=1 to 33 by size;
      input >pos number ib8.;
      print number;
   end;
```

record-directives
> are used to advance to a new record. *Record-directives* are the following:

holding @ sign	is used at the end of an INPUT statement to instruct IML to hold the current record so that you can continue to read from the record with later INPUT statements. Otherwise, IML automatically goes to the next record for the next INPUT statement.
/	advances to the next record.
> *operand*	specifies that the next record to be read starts at the indicated byte position in the file (for RECFM= N files only). The *operand* is a literal number, a variable name, or an expression in parentheses.
< *operand*	instructs IML to read the indicated number of bytes as the next record. The record directive must be specified for binary files (RECFM=N). The *operand* is a literal number, a variable name, or an expression in parentheses.

positionals
> instruct PROC IML to go to a specific column on the record. The *positionals* are the following:

@ *operand*	instructs IML to go to the indicated column, where *operand* is a literal number, a variable name, or an expression in parentheses. For example, @30 means to go to column 30. The operand can also be a character operand when pattern searching is needed. For more information, see Chapter 7, "File Access."
+ *operand*	specifies to skip the indicated number of columns. The *operand* is a literal number, a variable name, or an expression in parentheses.

INSERT Function

inserts one matrix inside another

returns the resulting matrix

Syntax

INSERT(*x,y,row*<,*column*>)

where

x	is the target matrix. It can be either numeric or character.
y	is the matrix to be inserted into the target. It can be either numeric or character, depending on the type of the target matrix.
row	is the row where the insertion is to be made.
column	is the column where the insertion is to be made.

Description

The INSERT function returns the result of inserting the matrix *y* inside the matrix *x* at the place specified by the *row* and *column* arguments. This is done by splitting *x* either horizontally or vertically before the row or column specified and concatenating *y* between the two pieces. Thus, if *x* is *m* rows by *n* columns, *row* can range from 0 to *m*+1 and *column* can range from 0 to *n*+1. However, it is not possible to insert in both dimensions simultaneously, so either *row* or *column* must be 0, but not both. The *column* argument is optional and defaults to 0. Also, the matrices must conform in the dimension in which they are joined.

For example, the statements

```
a={1 2, 3 4};
b={5 6, 7 8};
c=insert(a, b, 2, 0);
d=insert(a, b, 0, 3);
```

produce the result

```
       C        4 rows     2 cols    (numeric)

                             1          2
                             5          6
                             7          8
                             3          4


       D        2 rows     4 cols    (numeric)

                    1        2         5          6
                    3        4         7          8
```

C shows the result of insertion in the middle, while **D** shows insertion on an end.

INT Function

truncates a value

returns the integer portion

Syntax

INT(*matrix*)

where *matrix* is a numeric matrix or literal.

Description

The INT function truncates the decimal portion of the value of the argument. The integer portion of the value of the argument remains. The INT function takes the integer value of each element of the argument matrix. An example using the INT function follows:

```
c=2.8;
b=int(c);
```

```
        B           1 row        1 col      (numeric)

                                    2
```

In the next example, notice that a value of 11 is returned. This is because of the maximal machine precision. If the difference is less than $1E-12$, the INT function rounds up.

```
x={12.95  10.9999999999999,
   -30.5   1e-6};
b=int(x);
```

```
        B           2 rows       2 cols     (numeric)

                                  12          11
                                 -30           0
```

INV Function

computes the matrix inverse

returns the inverse of a matrix

Syntax

INV(*matrix*)

where *matrix* is a square nonsingular matrix.

Description

The INV function produces a matrix that is the inverse of *matrix*, which must be square and nonsingular.

For $G = INV(A)$ the inverse has the properties

$$GA = AG = \text{identity} \quad .$$

To solve a system of linear equations $AX=B$ for X, you can use the statement

```
x=inv(a)*b;
```

However, the SOLVE function is more accurate and efficient for this task.

The INV function uses an LU decomposition followed by backsubstitution to solve for the inverse, as described in Forsythe, Malcolm, and Moler (1967).

INVUPDT Function

updates a matrix inverse

returns an updated matrix inverse

Syntax

INVUPDT(*matrix,vector*<*,scalar*>)

where

matrix   is an $n \times n$ positive definite matrix.

vector   is an $n \times 1$ or $1 \times n$ vector.

scalar   is a numeric scalar.

Description

The INVUPDT function updates a matrix inverse. For example, let

```
r=invupdt(a,x,w);
```

where **R** is an $n \times w$ matrix; **A** is an $n \times n$ positive definite matrix; **X** is an $n \times 1$ or $1 \times n$ vector; and w is an optional scalar (if not specified, w has default value 1).

The INVUPDT function computes the matrix expression

$$\mathbf{R} = \mathbf{A} - w\mathbf{AX}(1 + w\mathbf{X}'\mathbf{AX})^{-1}\mathbf{X}'\mathbf{A}^{-1}\mathbf{X}'\mathbf{A}$$

or, in matrix language,

```
r=a-w*a*x*inv(1+w*x`*a*x)*x`*a;
```

The INVUPDT function is used primarily to update a matrix inverse because the function has the property

$$\text{INVUPDT}(\mathbf{B}^{-1},\mathbf{X},w) = (\mathbf{B} + w\mathbf{XX}')^{-1} \quad .$$

If **Z** is a design matrix and **X** is a new observation to be used in estimating the parameters of a linear model, then the inverse crossproducts matrix that includes the new observation can be updated from the old inverse by

```
c2=invupdt(c,x);
```

where **C**=INV(**Z**'***Z**). Note that

```
c2=inv((z//x)`*(z//x));
```

If w is 1, the function adds an observation to the inverse; if w is -1, the function removes an observation from the inverse. If weighting is used, w is the weight.

To perform the computation, the INVUPDT function uses about $2n^2$ multiplications and additions, where n is the row dimension of the positive definite argument matrix.

IPF Call

performs an iterative proportional fit

Syntax

CALL IPF(*fit,status,dim,table,config*<*,initab*> <*,mod*>);

where

fit   is a returned matrix. The argument *fit* specifies an array of the estimates of the expected number in each cell under the model specified in *config*. This matrix conforms to *table*.

IPF Call *continued*

status is a returned matrix. The *status* argument specifies a row vector of length 3. If you specify STATUS= {*error, obs-maxdev, no-iterate*}, then *error* is 0 if there is convergence to the desired accuracy and is 3 if there is no convergence to the desired accuracy; *obs-maxdev* is the maximum difference between estimates of the last two iterations; and *no-iterate* is the number of iterations performed.

dim is an input matrix. The *dim* argument is a vector specifying the number of variables and the number of their possible levels in a contingency table. If *dim* is $1 \times v$, then there are v variables, and the value of the ith element is the number of levels of the ith variable.

table is an input matrix. The *table* argument specifies an array of the number of observations at each level of each variable. Variables are nested across columns and then across rows.

config an input matrix. The *config* argument gives an array specifying which marginal totals to fit. Each column specifies a distinct marginal in the model under consideration. Because the model is hierarchical, all subsets of specified marginals are included in fitting.

initab is an input matrix. The *initab* argument is an array of initial values for the iterative procedure. If you do not specify values, 1s are used. For incomplete tables, *initab* is set to 1 if the cell is included in the design, and 0 if it is not.

mod is an input matrix. The *mod* argument is a two-element vector specifying the stopping criteria. If *mod*= {*maxdev, maxit*}, then the procedure iterates until either *maxit* iterations are completed or the maximum difference between estimates of the last two iterations is less than *maxdev*. Default values are *maxdev*= .25 and *maxit*= 15.

Description

The IPF subroutine performs an iterative proportional fit of the marginal totals of a contingency table. The arguments used with the IPF function can be matrix names or literals.

The matrix *table* must conform in size to the contingency table as specified in *dim*. In particular, if *table* is $n \times m$, the product of the entries in *dim* must equal nm. Furthermore, there must be some integer k such that the product of the first k entries in *dim* equals m. If you specify *initab*, then it must be the same size as *table*.

For example, consider the no–three-factor-effect model for interpreting Bartlett's data as described in Bishop, Fienberg, and Holland (1975):

```
dim={2 2 2};
table={156  84  84  156,
       107 133  31  209};
config={1  1  2,
        2  3  3};
call ipf(fit,status,dim,table,config);
```

The result is

```
        FIT
161.062    78.938   78.907   161.093
101.905   138.095   36.119   203.881
```

```
        STATUS
        0    .166966      4
```

Equivalent results are obtained by the statement

```
table={156  84,
        84 156,
       107 133,
        31 209};
```

or the statement

```
table={156 84 84 156 107 133 31 209};
```

In the first specification, TABLE is interpreted as

variable 3	variable 2: variable 1:	1		2	
		1	2	1	2
1		156	84	84	156
2		107	133	31	209

IPF Call *continued*

In the second specification, TABLE is interpreted as

variable 3	variable 2	variable 1:	1	2
1	1		156	84
	2		84	156
2	1		107	133
	2		31	209

And in the third specification, TABLE is interpreted as

variable 3:		1				2		
variable 2:	1		2		1		2	
variable 1:	1	2	1	2	1	2	1	2
	156	84	84	156	107	133	31	209 .

J Function

creates a matrix of identical values

returns a matrix

Syntax

J(*nrow*<,*ncol*<,*value*>>)

where

nrow   is a numeric matrix or literal giving the number of rows.

ncol   is a numeric matrix or literal giving the number of columns.

value is a numeric or character matrix or literal for filling the rows and
columns of the matrix.

Description

The J function creates a matrix with *nrow* rows and *ncol* columns with all
elements equal to *value*. If *ncol* is not specified, it defaults to *nrow*. If *value* is not
specified, it defaults to 1. The REPEAT and SHAPE functions can also perform
this operation, and they are more general.

Examples of the J function are shown below:

```
b=j(3);
```

B   3 rows   3 cols   (numeric)

                        1   1   1
                        1   1   1
                        1   1   1

```
r=j(5,2,'xyz');
```

R   5 rows   2 cols   (character, size 3)

                                    xyz xyz
                                    xyz xyz
                                    xyz xyz
                                    xyz xyz
                                    xyz xyz

LCP Call

solves the linear complementarity problem

Syntax

CALL LCP(*rc,w,z,m,q,<epsilon>*);

where

m	is an $m \times m$ matrix.
q	is an $m \times 1$ matrix.
epsilon	is a scalar defining virtual zero. The default value of *epsilon* is 1.0E−8.
rc	returns one of the following scalar return codes:

      0   solution found

      1   no solution possible

      5   solution is numerically unstable

      6   subroutine could not obtain enough memory.

w and *z*	return the solution in an *m*-element column vector.

LCP Call *continued*

Description

The LCP subroutine solves the linear complementarity problem:

$$\mathbf{w} = \mathbf{Mz} + \mathbf{q}$$
$$\mathbf{w'z} = 0$$
$$\mathbf{w,z} \geq 0$$

Consider the following example:

```
q={1, 1};
m={1 0,
   0 1};
call lcp(rc,w,z,m,q);
```

The result is

```
        RC          1 row      1 col    (numeric)

                                 0


         W          2 rows     1 col    (numeric)

                                 1
                                 1


         Z          2 rows     1 col    (numeric)

                                 0
                                 0
```

The next example shows the relationship between quadratic programming and the linear complementarity problem. Consider the linearly constrained quadratic program:

$$\min \mathbf{c'x} + \mathbf{x'Hx}/2$$

$$\text{st. } \mathbf{Gx} \geq \mathbf{b} \qquad \text{(QP)}$$

$$\mathbf{x} \geq 0 \quad .$$

If **H** is positive semidefinite, then a solution to the Kuhn-Tucker conditions solves QP. The Kuhn-Tucker conditions for QP are

$$\mathbf{c} + \mathbf{Hx} = \mu + \mathbf{G'\lambda}$$
$$\lambda'(\mathbf{Gx} - \mathbf{b}) = 0$$
$$\mu'\mathbf{x} = 0$$
$$\mathbf{Gx} \geq \mathbf{b}$$

$$\mathbf{x}, \mu, \lambda \geq 0 \quad .$$

In the linear complementarity problem, let

$$\mathbf{M} = \begin{bmatrix} \mathbf{H} & -\mathbf{G'} \\ \mathbf{G} & 0 \end{bmatrix}$$

$$\mathbf{w'} = (\mu'\mathbf{s'})$$
$$\mathbf{z'} = (\mathbf{x'}\lambda')$$
$$\mathbf{q'} = (\mathbf{c'} - \mathbf{b}) \quad .$$

Then the Kuhn-Tucker conditions are expressed as finding **w** and **z** that satisfy

$$\mathbf{w} = \mathbf{Mz} + \mathbf{q}$$
$$\mathbf{w'z} = 0$$
$$\mathbf{w,z} \geq 0 \quad .$$

From the solution **w** and **z** to this linear complementarity problem, the solution to QP is obtained; namely, **x** is the primal structural variable, **s**= **Gx**−**b** the surpluses, and μ and λ are the dual variables. Consider a quadratic program with the following data:

$$C' = (1\ 2\ 4\ 5) \quad B' = (\ 1\ 1\ )$$

$$H = \begin{bmatrix} 100 & 10 & 1 & 0 \\ 10 & 100 & 10 & 1 \\ 1 & 10 & 100 & 10 \\ 0 & 1 & 10 & 100 \end{bmatrix}$$

$$G = \begin{bmatrix} 1 & 2 & 3 & 4 \\ 10 & 20 & 30 & 40 \end{bmatrix}$$

LCP Call *continued*

This problem is solved using the LCP subroutine in PROC IML as follows:

```
    /*---- Data for the Quadratic Program -----*/
c={1,2,3,4};
h={100 10 1 0, 10 100 10 1, 1 10 100 10, 0 1 10 100};
g={1 2 3 4, 10 20 30 40 };
b={1, 1};

    /*----- Express the Kuhn-Tucker Conditions as an LCP ----*/
m=h||-g';
m=m//(g||j(nrow(g),nrow(g),0));
q=c//-b ;

    /*----- Solve for a Kuhn-Tucker Point --------*/
call lcp(rc,w,z,m,q);

    /*------ Extract the Solution to the Quadratic Program ----*/
x=z[1:nrow(h)];
print rc x;
```

The printed solution is

```
              RC         1 row      1 col     (numeric)

                                       0

              X          4 rows     1 col     (numeric)

                                   0.0307522
                                   0.0619692
                                   0.0929721
                                   0.1415983
```

LENGTH Function

finds the lengths of character matrix elements

returns a numeric matrix containing string lengths

Syntax

LENGTH(*matrix*)

where *matrix* is a character matrix or quoted literal.

Description

The LENGTH function takes a character matrix as an argument and produces a numeric matrix as a result. The result matrix has the same dimensions as the argument and contains the lengths of the corresponding string elements in *matrix*. The length of a string is equal to the position of the rightmost nonblank character in the string. If a string is entirely blank, its length value is set to 1. An example of the LENGTH function is shown below:

```
c={'Hello' 'My name is Jenny'};
b=length(c);
```

```
         B           1 row        2 cols     (numeric)

                        5            16
```

See also the description of the NLENG function.

LINK and RETURN Statements

jump to another statement

Syntax

LINK *label*;
     statements
label:statements
RETURN;

Description

The LINK statement, like the GOTO statement, directs IML to jump to the statement with the specified label. Unlike the GOTO statement, however, IML remembers where the LINK statement was issued and returns to that point when a RETURN statement is executed. This statement can only be used inside modules and DO groups.

The LINK statement provides a way of calling sections of code as if they were subroutines. The LINK statement calls the routine. The routine begins with the label and ends with a RETURN statement. LINK statements can be nested within other LINK statements to any level. A RETURN statement without a LINK statement is executed the same as the STOP statement.

Any time you use a LINK statement, you may consider using a RUN statement and a module defined using the START and FINISH statements instead.

LINK and RETURN Statements *continued*

An example using the LINK statement is shown below:

```
start a;
   x=1;
   y=2;
   link sum1;
   print z;
   stop;
   sum1:
       z=x+y;
   return;
finish a;
run a;
```

```
              Z              1 row       1 col      (numeric)

                                      3
```

LIST Statement

displays observations of a data set

Syntax

LIST <*range*> <VAR *operand*> <WHERE(*expression*)>;

where

range	specifies a range of observations
operand	specifies a set of variables
expression	is an expression evaluated to be true or false.

Description

The LIST statement prints selected observations of a data set. If all data values for variables in the VAR clause fit on a single line, values are displayed in columns headed by the variable names. Each record occupies a separate line. If the data values do not fit on a single line, values from each record are grouped into paragraphs. Each element in the paragraph has the form *name=value*.

You can specify a *range* of observations with a keyword or by record number using the POINT option. You can use any of the following keywords to specify a *range*:

ALL	all observations
CURRENT	the current observation (this is the default for the LIST statement)
NEXT <*number*>	the next observation or the next *number* of observations
AFTER	all observations after the current one
POINT *operand*	observations specified by number, where *operand* can be one of the following:

Operand	Example
a single record number	`point 5`
a literal giving several record numbers	`point {2 5 10}`
the name of a matrix containing record numbers	`point p`
an expression in parentheses	`point (p+1)`

If the current data set has an index in use, the POINT option is invalid.

You can specify a set of variables to use with the VAR clause. The *operand* can be specified as one of the following:

□ a literal containing variable names

□ the name of a matrix containing variable names

□ an expression in parentheses yielding variable names

□ one of the keywords described below:

ALL	for all variables
CHAR	for all character variables
NUM	for all numeric variables.

Below are examples showing each possible way you can use the VAR clause:

```
var {time1 time5 time9};    /* a literal giving the variables  */
var time;                   /* a matrix containing the names   */
var('time1':'time9');       /* an expression                   */
var _all_;                  /* a keyword                       */
```

LIST Statement *continued*

The WHERE clause conditionally selects observations, within the *range* specification, according to conditions given in the clause. The general form of the WHERE clause is

WHERE(*variable comparison-op operand*)

where

variable   is a variable in the SAS data set.

comparison-op   is any one of the following comparison operators:

<	less than
<=	less than or equal to
=	equal to
>	greater than
>=	greater than or equal to
^=	not equal to
?	contains a given string
^?	does not contain a given string
= :	begins with a given string
= *	sounds like or is spelled similar to a given string.

operand   is a literal value, a matrix name, or an expression in parentheses.

WHERE comparison arguments can be matrices. For the following operators, the WHERE clause succeeds if *all* the elements in the matrix satisfy the condition:

    ^= ^? < <= > >=

For the following operators, the WHERE clause succeeds if *any* of the elements in the matrix satisfy the condition:

    = ? =: =*

Logical expressions can be specified within the WHERE clause using the AND (&) and OR (|) operators. The general form is

clause&*clause*   (for an AND clause)
clause | *clause*   (for an OR clause)

where *clause* can be a comparison, a parenthesized clause, or a logical expression clause that is evaluated using operator precedence.

Note: The expression on the left-hand side refers to values of the data set variables and the expression on the right-hand side refers to matrix values. Below are several examples on using the LIST statement:

```
list all;                        /* lists whole data set */
list;                        /* lists current observation */
list var{name addr};        /* lists NAME and ADDR in current obs */
list all where(age>30);  /* lists all obs where condition holds */
list next;                        /* lists next observation */
list point 24;                        /* lists observation 24 */
list point (10:15);        /* lists observations 10 through 15 */
```

LOAD Statement

loads modules and matrices from library storage

Syntax

LOAD <MODULE=(*module-list*)> <*matrix-list*>;

where

module-list	is a list of modules.
matrix-list	is a list of matrices.

Description

The LOAD statement loads modules or matrix values from the current library storage into the current workspace. For example, to load three modules A, B, and C and one matrix X, specify the statement

```
load module=(A B C) X;
```

The special operand _ALL_ can be used to load all matrices or all modules. For example, if you want to load all matrices, specify

```
load _all_;
```

If you want to load all modules, specify

```
load module=_all_;
```

To load all matrices and modules stored in the library storage, you can enter the LOAD command without any arguments:

```
load;
```

LOAD Statement *continued*

The storage library can be specified using a RESET STORAGE command. The default library is SASUSER.IMLSTOR. For more information, see Chapter 12, "Storage Features," and the descriptions of the STORE, REMOVE, RESET, and SHOW statements.

LOC Function

finds nonzero elements of a matrix

returns a row vector

Syntax

LOC(*matrix*)

where *matrix* is a numeric matrix or literal.

Description

The LOC function creates a $1 \times n$ row vector, where n is the number of nonzero elements in the argument. Missing values are treated as zeros. The values in the resulting row vector are the locations of the nonzero elements in the argument (in row-major order, like subscripting). For example, the statements

```
a={1 0 2 3 0};
b=loc(a);
```

result in the row vector

```
            B           1 row      3 cols     (numeric)

                          1          3          4
```

since the first, third, and fourth elements of **A** are nonzero. If every element of the argument vector is 0, the result is empty; that is, **B** has zero rows and zero columns.

The LOC function is useful for subscripting parts of a matrix that satisfy some condition. For example, suppose you want to create a matrix **Y** containing the rows of **X** that have a positive element in the diagonal of **X**. Specify the statements

```
x={1  1  0,
   0 -2  2,
   0  0  3};
y=x[loc(vecdiag(x)>0),];
```

The result is

Y = **X**[{1 3},]

or the matrix

Y	2 rows	3 cols	(numeric)
1	1	0	
0	0	3	

since the first and third rows of **X** have positive elements on the diagonal of **X**.
The next example selects all positive elements of a column vector **A**:

```
a={0,
  -1,
   2,
   0};
y=a[loc(a>0),];
```

The result is

Y = **A**[3,]

or the scalar

Y	1 row	1 col	(numeric)
2			

LOG Function

takes the natural logarithm

returns a matrix of logarithms

Syntax

LOG(*matrix*)

where *matrix* is a numeric matrix or literal.

Description

The LOG function is the scalar function that takes the natural logarithm of each element of the argument matrix. An example of a valid statement is shown below:

```
b=log(c);
```

LP Call

solves the linear programming problem

Syntax

CALL LP(*rc,x,dual,a,b,<cntl>,<u>,<l>,<basis>*);

where

a	is an $m \times n$ vector specifying the technological coefficients, where m is less than or equal to n.
b	is an $m \times 1$ vector specifying the right-side vector.
cntl	is an optional row vector with 1 to 5 elements. If CNTL= (*indx,nprimal,ndual,epsilon,infinity*),

then

indx	is the subscript of nonzero objective coefficient.
nprimal	is the maximum number of primal iterations.
ndual	is the maximum number of dual iterations.

epsilon   is the value of virtual zero.

infinity   is the value of virtual infinity.

The default values are as follows: *indx* equals *n*, *nprimal* equals 999999, *ndual* equals 999999, *epsilon* equals 1.0E−8, and *infinity* is machine dependent. If you specify *ndual* or *nprimal* or both, then on return they contain the number of iterations actually performed.

u   is an optional array of dimension *n* specifying upper bounds on the decision variables. If you do not specify *u*, the upper bounds are assumed to be *infinity*.

l   is an optional array of dimension *n* specifying lower bounds on the decision variables. If *l* is not given, then the lower bounds are assumed to be 0 for all the decision variables. This includes the decision variable associated with the objective value, which is specified by the value of *indx*.

basis   is an optional array of dimension *n* specifying the current basis. This is given by identifying which columns are explicitly in the basis and which columns are at their upper bound, as given in *u*. The absolute value of the elements in this vector is a permutation of the column indices. The columns specified in the first *m* elements of *basis* are considered the explicit basis. The absolute value of the last *n*−*m* elements of *basis* are the indices of the nonbasic variables. Any of the last *n*−*m* elements of *basis* that are negative indicate that that nonbasic variable is at its upper bound. On return from the LP subroutine, the *basis* vector contains the final basis encountered. If you do not specify *basis*, then the subroutine assumes that an initial basis is in the last *m* columns of **A** and that no nonbasic variables are at their upper bound.

rc   returns one of the following scalar return codes:

   0   solution is optimal

   1   solution is primal infeasible and dual feasible

   2   solution is dual infeasible and primal feasible

   3   solution is neither primal nor dual feasible

   4   singular basis encountered

   5   solution is numerically unstable

   6   subroutine could not obtain enough memory

   7   number of iterations exceeded.

x   returns the current primal solution in a column vector of length *n*.

dual   returns the current dual solution in a row vector of length *m*.

LP Call *continued*

Description

The LP subroutine solves the linear program:

max (0,...,0,1,0,...,0) **x**

st. **Ax = b**

l ≤ **x** ≤ **u**

The subroutine first inverts the initial basis. If the **BASIS** vector is given, then the initial basis is the $m \times m$ submatrix identified by the first m elements in **BASIS**; otherwise, the initial basis is defined by the last m columns of **A**. If the initial basis is singular, the subroutine returns with RC=4. If the basis is nonsingular, then the current dual and primal solutions are evaluated. If neither is feasible, then the subroutine returns with RC=3. If the primal solution is feasible, then the primal algorithm iterates until either a dual feasible solution is encountered or the number of NPRIMAL iterations is exceeded. If the dual solution is feasible, then the dual algorithm iterates until either a primal feasible solution is encountered or the number of NDUAL iterations is exceeded. When a basis is identified that is both primal and dual feasible, then the subroutine returns with RC=0.

Note that care must be taken when solving a sequence of linear programs and using the NPRIMAL or NDUAL control parameters or both. Because the LP subroutine resets the NPRIMAL and NDUAL parameters to reflect the number of iterations executed, subsequent invocations of the LP subroutine will have the number of iterations limited to the number used by the last LP subroutine executed. In these cases you should consider resetting these parameters prior to each LP call.

Consider the following example to maximize X_1 subject to the constraints $X_1 - X_2 = 10$ and $X_1 \geq 0$. The problem is solved as follows:

```
    /* the problem data */
obj={1 0};
coef={1 1};
b={0, 10};

    /* embed the objective function */
    /* in the coefficient matrix    */
a=obj//coef;
a=a||{-1, 0};

    /* solve the problem */
call lp(rc,x,dual,a,b);
```

The result is

RC	1 row	1 col	(numeric)

0

X	3 rows	1 col	(numeric)
		10	
		0	
		10	

DUAL	1 row	2 cols	(numeric)
	-1	1	

MARG Call

evaluates marginal totals in a multiway contingency table

Syntax

CALL MARG(*locmar,marginal,dim,table,config*);

where

dim	specifies a vector containing the number of variables and the number of their possible levels in a contingency table. If *dim* is $1 \times v$, then there are v variables, and the value of the *i*th element is the number of levels of the *i*th variable.
table	specifies an array containing the number of observations at each level of each variable. Variables are nested across columns and then across rows.
config	specifies an array containing the marginal totals to be evaluated. Each column specifies a distinct marginal.
locmar	returns a vector of indices to each new set of marginal totals specified by *config*. A marginal total is exhibited for each level of the specified marginal. These indices help locate particular totals.
marginal	returns a vector of marginal totals.

Description

The matrix *table* must conform in size to the contingency table specified in *dim*. In particular, if *table* is $n \times m$, the product of the entries in the *dim* vector must equal nm. In addition, there must be some integer k such that the product of the first k entries in *dim* equals m. See the description of the IPF function for more information on specifying *table*.

MARG Call *continued*

For example, consider the no–three-factor-effect model for Bartlett's data as described in Bishop, Fienberg, and Holland (1975):

```
dim={2 2 2};
table={156   84   84   156,
       107  133   31   209};
config={1 1 2,
        2 3 3};
call marg(locmar,marginal,dim,table,config);
```

These statements return

	LOCMAR	1 row	3 cols	(numeric)		
		1	5	9		
	MARGINAL	1 row	12 cols	(numeric)		
	263	217	115	365	240	240	138
:	342	240	240	240	240		

MATTRIB Statement

associates printing attributes with matrices

Syntax

MATTRIB *name* <ROWNAME=*row-name*> <COLNAME=*column-name*>
      <LABEL=*label*> <FORMAT=*format*>;

where

name	is a character matrix or quoted literal giving the name of a matrix.
row-name	is a character matrix or quoted literal specifying row names.
column-name	is a character matrix or quoted literal specifying column names.
label	is a character matrix or quoted literal associating a label with the matrix. The *label* argument has a maximum length of 40 characters.
format	is a valid SAS format.

Description

The MATTRIB statement associates printing attributes with matrices. Each matrix can be associated with a ROWNAME= matrix and a COLNAME= matrix, which is used whenever the matrix is printed to label the rows and columns, respectively. The statement is written as the keyword MATTRIB followed by a list of one or more names and attribute associations. It is not necessary to specify all attributes. The attribute associations are applied to the previous *name*. Thus, the following statement gives a row name RA and a column name CA to **A**, and a column name CB to **B**:

```
mattrib a rowname=ra colname=ca b colname=cb;
```

You cannot group names; although the following statement is valid, it does not associate anything with **A**:

```
mattrib a b rowname=n;
```

The values of the associated matrices are not looked up until they are needed. Thus, they need not have values at the time the MATTRIB statement is specified. They can be specified later when the object matrix is printed. The attributes continue to bind with the matrix until reassigned with another MATTRIB statement. To eliminate an attribute, specify EMPTY as the name, for example, ROWNAME=EMPTY. Labels can be up to 40 characters long. Longer labels are truncated. Use the SHOW NAMES statement to view current matrix attributes.

An example using the MATTRIB statement follows:

```
rows='xr1':'xr5';
print rows;
```

```
                              ROWS
                              xr1 xr2 xr3 xr4 xr5
```

```
cols='cl1':'cl5';
print cols;
```

```
                              COLS
                              cl1 cl2 cl3 cl4 cl5
```

```
x={1 1 1 1,2 2 2 2,3 3 3 3};
mattrib x rowname=(rows [1:3 ])
          colname=(cols [1:4])
          label={'matrix,x'}
          format=5.2;
print x;
```

```
                    matrix,x   cl1   cl2   cl3   cl4

                       xr1    1.00  1.00  1.00  1.00
                       xr2    2.00  2.00  2.00  2.00
                       xr3    3.00  3.00  3.00  3.00
```

MAX Function

finds the maximum value of matrix

returns a scalar (or character string value)

Syntax

MAX(*matrix1* <,*matrix2*,...,*matrix15*>)

where *matrix* is a numeric or character matrix or literal.

Description

The MAX function produces a single numeric value (or a character string value) that is the largest element (or highest character string value) in all arguments. There can be as many as 15 argument matrices. The function checks for missing numeric values and does not include them in the result. If all arguments are missing, then the machine's most negative representable number is the result.

If you want to find the elementwise maximums of the corresponding elements of two matrices, use the maximum operator (<>).

For character arguments, the size of the result is the size of the largest of all arguments.

An example using the MAX function is shown below:

```
b=max(c);
```

MIN Function

finds the smallest element of a matrix

returns a scalar (or character string value)

Syntax

MIN(*matrix1* <,*matrix2*,...,*matrix15*>)

where *matrix* is a numeric or character matrix or literal.

Description

The MIN function produces a single numeric value (or a character string value) that is the smallest element (lowest character string value) in all arguments. There can be as many as 15 argument matrices. The function checks for missing numeric values and excludes them from the result. If all arguments are missing, then the machine's largest representable number is the result.

If you want to find the elementwise minimums of the corresponding elements of two matrices, use the element minimum operator (><).

For character arguments, the size of the result is the size of the largest of all arguments.

An example using the MIN function is shown below.

```
b=min(c);
```

MOD Function

computes the modulo (remainder)

returns the remainder

Syntax

MOD(*value,divisor*)

where

value   is a numeric matrix or literal giving the dividend.

divisor   is a numeric matrix or literal giving the divisor.

Description

The MOD function is the scalar function returning the remainder of the division of elements of the first argument by elements of the second argument. An example of a valid statement follows.

```
b=mod(4,1);
```

NAME Function

lists the names of arguments

returns a character column vector

Syntax

NAME(*arguments*);

where *arguments* are the names of existing matrices.

NAME Function *continued*

Description

The NAME function returns the names of the arguments in a column vector. In the following example, **N** is a 3×1 character matrix of element size 8 containing the character values A, B, and C:

```
n=name(a,b,c);
```

The main use of the NAME function is with macros when you want to use an argument for both its name and its value.

NCOL Function

finds the number of columns of a matrix

returns a scalar containing the number of columns of a matrix

Syntax

NCOL(*matrix*)

where *matrix* is a numeric or character matrix.

Description

The NCOL function returns a single numeric value that is the number of columns in *matrix*. If the matrix has not been given a value, the NCOL function returns a value of 0.

For example, to let B contain the number of columns of matrix **S**, use the statement

```
b=ncol(s);
```

NLENG Function

finds the size of an element

returns a scalar

Syntax

NLENG(*matrix*)

where *matrix* is a numeric or character matrix.

Description

The NLENG function returns a single numeric value that is the size in bytes of each element in *matrix*. All matrix elements have the same size. If the matrix does not have a value, then the NLENG function returns a value of 0. This function is different from the LENGTH function, which returns the size of each element of a character matrix, omitting the trailing blanks.

The following statement returns the value 7:

```
a=nleng(("ab " "ijklm ",
         "x"    " "));
```

NORMAL Function

generates a pseudo-random normal deviate

returns a pseudo-random number

Syntax

NORMAL(*seed*)

where *seed* is a numeric matrix or literal. The *seed* argument can be any integer value up to $2^{31}-1$.

Description

The NORMAL function is a scalar function that returns a pseudo-random number having a normal distribution with a mean of 0 and a standard deviation of 1. The NORMAL function returns a matrix with the same dimensions as the argument. The first argument on the first call is used for the seed (or if that is 0, the system clock is used for the seed). This function is synonymous with the DATA step function RANNOR. The Box-Muller transformation of the UNIFORM function deviates is used to generate the numbers.

NROW Function

finds the number of rows of a matrix

returns a scalar containing the number of rows of a matrix

Syntax

NROW(*matrix*)

where *matrix* is a numeric or character matrix.

NROW Function *continued*

Description

The NROW function returns a single numeric value that is the number of rows in matrix. If the matrix has not been given a value, the NROW function returns a value of 0.

For example, to let J contain the number of rows of the matrix **S**, use the statement

```
j=nrow(s);
```

NUM Function

produces a numeric representation of a character matrix

returns a numeric matrix

Syntax

NUM(*matrix*)

where *matrix* is a character matrix or a quoted literal.

Description

The NUM function takes as an argument a character matrix whose elements are character numerics and produces a numeric matrix whose dimensions are the same as the dimensions of the argument and whose elements are the numeric representations (double-precision floating-point) of the corresponding elements of the argument. An example using the NUM function is shown below:

```
c={'1' '2' '3'};
j=num(c);
```

```
        C           1 row      3 cols    (character, size 1)

                                1 2 3


        J           1 row      3 cols    (numeric)
                        1          2          3
```

See also the description of the CHAR function, which does the reverse conversion.

OPSCAL Function

rescales qualitative data to be a least-squares fit to qualitative data

returns the optimal scaling transformation

Syntax

OPSCAL(*mlevel,quanti* <*,qualit*>)

where

mlevel   specifies a scalar that has one of two values. When *mlevel* is 1 the *qualit* matrix is at the nominal measurement level; when *mlevel* is 2 it is at the ordinal measurement level.

quanti   specifies an $m \times n$ matrix of quantitative information assumed to be at the interval level of measurement.

qualit   specifies an $m \times n$ matrix of qualitative information whose level of measurement is specified by *mlevel*. When *qualit* is omitted, *mlevel* must be 2. When omitted, a temporary *qualit* is constructed that contains the integers from 1 to n in the first row, from $n+1$ to $2n$ in the second row, from $2n+1$ to $3n$ in the third row, and so forth, up to the integers $(m-1)n$ to mn in the last (*m*th) row. Note that you cannot specify *qualit* as a character matrix.

Description

The result of the OPSCAL function is the optimal scaling transformation of the qualitative (nominal or ordinal) data in *qualit*. The optimal scaling transformation result

□   is a least-squares fit to the quantitative data in *quanti*

□   preserves the qualitative measurement level of *qualit*.

The OPSCAL function performs as a function or call. When used as a call, the first argument of the call is the matrix to contain the result returned.

   When *qualit* is at the nominal level of measurement, the optimal scaling transformation result is a least-squares fit to *quanti*, given the restriction that the category structure of *qualit* must be preserved. If element i of *qualit* is in category c, then element i of the optimum scaling transformation result is the mean of all those elements of *quanti* that correspond to elements of *qualit* that are in category c.

OPSCAL Function *continued*

For example, consider these statements:

```
quanti=(5  4  6  7  4  6  2  4  8  6);
qualit=(6  6  2 12  4 10  4 10  8  6);
os=opscal(1,quanti,qualit);
```

The resulting vector **OS** has the following values:

```
        OS              1 row      10 cols    (numeric)

       5       5        6       7       3       5       3
 :     5       8        5
```

The optimal scaling transformation result is said to preserve the nominal measurement level of *qualit* because wherever there was a *qualit* category *c*, there is now a result category label *v*. The transformation is least squares because the result element *v* is the mean of appropriate elements of *quanti*. This is Young's (1981) discrete-nominal transformation.

When *qualit* is at the ordinal level of measurement, the optimal scaling transformation result is a least-squares fit to *quanti*, given the restriction that the ordinal structure of *qualit* must be preserved. This is done by determining blocks of elements of *qualit* so that if element *i* of *qualit* is in block *b*, then element *i* of the result is the mean of all those *quanti* elements corresponding to block *b* elements of *qualit* so that the means are (weakly) in the same order as the elements of *qualit*. For example, consider these statements:

```
quanti=(5  4  6  7  4  6  2  4  8  6);
qualit=(6  6  2 12  4 10  4 10  8  6);
os=opscal(2,quanti,qualit);
```

The resulting vector **OS** has the following values:

```
        OS              1 row      10 cols    (numeric)

       5       5        4       7       4       6       4
 :     6       6        5
```

This transformation preserves the ordinal measurement level of *qualit* because the elements of *qualit* and the result are (weakly) in the same order. It is least-squares because the result elements are the means of appropriate elements of *quanti*. By comparing this result to the nominal one, you see that categories whose means are incorrectly ordered have been merged together to form correctly ordered blocks. This is known as Kruskal's (1964) least-squares monotonic transformation.

Consider the following statements:

```
quanti=(5  3  6  7  5  7  8  6  7  8);
os=opscal(2,quanti);
```

These statements imply that

```
qualit={1  2  3  4  5  6  7  8  9 10 };
```

which means that the resulting vector has the values

```
            OS            1 row     10 cols    (numeric)

        4        4        6        6        6        7        7
    :   7        7        8
```

ORPOL Function

generates orthogonal polynomials

returns a matrix with orthogonal columns

Syntax

ORPOL(*vector*, <*maxdegree*<,*weights*>>)

where

vector   is an $n \times 1$ (or $1 \times n$) vector of values over which the polynomials are to be defined.

maxdegree   specifies the maximum degree polynomial to be computed. Note that the number of columns in the computed result is $1 + maxdegree$, whether *maxdegree* is specified or the default value is used. If *maxdegree* is omitted, IML uses the default value $\min(n, 19)$.
If *weights* is specified, *maxdegree* must also be specified.

weights   specifies an $n \times 1$ (or $1 \times n$) vector of nonnegative weights to be used in defining orthogonality:

$$\mathbf{P}'^{*}\text{DIAG}(weights)^{*}\mathbf{P} = \mathbf{I} \quad .$$

If you specify *weights*, you *must* also specify *maxdegree*. If *maxdegree* is not specified or is specified incorrectly, the default weights (all weights are 1) are used.

Description

The ORPOL matrix function generates orthogonal polynomials. The result is a column-orthonormal matrix **P** with the same number of rows as the vector and with *maxdegree*+1 columns:

$$\mathbf{P}'^{*}\text{DIAG}(weights)^{*}\mathbf{P} = \mathbf{I} \quad .$$

The result is computed such that $\mathbf{P}[i,j]$ is the value of a polynomial of degree $j-1$ evaluated at the ith element of the vector.

ORPOL Function *continued*

The maximum number of nonzero orthogonal polynomials (r) that can be computed from the vector and the weights is the number of distinct values in the vector, ignoring any value associated with a zero weight.

The polynomial of maximum degree has degree of $r-1$. If the value of *maxdegree* exceeds $r-1$, then columns $r+1$, $r+2$, ..., *maxdegree*$+1$ of the result are set to 0. In this case,

$$\mathbf{P}' * \text{DIAG(weights)}^*\mathbf{P} = \begin{bmatrix} \text{I}(r) & 0 \\ 0 & 0^*\mathbf{J}(maxdegree + 1 - r) \end{bmatrix}$$

The statement below results in a matrix with 3 orthogonal columns:

```
orpol=orpol(1:5,2);
```

```
ORPOL          5 rows      3 cols      (numeric)

              0.4472136 -0.632456 0.5345225
              0.4472136 -0.316228 -0.267261
              0.4472136         0 -0.534522
              0.4472136 0.3162278 -0.267261
              0.4472136 0.6324555 0.5345225
```

See "Acknowledgments" in the front of this book for authorship of the ORPOL function.

PARSE Statement

parses matrix elements as statements

Syntax

PARSE *matrices* <(*matrix-names*)>;

where

matrices   are character matrices containing IML module statements.

(*matrix-names*)   are character matrices whose elements are the names of character matrices containing IML module statements.

Description

Use the PARSE statement to parse the elements of a character matrix containing IML module statements. For example, the following statement parses the elements (rows) of matrix **A** as lines of code:

```
parse a;
```

You can parse several matrices with one PARSE statement either by listing all of their names in the statement or by first creating a character matrix, say **N**, containing their names as elements and then parsing **N**. Each element of **N** is the name of a matrix containing IML module statements. In this case, enclose **N** in parentheses in the PARSE statement to indicate the indirect references to the elements of **N**.

For example, the statements

```
a={"start mod1;",
   "x={1 2 3};",
   "print x;",
   "finish;"};
parse a;
run mod1;
```

produce the result

```
NOTE: Module MOD1 defined.
```

```
          X
          1         2         3
```

Alternatively, you can use the following statements to obtain the same result:

```
a={"start mod1;",
   "x={1 2 3};",
   "print x;",
   "finish;"};
c={a};
parse (c);
run mod1;
```

PAUSE Statement

interrupts module execution

Syntax

PAUSE <*expression*> <***>;

where

expression   is a character matrix or quoted literal giving a message to print.

*   suppresses any messages.

Description

The PAUSE statement stops execution of a module, saves the calling chain so that execution can resume later (by a RESUME statement), prints a pause

PAUSE Statement *continued*

message that you can specify, and puts you in immediate mode so you can enter more statements.

You can specify an operand in the PAUSE statement to supply a message to be printed for the pause prompt. If no operand is specified, the default message,

```
paused in module XXX
```

is printed, where *XXX* is the name of the module containing the pause. If you want to suppress all messages in a PAUSE statement, use an asterisk as the operand:

```
pause *;
```

The PAUSE statement should only be specified in modules. It generates a warning if executed in immediate mode.

When an error occurs while executing inside a module, IML automatically behaves as though a PAUSE statement was issued. PROC IML prints a note saying

```
paused in module
```

and IML puts you in immediate mode within the module environment, where you can correct the error. You can then resume execution from the statement following the one where the error occurred by issuing a RESUME command.

IML supports pause processing of both subroutine and function modules. See also the description of the SHOW statement with the PAUSE option.

PGRAF Call

produces scatterplots

Syntax

CALL PGRAF(*xy*<,*id*><,*xlabel*><,*ylabel*><,*title*>);

where

xy   is an $n \times 2$ matrix of (x,y) points.

id   is an $n \times 1$ character matrix of labels for each point. The PGRAF subroutine uses up to 8 characters per point. If *id* is a scalar (1×1), then the same label is used for all of the points. The label is centered over the actual point location. If you do not specify *id*, **x** is the default character for labeling the points.

xlabel   is a character scalar or quoted literal that labels the *x* axis (centered below the *x* axis) .

ylabel   is a character scalar or quoted literal that labels the *y* axis (printed vertically to the left of the *y* axis).

title   is a character scalar or quoted literal printed above the graph.

Description

The PGRAF subroutine produces a scatterplot suitable for display on a line printer or similar device. The statements below specify a plotting symbol, axis labels, and a title to produce the plot shown:

```
xy={1 2, 3 3, 5 4, 6 2};
call pgraf(xy,'*','X','Y','Plot of X vs Y');
```

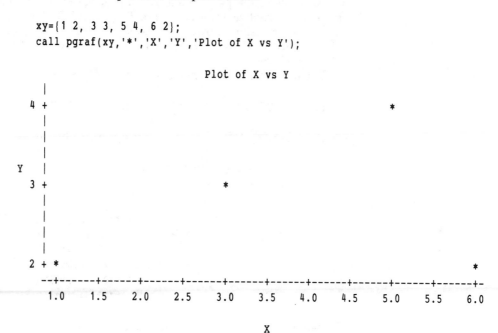

POLYROOT Function

finds zeros of a real polynomial

returns a matrix of roots

Syntax

POLYROOT(*vector*)

where *vector* is an $n \times 1$ (or $1 \times n$) vector containing the coefficients of an $(n-1)$ degree polynomial with the coefficients arranged in order of decreasing powers. The POLYROOT function returns the array *r*, which is an $(n-1) \times 2$ matrix containing the roots of the polynomial. The first column of *r* contains the real part of the complex roots and the second column contains the imaginary part. If a root is real, the imaginary part will be 0.

POLYROOT Function *continued*

Description

The POLYROOT function finds the real and complex roots of a polynomial with real coefficients.

The POLYROOT function uses an algorithm proposed by Jenkins and Traub (1970) to find the roots of the polynomial. The algorithm is not guaranteed to find all roots of the polynomial. An appropriate warning message is issued when one or more roots cannot be found. The POLYROOT algorithm produces roots within the precision allowed by the hardware. If r is given as a root of the polynomial $P(x)$, then $1 + P(R) = 1$ based on the roundoff error of the computer that is employed.

For example, to find the roots of the polynomial

$$P(x) = 0.2567x^4 + 0.1570x^3 + 0.0821x^2 - 0.3357x + 1$$

use the following IML code to produce the result shown:

```
p={0.2567 0.1570 0.0821 -0.3357 1};
r=polyroot(p);
```

```
          R           4 rows      2 cols      (numeric)

                  0.8383029 0.8514519
                  0.8383029 -0.851452
                  -1.144107 1.1914525
                  -1.144107 -1.191452
```

The polynomial has two conjugate pairs of roots that, within machine precision, are given by $r = 0.8383029 \pm 0.8514519i$ and $r = -1.144107 \pm 1.1914525i$.

PRINT Statement

prints matrix values

Syntax

PRINT *<matrices>* *<(expression)>* *<"message">* *<pointer-controls>*
    <[options]>;

where

matrices	are the names of matrices.
(expression)	is an expression in parentheses that is evaluated. The result of the evaluation is printed. The evaluation of a subscripted matrix used as an expression results in printing the submatrix.

"*message*"   is a message in quotes.

pointer-controls   control the pointer for printing. For example, using a
                         comma (,) skips a single line and using a slash (/) skips to a
                         new page.

[*options*]   are described below.

Description

The PRINT statement prints the specified matrices or message. The options
below can appear in the PRINT statement. They are specified in brackets after
the matrix name to which they apply.

COLNAME=*matrix*

   specifies the name of a character matrix whose first *ncol* elements are to be
   used for the column labels of the matrix to be printed, where *ncol* is the
   number of columns in the matrix. (You can also use the RESET AUTONAME
   statement to automatically label columns as COL1, COL2, and so on.)

FORMAT=*format*

   specifies a valid SAS or user-defined format to use in printing the values of
   the matrix, for example,

```
print x[format=5.3];
```

ROWNAME=*matrix*

   specifies the name of a character matrix whose first *nrow* elements are to be
   used for the row labels of the matrix to be printed, where *nrow* is the
   number of rows in the matrix and where the scan to find the first *nrow*
   elements goes across row 1, then across row 2, and so forth through row *n*.
   (You can also use the RESET AUTONAME statement to automatically label
   rows as ROW1, ROW2, and so on.)

```
reset autoname;
```

For example, you can use the statement below to print a matrix called **X** in
format 12.2 with columns labeled AMOUNT and NET PAY, and rows labeled
DIV A and DIV B:

```
x={45.125 50.500,
   75.375 90.825};
r={"DIV A"  "DIV B"};
c={"AMOUNT" "NET PAY"};

print x[rowname=r colname=c format=12.2];
```

The output is

```
        X        AMOUNT    NET PAY

      DIV A       45.13      50.50
      DIV B       75.38      90.83
```

PRINT Statement *continued*

To permanently associate the above options with a matrix name, refer to the description of the MATTRIB statement.

If there is not enough room to print all the matrices across the page, then one or more matrices are printed out in the next group. If there is not enough room to print all the columns of a matrix across the page, then the columns are folded, with the continuation lines identified by a colon(:). The spacing between adjacent matrices can be controlled by the SPACES= option of the RESET statement. The FW= option of the RESET statement can be used to control the number of print positions used to print each numeric element. For more print-related options, see the description of the RESET statement. The example below shows how to print part of a matrix:

```
y=1:10;
    /* prints first five elements of y*/
print (y[1:5]) [format=5.1];
```

PRODUCT Function

multiplies matrices of polynomials

returns a matrix of polynomial products

Syntax

PRODUCT(a,b<,dim>)

where

a   is an $m \times (n*s)$ numeric matrix. The first $m \times n$ submatrix contains the constant terms of the polynomials, the second $m \times n$ submatrix contains the first order terms, and so on.

b   is an $n \times (p*t)$ matrix. The first $n \times p$ submatrix contains the constant terms of the polynomials, the second $n \times p$ submatrix contains the first order terms, and so on.

dim   is a 1×1 matrix, with value $p > 0$. The value of this matrix is used to set p above. If omitted, the value of p is set to 1.

Description

The PRODUCT function multiplies matrices of polynomials. The value returned is the $m \times (p*(s+t-1))$ matrix of the polynomial products. The first $m \times p$ submatrix contains the constant terms, the second $m \times p$ submatrix contains the first order terms, and so on.

Note: The PRODUCT function can be used to multiply the matrix operators employed in a multivariate time-series model of the form

$$\Phi_1(B)\Phi_2(B)Y_t = \Theta_1(B)\,\Theta_2(B)\varepsilon_t$$

where $\Phi_1(B)$, $\Phi_2(B)$, $\Theta_1(B)$, and $\Theta_2(B)$ are matrix polynomial operators whose first matrix coefficients are identity matrices. Often $\Phi_2(B)$ and $\Theta_2(B)$ represent seasonal components that are isolated in the modeling process but multiplied with the other operators when forming predictors or estimating parameters. The RATIO function is often employed in a time series context as well.

For example, the statement

```
r=product(( 1  2  3  4,
            5  6  7  8),
          ( 1  2  3,
            4  5  6), 1);
```

produces the result

```
           R              2 rows      4 cols     (numeric)

                     9        31         41         33
                    29        79        105         69
```

PURGE Statement

removes observations marked for deletion and renumbers records

Syntax

PURGE;

Description

The PURGE data processing statement is used to remove observations marked for deletion and to renumber the remaining observations. This closes the gaps created by deleted records. Execution of this statement may be time-consuming because it involves rewriting the entire data set.

▶ *Caution* .... *Any indexes associated with the data set are lost after a purge.* ▲

IML does not do an automatic purge for you at quit time.

In the example that follows, a data set named A is created. Then, you begin an IML session and edit A. You delete the fifth observation, list the data set, and

PURGE Statement *continued*

issue a PURGE statement to delete the fifth observation and renumber the remaining observations.

```
data a;
   do i=1 to 10;
      output;
   end;
run;

proc iml;
   edit a;
   delete point 5;
   list all;
   purge;
   list all;
```

PUSH Call

pushes SAS statements into the command input stream

Syntax

CALL PUSH(*argument1* <,*argument2*,...,*argument15*>);

where *argument* is a character matrix or quoted literal containing valid SAS statements.

Description

The PUSH subroutine pushes character arguments containing valid SAS statements (usually SAS/IML statements or global statements) to the input command stream. You can specify up to 15 arguments. Any statements pushed to the input command queue get executed when the module is put in a hold state. This is usually induced by one of the following:

□   an execution error within a module

□   an interrupt

□   a pause command.

The string pushed is read before any other lines of input. If you call the PUSH subroutine several times, the strings pushed each time are ahead of the less recently pushed strings. If you would rather place the lines after others in the input stream, then use the QUEUE command instead.

The strings you push do not appear on the log.

▶ *Caution* . . . . *Do not push too much code at one time.*

Pushing too much code at one time, or getting into infinite loops of pushing, causes problems that may result in SAS exiting. ▲

For details, see Chapter 13, "Using SAS/IML Software to Generate IML Statements."

An example using the PUSH subroutine is shown below:

```
start;
   code='reset pagesize=25;';
   call push(code,'resume;');
   pause;
      /* show that pagesize was set to 25 during  */
      /* a PAUSE state of a module                */
   show options;
finish;
run;
```

PUT Statement

writes data to an external file

Syntax

PUT <*operand*> <*record-directives*> <*positionals*> <*format*>;

where

operand

specifies the value you want to output to the current position in the record. The *operand* can be either a variable name, a literal value, or an expression in parentheses. The *operand* can be followed immediately by an output format specification.

record-directives

start new records. There are three types:

holding @   at the end of a PUT statement, instructs IML to put a hold on the current record so that IML can write more to the record with later PUT statements. Otherwise, IML automatically begins the next record for the next PUT statement.

/   writes out the current record and begins forming a new record.

> *operand*   specifies that the next record written will start at the indicated byte position in the file (for RECFM=N files only). The *operand* is a literal number, a variable name, or an expression in parentheses, for example,

```
put >3 x 3.2;
```

PUT Statement *continued*

positionals

specify the column on the record to which the PUT statement should go. There are two types of positionals:

@ *operand*   specifies to go to the indicated column, where *operand* is a literal number, a variable name, or an expression in parentheses. For example, @30 means to go to column 30.

+ *operand*   specifies that the indicated number of columns are to be skipped, where *operand* is a literal number, a variable name, or an expression in parentheses.

format

specifies a valid SAS or user-defined output format. These are of the form *w.d* or $w.$ for standard numeric and character formats, respectively, where *w* is the width of the field and *d* is the decimal parameter, if any. They can also be a named format of the form *NAMEw.d*, where *NAME* is the name of the format. If the width is unspecified, then a default width is used; this is 9 for numeric variables.

Description

The PUT statement writes to the file specified in the previously executed FILE statement, putting the values from IML variables. The statement is described in detail in Chapter 7, "File Access." The PUT statement is a sequence of positionals and record directives, variables, and formats. An example using the PUT statement is shown below:

```
/* output variable A in column 1 using SAS format 6.4.  */
/* Skip 3 columns and output X using format 8.4         */

put @1 a 6.4 +3 x 8.4;
```

QUEUE Call

queues SAS statements into the command input stream

Syntax

CALL QUEUE(*argument1* <,*argument2*,...,*argument15*>);

where *argument* is a character matrix or quoted literal containing valid SAS statements.

Description

The QUEUE subroutine places character arguments containing valid SAS statements (usually SAS/IML statements or global statements) at the end of the input command stream. You can specify up to 15 arguments. The string queued is read after other lines of input already on the queue. If you want to push the lines in front of other lines already in the queue, use the PUSH subroutine instead. Any statements queued to the input command queue get executed when the module is put in a hold state. This is usually induced by one of the following:

□ an execution error within a module

□ an interrupt

□ a pause command.

The strings you queue do not appear on the log.

▶ *Caution* . . . . *Do not queue too much code at one time.*
Queuing too much code at one time, or getting into infinite loops of queuing, causes problems that may result in SAS exiting. ▲

For more examples, consult Chapter 13, "Using SAS/IML Software to Generate IML Statements." An example using the QUEUE subroutine follows:

```
start mod(x);
   code='x=0;';
   call queue (code,'resume;');
   pause;
finish;
x=1;
run mod(x);
print(x);
```

produces

X

0

QUIT Statement

exits from IML

Syntax

QUIT;

Description

Use the QUIT statement to exit IML. If a DATA or PROC statement is encountered, QUIT is implied. The QUIT statement is executed immediately; therefore, you cannot use QUIT as an executable statement, that is, as part of a module or conditional clause. (See the description of the ABORT statement.)

   PROC IML closes all open data sets and files when a QUIT statement is encountered. Workspace and symbol spaces are freed up. If you need to use any matrix values or any module definitions in a later session, you must store them in a storage library before you quit.

RANK Function

ranks elements of a matrix

returns a matrix containing the ranks of the elements

Syntax

RANK(*matrix*)

where *matrix* is a numeric matrix or literal.

Description

The RANK function creates a new matrix whose elements are the ranks of the corresponding elements of *matrix*. The ranks of tied values are assigned arbitrarily rather than averaged. (See the description of the RANKTIE function.)

   For example, the statements

```
x={2 2 1 0 5};
y=rank(x);
```

produce the vector

```
                            Y

              3       4       2       1       5
```

The RANK function can be used to sort a vector **x**:

```
b=x;
x[,rank(x)]=b;
```

```
              X
       0      1      2      2      5
```

The RANK function can also be used to find anti-ranks of **x**:

```
r=rank(x);
i=r;
i[,r]=1:ncol(x);
```

```
              I
       4      3      1      2      5
```

IML does not have a function that directly computes the rank of a matrix. You can use the following technique to compute the rank of matrix A:

```
rank=round(trace(ginv(a)*a));
```

RANKTIE Function

ranks matrix elements using tie-averaging

returns a matrix containing the ranks of the elements

Syntax

RANKTIE(*matrix*)

where *matrix* is a numeric matrix or literal.

Description

The RANKTIE function creates a new matrix whose elements are the ranks of the corresponding elements of *matrix*. The ranks of tied values are averaged.

For example, the statements

```
x={2 2 1 0 5};
y=ranktie(x);
```

produces the vector

```
              Y
      3.5    3.5     2      1      5
```

The RANKTIE function differs from the RANK function in that RANKTIE averages the ranks of tied values, whereas RANK breaks ties arbitrarily.

RATIO Function

divides matrix polynomials

returns a matrix containing the terms of $\Phi(B)^{-1}\Theta(B)$ considered as a matrix of rational functions in B that have been expanded as power series

Syntax

RATIO(*ar,ma,terms,<dim>*)

where

ar	is an $n \times (n^*s)$ matrix representing a matrix polynomial generating function, $\Phi(B)$, in the variable B. The first $n \times n$ submatrix represents the constant term and must be nonsingular, the second $n \times n$ submatrix represents the first order coefficients, and so on.
ma	is an $n \times (m^*t)$ matrix representing a matrix polynomial generating function, $\Theta(B)$, in the variable B. The first $n \times m$ submatrix represents the constant term, the second $n \times m$ submatrix represents the first order term, and so on.
terms	is a scalar containing the number of terms to be computed, denoted by r in the discussion below. This value must be positive.
dim	is a scalar containing the value of m above. The default value is 1.

Description

The RATIO function multiplies a matrix of polynomials by the inverse of another matrix of polynomials. It is useful for expressing univariate and multivariate ARMA models in pure moving-average or pure autoregressive forms.

Note that the order of the first two arguments is reversed from the corresponding PROC MATRIX function.

The value returned is an $n \times (m^*r)$ matrix containing the terms of $\Phi(B)^{-1}\Theta(B)$ considered as a matrix of rational functions in B that have been expanded as power series.

Note: The RATIO function can be used to consolidate the matrix operators employed in a multivariate time-series model of the form

$$\Phi(B)\mathbf{Y}_t = \Theta(B)\varepsilon_t$$

where $\Phi(B)$ and $\Theta(B)$ are matrix polynomial operators whose first matrix coefficients are identity matrices. The RATIO function can be used to compute a truncated form of $\Psi(B) = \Phi(B)^{-1}\Theta(B)$ for the equivalent infinite order model

$$\mathbf{Y}_t = \Psi(B)\varepsilon_t \quad .$$

The RATIO function can also be employed for simple scalar polynomial division, giving a truncated form of $\theta(x)/\varphi(x)$ for two scalar polynomials $\theta(x)$ and $\varphi(x)$.

The cumulative sum of the elements of a column vector **x** can be obtained using

```
ratio({1 -1},x,ncol(x));
```

Consider the following example for multivariate ARMA(1,1):

```
ar={1 0 -.5  2,
    0 1   3 -.8};
ma={1 0 .9  .7,
    0 1   2 -.4};
psi=ratio(ar,ma,4,2);
```

The matrix produced in

```
       PSI
         1        0      1.4     -1.3      2.7    -1.45    11.35
    :  -9.165

         0        1       -1      0.4       -5     4.22    -12.1
    :   7.726
```

READ Statement

reads observations from a data set

Syntax

READ <*range*> <VAR *operand*> <WHERE(*expression*)>
     <INTO *name* <[ROWNAME=*row-name* COLNAME=*column-name*>] >>;

where

range	specifies a range of observations.
operand	selects a set of variables.
expression	is evaluated for being true or false.
name	is the name of the target matrix.
row-name	is a character matrix or quoted literal giving descriptive row labels.
column-name	is a character matrix or quoted literal giving descriptive column labels.

The clauses and options are explained below.

Description

Use the READ statement to read variables or records from the current SAS data set into column matrices of the VAR clause or into the single matrix of the INTO clause. When the INTO clause is used, each variable in the VAR clause becomes a column of the target matrix, and all variables in the VAR clause must be of the same type. If you specify no VAR clause, the default variables for the INTO clause are all numeric variables. Read all character variables into a target matrix by using VAR _CHAR_.

READ Statement *continued*

You can specify a *range* of observations with a keyword or by record number using the POINT option. You can use any of the following keywords to specify a range:

ALL	all observations
CURRENT	the current observation
NEXT *<number>*	the next observation or the next *number* of observations
AFTER	all observations after the current one
POINT *operand*	observations specified by number, where *operand* can be one of the following:

Operand	Example
a single record number	`point 5`
a literal giving several record numbers	`point {2 5 10}`
the name of a matrix containing record numbers	`point p`
an expression in parentheses	`point (p+1)`

If the current data set has an index in use, the POINT option is invalid.

You can specify a set of variables to use with the VAR clause.

The *operand* in the VAR clause can be one of the following:

□ a literal containing variable names

□ the name of a matrix containing variable names

□ an expression in parentheses yielding variable names

□ one of keywords described below:

ALL	for all variables
CHAR	for all character variables
NUM	for all numeric variables.

Below are examples showing each possible way you can use the VAR clause:

```
var {time1 time5 time9};    /* a literal giving the variables   */
var time;                   /* a matrix containing the names    */
var('time1':'time9');       /* an expression                    */
var _all_;                  /* a keyword                        */
```

The WHERE clause conditionally selects observations, within the *range* specification, according to conditions given in the clause.

The general form of the WHERE clause is

WHERE(*variable comparison-op operand*)

where

variable	is a variable in the SAS data set.
comparison-op	is one of the following comparison operators:

<	less than
<=	less than or equal to
=	equal to
>	greater than
>=	greater than or equal to
^=	not equal to
?	contains a given string
^?	does not contain a given string
=:	begins with a given string
=*	sounds like or is spelled similar to a given string.

operand	is a literal value, a matrix name, or an expression in parentheses.

WHERE comparison arguments can be matrices. For the following operators, the WHERE clause succeeds if *all* the elements in the matrix satisfy the condition:

   ^= ^? < <= > >=

For the following operators, the WHERE clause succeeds if *any* of the elements in the matrix satisfy the condition:

   = ? =: =*

Logical expressions can be specified within the WHERE clause using the AND (&) and OR (|) operators. The general form is

clause&*clause* (for an AND clause)
clause | *clause* (for an OR clause)

where *clause* can be a comparison, a parenthesized clause, or a logical expression clause that is evaluated using operator precedence.

   Note: The expression on the left-hand side refers to values of the data set variables, and the expression on the right-hand side refers to matrix values.

   You can specify ROWNAME= and COLNAME= matrices as part of the INTO clause. The COLNAME= matrix specifies the name of a new character matrix to be created. This COLNAME= matrix is created in addition to the target matrix of the INTO clause and contains variable names from the input data set corresponding to columns of the target matrix. The COLNAME= matrix has dimension $1 \times nvar$, where *nvar* is the number of variables contributing to the target matrix.

READ Statement *continued*

The ROWNAME= option specifies the name of a character variable in the input data set. The values of this variable are put in a character matrix with the same name as the variable. This matrix has the dimension *nobs*×1, where *nobs* is the number of observations in the range of the READ statement. The *range*, VAR, WHERE, and INTO clauses are all optional and can be specified in any order.

Row names created via a READ statement are permanently associated with the INTO matrix. You do not need to use a MATTRIB statement to get this association.

For example, to read all observations from the data set variables NAME and AGE, use a READ statement with the VAR clause and the keyword ALL for the *range* operand. This creates two IML variables with the same names as the data set variables.

```
read all var{name age};
```

To read all variables for the 23rd observation only, use the statement

```
read point 23;
```

To read the data set variables NAME and ADDR for all observations with a STATE value of **NJ**, use the statement

```
read all var{name addr} where(state="NJ");
```

See Chapter 6, "Working with SAS Data Sets," for further information.

REMOVE Function

discards elements from a matrix

returns a row vector with elements removed

Syntax

REMOVE(*matrix,indices*)

where

matrix   is a numeric or character matrix or literal.

indices   refers to a matrix containing the indices of elements that are removed from *matrix*.

Description

The REMOVE function returns as a row vector elements of the first argument, with elements corresponding to the indices in the second argument discarded and the gaps removed. The first argument is indexed in row-major order, as in subscripting, and the indices must be in the range 1 to the number of elements in the first argument. Non-integer indices are truncated to their integer part. You can repeat the indices, and you can give them in any order. If all elements are removed, the result is a null matrix (zero rows and zero columns).

Thus, the statement

```
a=remove({5 6, 7 8}, 3);
```

removes the third element, producing the result shown:

```
      A
5     6     8
```

The statement

```
a=remove({5 6 7 8}, {3 2 3 1});
```

causes all but the fourth element to be removed, giving the result shown:

```
A
8
```

REMOVE Statement

removes matrices from storage

Syntax

REMOVE <MODULE=(*module-list*) <*matrix-list*>>;

where

module-list   specifies a module or modules to remove from storage.

matrix-list   specifies a matrix or matrices to remove from storage.

Description

The REMOVE statement removes matrices or modules or both from the current library storage. For example, the statement below removes the three modules A, B, and C and the matrix X:

```
remove module=(A B C) X;
```

The special operand _ALL_ can be used to remove all matrices or all modules or both. For example, the following statement removes everything:

```
remove _all_ module=_all_;
```

See Chapter 12, "Storage Features," and also the descriptions of the LOAD, STORE, RESET, and SHOW statements for related information.

RENAME Call

renames a SAS data set

Syntax

CALL RENAME(*<libname,>* *member-name*, *new-name*);

where

libname	is a character matrix or quoted literal containing the name of the SAS data library.
member-name	is a character matrix or quoted literal containing the current name of the data set.
new-name	is a character matrix or quoted literal containing the new data set name.

Description

The RENAME subroutine renames a SAS data set in the specified library. All of the arguments can directly be specified in quotes, although quotes are not required. If a one-word data set name is specified, the libname specified by the RESET DEFLIB statement is used. Examples of valid statements follow:

```
call rename('a','b');
call rename(a,b);
call rename(work,a,b);
```

REPEAT Function

creates a new matrix of repeated values

returns a matrix of repeated values

Syntax

REPEAT(*matrix,nrow,ncol*)

where

matrix	is a numeric matrix or literal.
nrow	gives the number of times *matrix* is repeated across rows.
ncol	gives the number of times *matrix* is repeated across columns.

Description

The REPEAT function creates a new matrix by repeating the values of the argument matrix *nrow*ncol* times, *ncol* times across the rows, and *nrow* times down the columns. The *matrix* argument can be numeric or character. For example, the following statements result in the matrix **Y**, repeating the **X** matrix twice down and three times across:

```
x=(1 2 ,
   3 4);
y=repeat(x,2,3);
```

```
        Y
        1       2       1       2       1       2
        3       4       3       4       3       4
        1       2       1       2       1       2
        3       4       3       4       3       4
```

REPLACE Statement

replaces values in observations and updates observations

Syntax

REPLACE < *range* > <VAR *operand* > <WHERE(*expression*)>;

where

range	specifies a range of observations.
operand	selects a set of variables.
expression	is evaluated for being true or false.

Description

The REPLACE statement replaces the values of observations in a SAS data set with current values of IML matrices with the same name. Use the *range*, VAR, and WHERE arguments to limit replacement to specific variables and observations. Replacement matrices should be the same type as the data set variables. The REPLACE statement uses matrix elements in row order replacing the value in the *i*th observation with the *i*th matrix element. If there are more observations in *range* than matrix elements, the REPLACE statement continues to use the last matrix element.

For example, the statements below cause all occurrences of **ILL** to be replaced by **IL** for the variable STATE:

```
state="IL";
replace all var(state) where(state="ILL");
```

REPLACE Statement *continued*

You can specify a *range* of observations with a keyword or by record number using the POINT option. You can use any of the following keywords to specify a range:

ALL	all observations
CURRENT	the current observation
NEXT <*number*>	the next observation or the next *number* of observations
AFTER	all observations after the current one
POINT *operand*	observations by number, where *operand* can be one of the following:

Operand	Example
a single record number	`point 5`
a literal giving several record numbers	`point {2 5 10}`
the name of a matrix containing record numbers	`point p`
an expression in parentheses	`point (p+1)`

If the current data set has an index in use, the POINT option is invalid.

You can specify a set of variables to use with the VAR clause. The *variables* argument can have the following values:

□ a literal containing variable names

□ the name of a matrix containing variable names

□ an expression in parentheses yielding variable names

□ one of the keywords described below:

ALL	for all variables
CHAR	for all character variables
NUM	for all numeric variables.

Below are examples showing each possible way you can use the VAR clause:

```
var {time1 time5 time9};    /* a literal giving the variables  */
var time;                   /* a matrix containing the names   */
var('time1':'time9');       /* an expression                   */
var _all_;                  /* a keyword                       */
```

The WHERE clause conditionally selects observations, within the range specification, according to conditions given in the clause. The general form of the WHERE clause is

WHERE(*variable comparison-op operand*)

where

variable	is a variable in the SAS data set.
comparison-op	is any one of the following comparison operators:

<	less than
<=	less than or equal to
=	equal to
>	greater than
>=	greater than or equal to
^=	not equal to
?	contains a given string
^?	does not contain a given string
=:	begins with a given string
=*	sounds like or is spelled similar to a given string.

operand	is a literal value, a matrix name, or an expression in parentheses.

WHERE comparison arguments can be matrices. For the following operators, the WHERE clause succeeds if *all* the elements in the matrix satisfy the condition:

```
^=  ^?  <  <=  >  >=
```

For the following operators, the WHERE clause succeeds if *any* of the elements in the matrix satisfy the condition:

```
=  ?  =:  =*
```

Logical expressions can be specified within the WHERE clause using the AND (&) and OR (|) operators. The general form is

clause&*clause* (for an AND clause)
clause | *clause* (for an OR clause)

where *clause* can be a comparison, a parenthesized clause, or a logical expression clause that is evaluated using operator precedence.

Note: The expression on the left-hand side refers to values of the data set variables, and the expression on the right-hand side refers to matrix values.

The code statement below replaces all variables in the current observation:

```
replace;
```

RESET Statement

sets processing options

Syntax

RESET <*options*>;

where the options are described below.

Description

The RESET statement sets processing options. The options described below are currently implemented options. Note that the prefix NO turns off the feature where indicated. For options that take operands, the operand should be a literal, a name of a matrix containing the value, or an expression in parentheses. The SHOW OPTIONS statement displays the current settings of all of the options.

AUTONAME
NOAUTONAME
>   specifies whether rows are automatically labeled ROW1, ROW2, and so on, and columns are labeled COL1, COL2, and so on, when a matrix is printed. Row-name and column-name attributes specified in the PRINT statement or associated via the MATTRIB statement override the default labels. The AUTONAME option causes the SPACES option to be reset to 4. The default is NOAUTONAME.

CENTER
NOCENTER
>   specifies whether output from the PRINT statement is centered on the page. The default is CENTER.

CLIP
NOCLIP
>   specifies whether SAS/IML graphs are automatically clipped outside the viewport; that is, any data falling outside the current viewport is not displayed. NOCLIP is the default.

DEFLIB=*operand*
>   specifies the default libname for SAS data sets when no other libname is given. This defaults to USER if a USER libname is set up, or WORK if not. The libname operand can be specified with or without quotes.

DETAILS
NODETAILS
>   specifies whether additional information is printed from a variety of operations, such as when files are opened and closed. The default is NODETAILS.

FLOW
NOFLOW
>   specifies whether operations are shown as executed. It is used for debugging only. The default is NOFLOW.

FUZZ<=*number*>
NOFUZZ

> specifies whether very small numbers are printed as zero rather than in scientific notation. If the absolute value of the number is less than the value specified in *number*, it will be printed as 0. The *number* argument is optional, and the default varies across hosts but is typically around $1E-12$. The default is NOFUZZ.

FW=*number*

> sets the field width for printing numeric values. The default field width is 9.

LINESIZE=*n*

> specifies the linesize for printing. The default is usually 78.

LOG
NOLOG

> specifies whether output is routed to the log file rather than to the print file. On the log, the results are interleaved with the statements and messages. The NOLOG option routes output to the OUTPUT window in display manager and to the listing file in batch modes. The default is NOLOG.

NAME
NONAME

> specifies whether the matrix name or label is printed with the value for the PRINT statement. The default is NAME.

PAGESIZE=*n*

> specifies the pagesize for printing. The default is usually 21.

PRINT
NOPRINT

> specifies whether the final results from assignment statements are printed automatically. NOPRINT is the default.

PRINTALL
NOPRINTALL

> specifies whether the intermediate and final results are printed automatically. The default is NOPRINTALL.

SPACES=*n*

> specifies the number of spaces between adjacent matrices printed across the page. The default is 1, except when AUTONAME is on. Then, the default is 4.

STORAGE=<*libname.*>*memname*;

> specifies the file to be the current library storage for STORE and LOAD statements. The default library storage is SASUSER.IMLSTOR. The *libname* argument is optional and defaults to SASUSER. It can be specified with or without quotes.

RESUME Statement

resumes execution

Syntax

RESUME;

Description

The RESUME statement allows you to continue execution from the line in the module where the most recent PAUSE statement was executed. PROC IML issues an automatic pause when an error occurs inside a module. If a module was paused due to an error, the RESUME statement resumes execution immediately after the statement that caused the error. The SHOW PAUSE statement displays the current state of all paused modules.

RETURN Statement

returns to caller

Syntax

RETURN<(*operand*)>;

where *operand* is the value of the function returned. Use *operand* only in function modules.

Description

The RETURN statement causes IML to return to the calling point in a program. If a LINK statement has been issued, IML returns to the statement following the LINK. If no LINK statement was issued, the RETURN statement exits a module. If not in a module, execution is stopped (as with a STOP statement), and IML looks for more statements to parse.

The RETURN statement with an *operand* is used in function modules that return a value. The *operand* can be a variable name or an expression. It is evaluated, and the value is returned.

See the description of the LINK statement. Also, see Chapter 5, "Programming," for details.

If you use a LINK statement, you need a RETURN statement at the place where you want to go back to the statement after LINK.

If you are writing a function, use a RETURN to return the value of the function. An example is shown below.

```
start sum1(a,b);
   sum=a+b;
   return(sum);
finish;
```

ROOT Function

performs the Cholesky decomposition of a matrix

returns an upper triangular matrix

Syntax

ROOT(*matrix*)

where *matrix* is a symmetric positive-definite matrix.

Description

The ROOT function performs the Cholesky decomposition of a matrix (for example, **A**) such that

$$U'U = A$$

where **U** is upper triangular. The matrix **A** must be symmetric and positive definite.

For example, the statements

```
xpx={4 15, 15 85};
a=root(xpx);
```

produce the result shown below:

```
          A        2 rows      2 cols     (numeric)

                       2        7.5
                       0   5.3619026
```

ROWCAT Function

concatenates rows without using blank compression

returns a one-column matrix with all row elements concatenated into a single string

Syntax

ROWCAT(*matrix*<,*rows*<,*columns*>>);

where

matrix   is a character matrix or quoted literal.

rows   select the rows of *matrix*.

columns   select the columns of *matrix*.

ROWCAT Function *continued*

Description

The ROWCAT function takes a character matrix or submatrix as its argument and creates a new matrix with one column whose elements are the concatenation of all row elements into a single string. If the argument has *n* rows and *m* columns, the result will have *n* rows and 1 column. The element length of the result will be *m* times the element length of the argument. The optional rows and columns arguments may be used to select which rows and columns are concatenated. For example, the statements

```
b={"ABC"  "D "  "EF ",
   " GH"  " I "  " JK"};
a=rowcat(b);
```

produce the 2×1 matrix:

```
        A              2 rows      1 col      (character, size 9)

                                   ABCD EF
                                   GH I JK
```

Quotes (") are needed only if you want to embed blanks or special characters or to maintain uppercase and lowercase distinctions.

The form

ROWCAT(*matrix,rows,columns*)

returns the same result as

ROWCAT(*matrix[rows,columns]*)

The form

ROWCAT(*matrix,rows*)

returns the same result as

ROWCAT(*matrix[rows,]*)

ROWCATC Function

concatenates rows using blank compression

returns a matrix

Syntax

ROWCATC(*matrix*<,*rows*<,*columns*>>);

where

matrix　　is a character matrix or quoted literal.

rows　　select the rows of *matrix*.

columns　　select the columns of *matrix*.

Description

The ROWCATC function works the same way as the ROWCAT function except that blanks in element strings are moved to the end of the concatenation. For example, the statements

```
b={"ABC"  "D "  "EF ",
   " GH"  " I "  " JK"};
a=rowcatc(b);
```

produce the matrix **A** as shown:

```
       A          2 rows     1 col     (character, size 9)

                             ABCDEF
                             GHIJK
```

Quotes (") are needed only if you want to embed blanks or special characters or to maintain uppercase and lowercase distinctions.

RUN Statement

executes statements in a module

Syntax

RUN <*name*> <(*arguments*)>;

where

name　　　　is the name of a user-defined module or an IML built-in subroutine.

RUN Statement *continued*

arguments   are arguments to the subroutine. Arguments can be both local and global.

Description

The RUN statement executes a user-defined module or invokes PROC IML's built-in subroutines.

The resolution order for the RUN statement is

1. A user-defined module

2. An IML built-in function or subroutine.

This resolution order need only be considered if you have defined a module that has the same name as an IML built-in subroutine. If a RUN statement cannot be resolved at resolution time, a warning is produced. If the RUN statement is still unresolved when executed and a storage library is open at the time, IML attempts to load a module from that storage. If no module is found, then the program is interrupted and an error message is generated. By default, the RUN statement tries to run the module named MAIN.

You will usually want to supply both a name and arguments, as in

```
run myf1(a,b,c);
```

See Chapter 5, "Programming," for further details.

SAVE Statement

saves data

Syntax

SAVE;

Description

The SAVE statement forces out any data residing in output buffers for all active output data sets and files to ensure that the data are written to disk. This is equivalent to closing and then reopening the files.

SETDIF Function

compares elements of two matrices

returns a row vector of unique element values

Syntax

SETDIF(*matrix1,matrix2*)

where

matrix1 is a reference matrix. Elements of *matrix1* not found in *matrix2* are returned in a vector. It can be either numeric or character.

matrix2 is the comparison matrix. Elements of *matrix1* not found in *matrix2* are returned in a vector. It can be either numeric or character, depending on the type of *matrix1*.

Description

The SETDIF function returns as a row vector the sorted set (without duplicates) of all element values present in *matrix1* but not in *matrix2*. If the resulting set is empty, the SETDIF function returns a null matrix (with zero rows and zero columns). The argument matrices and result can be either both character or both numeric. For character matrices, the element length of the result is the same as the element length of the *matrix1*. Shorter elements in the second argument are padded on the right with blanks for comparison purposes.

For example, the statements

```
a={1 2 4 5};
b={3 4};
c=setdif(a,b);
```

produce the result

```
C              1 row     3 cols    (numeric)

               1         2         5
```

SETIN Statement

makes a data set current for input

Syntax

SETIN *SAS-data-set* <NOBS *name*> <POINT *operand*>;

where

SAS-data-set   can be specified with a one-word name (for example, A) or a two-word name (for example, SASUSER.A). For more information on specifying SAS data sets, see Chapter 6, "SAS Files," in *SAS Language: Reference.*

name   is the name of a variable to contain the number of observations in the data set.

operand   specifies the current observation.

Description

The SETIN statement chooses the specified data set from among the data sets already opened for input by EDIT or USE statements. This data set becomes the current input data set for subsequent data management statements. The NOBS option is not required. If specified, the NOBS option returns the number of observations in the data set in the scalar variable *name*. The POINT option makes the specified observation the current one. It positions the data set to a particular observation. The SHOW DATASETS command lists data sets already opened for input.

In the example that follows, if the data set WORK.A has 20 observations, the variable SIZE is set to 20. Also, the current observation is set to 10.

```
setin work.a nobs size point 10;
list;                                    /* lists observation 10 */
```

SETOUT Statement

makes a data set current for output

Syntax

SETOUT *SAS-data-set* <NOBS *name*> <POINT *operand*>;

where

SAS-data-set   can be specified with a one-word name (for example, A) or a two-word name (for example, SASUSER.A). For more information on specifying SAS data sets, see Chapter 6, "SAS Files," in *SAS Language: Reference.*

name	is the name of a variable to contain the number of observations in the data set.
operand	specifies the observation to be made the current observation.

Description

The SETOUT statement chooses the specified data set from among those data sets already opened for output by EDIT or CREATE statements. This data set becomes the current output data set for subsequent data management statements. If specified, the NOBS option returns the number of observations currently in the data set in the scalar variable *name*. The POINT option makes the specified observation the current one.

In the example that follows, the data set WORK.A is made the current output data set and the fifth observation is made the current observation. The number of observations in WORK.A is returned in the variable SIZE.

```
setout work.a nobs size point 5;
```

SHAPE Function

reshapes and repeats values

returns a reshaped matrix

Syntax

SHAPE(*matrix*<,*nrow*<,*ncol*<,*pad-value*>>>)

where

matrix	is a numeric or character matrix or literal.
nrow	gives the number of rows of the new matrix.
ncol	gives the number of columns of the new matrix.
pad-value	is a fill value.

Description

The SHAPE function shapes a new matrix from a matrix with different dimensions; *nrow* specifies the number of rows, and *ncol* specifies the number of columns in the new matrix. The operator works for both numeric and character operands. The three ways of using the function are outlined below:

□ If only *nrow* is specified, the number of columns is determined as the number of elements in the object matrix divided by *nrow*. The number of elements must be exactly divisible; otherwise, a conformability error is diagnosed.

□ If both *nrow* and *ncol* are specified, but not *pad-value*, the result is obtained moving along the rows until the desired number of elements is obtained. The

SHAPE Function *continued*

operation cycles back to the beginning of the object matrix to get more elements, if needed.

□   If *pad-value* is specified, the operation moves the elements of the object matrix first and then fills in any extra positions in the result with the *pad-value*.

If *nrow* or *ncol* is specified as 0, the number of rows or columns, respectively, becomes the number of values divided by *ncol* or *nrow*.
For example, the statement

```
r=shape(12,3,4);
```

produces the result shown:

	R	3 rows	4 cols	(numeric)
	12	12	12	12
	12	12	12	12
	12	12	12	12

The next statement

```
r=shape(77,1,5);
```

produces the result matrix by moving along the rows until the desired number of elements is obtained, cycling back as necessary:

	R	1 row	5 cols	(numeric)	
77	77	77	77	77	

The statement below

```
r=shape({1 2, 3 4, 5 6),2);
```

has *nrow* specified and converts the 3×2 matrix into a 2×3 matrix.

	R	2 rows	3 cols	(numeric)
	1	2	3
	4	5	6

The statement

```
r=shape({99 31},3,3);
```

demonstrates the cycling back and repetition of elements in row-major order until the number of elements desired is obtained.

R	3 rows	3 cols	(numeric)
	99	31	99
	31	99	31
	99	31	99

SHOW Statement

prints system information

Syntax

SHOW *operands*;

where *operands* are any of the valid operands to the SHOW statement. These are given below.

Description

The SHOW statement prints system information. The following *operands* are available:

ALL	shows all the information included by OPTIONS, SPACE, DATASETS, FILES, and MODULES.
ALLNAMES	behaves like NAMES, but also shows names without values.
CONTENTS	shows the names and attributes of the variables in the current SAS data set.
DATASETS	shows all open SAS data sets.
FILES	shows all open files.
MEMORY	returns the size of the largest chunk of main memory available.
MODULES	shows all modules that exist in the current IML environment. A module already referenced but not yet defined is listed as undefined.
name	shows attributes of the specified matrix. If the name of a matrix is one of the SHOW keywords, then both the information for the keyword and the matrix are shown.
NAMES	shows attributes of all matrices having values. Attributes include number of rows, number of columns, data type, and size.

SHOW Statement *continued*

OPTIONS   shows current settings of all IML options (see the RESET statement).

PAUSE   shows the status of all paused modules that are pending resume.

SPACE   shows the workspace and symbolspace size and their current usage.

STORAGE   shows the modules and matrices in the current IML library storage.

WINDOWS   shows all active windows opened by WINDOW statements.

An example of a valid statement follows:

```
show all;
```

SOLVE Function

solves a system of linear equations

Syntax

SOLVE(*A,B*)

where

A   is an $n \times n$ nonsingular matrix.

b   is an $n \times p$ matrix.

Description

The SOLVE function solves the set of linear equations $\mathbf{AX} = \mathbf{B}$ for **X**. **A** must be square and nonsingular.

    $\mathbf{X} = \text{SOLVE}(\mathbf{A},\mathbf{B})$ is equivalent to using the INV function as $\mathbf{X} = \text{INV}(\mathbf{A})^*\mathbf{B}$. However, the SOLVE function is recommended over the INV function because it is more efficient and more accurate. An example follows:

```
x=solve(a,b);
```

The solution method used is discussed in Forsythe, Malcolm, and Moler (1967).

SORT Statement

sorts a SAS data set

Syntax

SORT <DATA=>*SAS-data-set* <OUT=*SAS-data-set*>
      BY <DESCENDING> *variables*;

where you can use the following clauses with the SORT statement:

DATA=*SAS-data-set*	names the SAS data set to be sorted. It can be specified with a one-word name (for example, A) or a two-word name (for example, SASUSER.A). For more information on specifying SAS data sets, see Chapter 6, "SAS Files," in *SAS Language: Reference*. Note that the DATA= portion of the specification is optional.
OUT=*SAS-data-set*	specifies a name for the output data set. If this clause is omitted, the DATA= data set is sorted and the sorted version replaces the original data set.
BY *variables*	specifies the variables to be sorted. A BY clause *must* be used with the SORT statement.
DESCENDING	specifies the variables are to be sorted in descending order.

Description

The SORT statement sorts the observations in a SAS data set by one or more variables, stores the resulting sorted observations in a new SAS data set, or replaces the original. As opposed to all other IML data processing statements, it is *mandatory* that the data set to be sorted be closed prior to the execution of the SORT statement.

The SORT statement first arranges the observations in the order of the first variable in the BY clause; then it sorts the observations with a given value of the first variable by the second variable, and so forth. Every variable in the BY clause can be preceded by the keyword DESCENDING to denote that the variable that follows is to be sorted in descending order. Note that the SORT statement in IML always retains the same relative positions of the observations with identical BY variable values.

For example, the IML statement

```
sort class out=sclass by descending age height;
```

sorts the SAS data set CLASS by the variables AGE and HEIGHT, where AGE is sorted in descending order, and all observations with the same AGE value are sorted by HEIGHT in ascending order. The output data set SCLASS contains the sorted observations. When a data set is sorted in place (without the OUT= clause) any indexes associated with the data set become invalid and are automatically deleted.

Note that all the clauses of the SORT statement must be specified in the order given above.

SOUND Call

produces a tone

Syntax

CALL SOUND(*freq*<*,dur*>);

where

freq   is a numeric matrix or literal giving the frequency in hertz.

dur   is a numeric matrix or literal giving the duration in seconds. Note that the *dur* argument differs from that in the DATA step.

Description

The SOUND subroutine generates a tone using *freq* for frequency (in hertz) and *dur* for duration (in seconds). Matrices may be specified for frequency and duration to produce multiple tones, but if both arguments are nonscalar, then the number of elements must match. The duration argument is optional and defaults to 0.25 (one quarter second).

For example, the following statements produce tones from an ascending musical scale, all with a duration of 0.2 seconds:

```
notes=400#(2##do(0, 1, 1/12));
call sound(notes,0.2);
```

SPLINE Call

evaluates points on the spline

returns a matrix containing fitted values

Syntax

CALL SPLINE(*fitted,data,*<*smooth*>,<*delta*>,<*nout*>,<*type*>,<*slope*>);

where

fitted   is a returned matrix. *fitted* is an $m \times 2$ matrix of fitted values.

data   is an $n \times 2$ or $n \times 3$ matrix of *(x,y)* points.

smooth   is a numeric scalar specifying the smoothing constant.

delta   is a numeric scalar specifying the resolution constant.

nout   is a numeric scalar specifying the number of fitted points. The default is 200.

type	is a character matrix or quoted literal giving the type of spline.
slope	is a 1×2 matrix of endpoint slopes given as angles in degrees.

Description

The SPLINE routine evaluates points on the spline given at least a matrix of *(x,y)* points. SPLINE returns a matrix of fitted values.

If *smooth* is omitted or has the value 0, a cubic interpolating spline is fit to the data. Otherwise, a cubic spline is used to smooth the data. The value of *smooth* must be greater than or equal to 0. Larger values of *smooth* generate more smoothing. If a smoothing spline is used, an optional weight column, column 3 of the data matrix, can be included to give differing weights to data points. If a weight is less than or equal to 0, the point is not used.

delta or *nout* control the number of fitted points. If *delta* is given, the fitted points are spaced by this much on the scale of the first column of *data* if a regular spline is used or on the scale of the curve length if a parametric spline is used. If *nout* is given, then *nout* equally spaced points are returned. If both are given, the *nout* overrides *delta*.

type controls the type of spline fit. In particular, *type* controls whether a parametric or a nonparametric spline is used and the endpoint conditions on the spline. The *type* is either a 1×1 or a 1×2 character matrix. The first element is either PERIODIC, to use periodic endpoints, or ZERO, to set second derivatives at ends to zero (this is the default), and controls the endpoint constraints unless the *slope* matrix is specified. If the *slope* matrix is specified, then the slopes at the ends are set to the two given values. If periodic endpoints are used, then the response values must be the same at the beginning and the end of the second column of *data* unless the smoothing spline is being used. If the response values are not the same, then an error message is issued and no spline is computed.

The second element of *type*, if given, is either NONPARAMETRIC, for a regular spline, or PARAMETRIC, for a parametric spline. The parametric spline first forms a parameter sequence (t_i) from the points (x_i, y_i) as

$$t_i = t_{i-1} + \sqrt{((x_i - x_{i-1})^2 + (y_i - y_{i-1})^2)}$$

with $t_1 = 0$. Splines are then fit to both the **x** and **y** vectors. The resulting splined values are paired to form the output array. Changing the relative scaling of the **x** and **y** vectors changes the output array because the above parameter sequence assumes a Euclidean distance.

The *slope* matrix is a 1×2 matrix of the slope angles at the two ends, given in degrees. If a parametric spline is used, the angle values are used modulo 360. Otherwise, the tangent of the angles is used to set the slopes (that is, the effective angles range from −90 to 90 degrees).

SPLINEC Call

evaluates points on the spline

returns matrices containing coefficients and endpoint slopes

Syntax

CALL SPLINEC(*fitted,coef,endval,data,<smooth>,<delta>,*
        <nout>,<type>,<slope>);

where

fitted	is a returned $m \times 2$ matrix of fitted values.
coeff	is a returned $n \times 5$ or $n \times 9$ matrix of spline coefficients.
endval	is a returned 1×2 matrix of endpoint slopes.
data	is an $n \times 2$ or $n \times 3$ matrix of *(x,y)* points.
smooth	is a numeric scalar specifying the smoothing constant.
delta	is a numeric scalar specifying the resolution constant.
nout	is a numeric scalar specifying the number of fitted points. The default is 200.
type	is a character matrix or quoted literal giving the type of spline.
slope	is a 1×2 matrix of endpoint slopes given as angles in degrees.

Description

The SPLINEC subroutine returns a matrix of spline coefficients and a matrix of endpoint slopes.

If *smooth* is omitted or has the value 0, a cubic interpolating spline is fit to the data. Otherwise, a cubic spline is used to smooth the data. The value of *smooth* must be greater than or equal to 0. Larger values of *smooth* generate more smoothing. If a smoothing spline is used, an optional weight column, column 3 of the data matrix, can be included to give differing weights to data points. If a weight is less than or equal to 0, the point is not used.

The *delta* argument or the *nout* argument controls the number of fitted points. If *delta* is given, the fitted points are spaced by this much on the scale of the first column of *data*, if a regular spline is used, or on the scale of the curve length, if a parametric spline is used. If *nout* is given, then *nout* equally spaced points are returned. If both are given, the *nout* argument overrides *delta*.

The *type* argument controls the type of spline fit. In particular, *type* controls whether a parametric or a nonparametric spline is used and what endpoint conditions are used on the spline. The *type* is either a 1×1 or a 1×2 character matrix. The first element is either PERIODIC, to use periodic endpoints, or ZERO, to set second derivatives at ends to 0. 0 is the default. It controls the endpoint constraints unless the *slope* matrix is specified. If the *slope* matrix is specified, then the slopes at the ends are set to the two given values. If periodic endpoints are used, then the response values must be the same at the beginning and the end of the second column of *data* unless the smoothing spline is being used. If the response values are not the same, then an error message is issued and no spline is computed.

The second element of *type*, if given, is either NONPARAMETRIC for a regular spline or PARAMETRIC for a parametric spline. The parametric spline first forms a parameter sequence (t_i) from the points (x_i, y_i) as

$$t_i = t_{i-1} + \sqrt{((\mathbf{x}_i - \mathbf{x}_{i-1})^2 + (\mathbf{y}_i - \mathbf{y}_{i-1})^2)}$$

with $t_1 = 0$. Splines are then fit to both the **x** and **y** vectors. The resulting splined values are paired to form the output array. Changing the relative scaling of the **x** and **y** vectors changes the output array because the above parameter sequence assumes a Euclidean distance.

The *slope* matrix is a 1×2 matrix of the slope angles at the two ends, given in degrees. If a parametric spline is used, the angle values are used modulo 360. Otherwise, the tangent of the angles is used to set the slopes (that is, the effective angles range from -90 to 90 degrees).

The *coeff* matrix returns the cubic polynomial coefficients for the spline for each interval. The first column gives the the left endpoint of the x interval for the regular spline and the left endpoint of the parameter for the parametric spline. The next column gives the constant coefficient. Subsequent columns give the linear, quadratic, and cubic coefficients. If a parametric spline is used, then columns 2 through 5 give the coefficients for the x component, and columns 6 through 9 give the coefficients for the y components of the curve. The coefficients for each interval are with respect to the variable $x - x_i$, where x_i is the left endpoint of the interval and x is the point of interest. The *coeff* matrix can be processed to yield the coefficients for the integral or the derivative of the spline. These in turn can be used with the SPLINEV function to evaluate the resulting curves.

The *endval* matrix returns the slopes of the two ends of the curve as angles expressed in degrees.

If the points are not arranged in order of the first column of *data* and the nonparametric spline is requested, an error message is printed. The data must be ordered by the value of the independent variable (the first column of *data*) or a parametric method must be used. See Stoer and Bulirsch (1980), Reinsch (1967), and Pizer (1975) for descriptions of the methods used to fit the spline.

SPLINEV Function

evaluates points on a spline

returns a matrix with fitted values

Syntax

SPLINEV(*coeff*<,*delta*> <,*nout*>)

where

coeff	is an $n \times 5$ or $n \times 9$ matrix of spline coefficients.
delta	is a numeric scalar specifying the resolution constant.

SPLINEV Function *continued*

> *nout*   is a numeric scalar specifying the number of fitted points. The default is 200.

Description

The SPLINEV function requires an input matrix of coefficients. It evaluates the spline using these coefficient values and returns a matrix of fitted values. Below is an example of a valid statement using this function:

```
fit=splinev(coef, delta, nout);
```

SQRSYM Function

> **converts a symmetric matrix to a square matrix**
>
> returns a square matrix

Syntax

SQRSYM(*matrix*)

where *matrix* is a symmetric numeric matrix.

Description

The SQRSYM function takes a matrix such as those generated by the SYMSQR function and transforms it back into a square matrix. The elements of the argument are unpacked into the lower triangle of the result and reflected across the diagonal into the upper triangle.

For example, the following statement

```
sqr=sqrsym(symsqr({1 2, 3 4}));
```

which is the same as

```
sqr=sqrsym({1, 3, 4});
```

produces the result

```
         SQR        2 rows      2 cols    (numeric)

                       1          3
                       3          4
```

SQRT Function

calculates the square root

returns a matrix containing the square roots of the elements

Syntax

SQRT(*matrix*)

where *matrix* is a numeric matrix or literal.

Description

The SQRT function is the scalar function returning the positive square roots of each element of the argument. An example of a valid statement follows:

```
a=sqrt(c);
```

SSQ Function

calculates the sum of squares of all elements

returns a scalar containing the sum of squares

Syntax

SSQ(*matrix1*<,*matrix2*,...,*matrix15*>)

where *matrix* is a numeric matrix or literal.

Description

The SSQ function returns as a single numeric value the (uncorrected) sum of squares for all the elements of all arguments. You can specify as many as 15 numeric argument matrices.

The SSQ function checks for missing arguments and does not include them in the accumulation. If all arguments are missing, the result is 0.

An example of a valid statement follows:

```
a={1 2 3, 4 5 6};
x=ssq(a);
```

START and FINISH Statements

define a module

Syntax

START <*name*> <*(arguments)*> <GLOBAL(*arguments*)>;
   module statements;
FINISH <*name*>;

where

name	is the name of a user-defined module.
arguments	are names of variable arguments to the module. Arguments can be either input variables or output (returned) variables. Arguments listed in the GLOBAL clause are treated as global variables. Otherwise, the arguments are local.
module statements	are statements making up the body of the module.

Description

The START statement instructs IML to enter a module-collect mode to collect the statements of a module rather than execute them immediately. The FINISH statement signals the end of a module. Optionally, the FINISH statement can take the module name as its argument. When no *name* argument is given in the START statement, the module name MAIN is used by default. If an error occurs during module compilation, the module is not defined. See Chapter 5, "Programming," for details.

    The example below defines a module named MYMOD that has two local variables (A and B) and two global variables (X and Y). The module creates the variable Y from the arguments A, B, and X.

```
start mymod(a,b) global(x,y);
   y=a*x+b;
finish;
```

STOP Statement

stops execution of statements

Syntax

STOP;

Description

The STOP statement stops the IML program, and no further matrix statements are executed. However, IML continues to execute if more statements are entered. See also the descriptions of the RETURN and ABORT statements.

If IML execution was interrupted by a PAUSE statement or by a break, the STOP statement clears all the paused states and returns to immediate mode.

IML supports STOP processing of both regular and function modules.

STORAGE Function

lists names of matrices and modules in storage

returns a character vector containing names of modules and matrices in storage

Syntax

STORAGE();

Description

The STORAGE function returns a matrix of the names of all of the matrices and modules in the current storage library. The result is a character vector with each matrix or module name occupying a row. Matrices are listed before modules. The SHOW STORAGE command separately lists all of the modules and matrices in storage.

For example, the following statements reset the current library storage to MYLIB and then print a list of the modules and matrices in storage:

```
reset storage="MYLIB";
```

Then issue the command below to get the resulting matrix:

```
a=storage();
print a;
```

STORE Statement

stores matrices and modules in library storage

Syntax

STORE <MODULE=*(module-list)*> <*matrix-list*>;

where

module-list   is a list of module names.

matrix-list   is a list of matrix names.

Description

The STORE statement stores matrices or modules in the storage library. For example, the following statement stores the modules A, B, and C and the matrix X:

```
store module=(A B C) X;
```

The special operand _ALL_ can be used to store all matrices or all modules. For example, the following statement stores all matrices and modules:

```
store _all_ module=_all_;
```

The storage library can be specified using the RESET STORAGE command and defaults to SASUSER.IMLSTOR. The SHOW STORAGE command lists the current contents of the storage library. The following statement stores all matrices:

```
store;
```

See Chapter 12, "Storage Features," and also the descriptions of the LOAD, REMOVE, RESET, and SHOW statements for related information.

SUBSTR Function

takes substrings of matrix elements

returns a character matrix

Syntax

SUBSTR(*matrix,position*<*,length*>)

where

matrix   is a character matrix or quoted literal.

position   is a numeric matrix or scalar giving the starting position.

length   is a numeric matrix or scalar giving the length of the substring.

Description

The SUBSTR function takes a character matrix as an argument along with starting positions and lengths and produces a character matrix with the same dimensions as the argument. Elements of the result matrix are substrings of the corresponding argument elements. Each substring is constructed using the starting *position* supplied. If a *length* is supplied, this length is the length of the substring. If no *length* is supplied, the remainder of the argument string is the substring.

The *position* and *length* arguments can be scalars or numeric matrices. If *position* or *length* is a matrix, its dimensions must be the same as the dimensions of the argument matrix or submatrix. If either one is a matrix, its values are applied to the substringing of the corresponding elements of the *matrix*. If *length* is supplied, the element length of the result is MAX(*length*); otherwise, the element length of the result is

$$NLENG(matrix) - MIN\ (position) + 1\quad.$$

The statements

```
B={abc def ghi, jkl mno pqr};
a=substr(b,3,2);
```

return the matrix

```
        A          2 rows      3 cols     (character, size 2)

                               C  F  I
                               L  O  R
```

The element size of the result is 2; the elements are padded with blanks.

SUM Function

sums all elements

returns a scalar

Syntax

SUM(*matrix1*<,*matrix2*,...,*matrix15*>)

where *matrix* is a numeric matrix or literal.

Description

The SUM function returns as a single numeric value the sum of all the elements in all arguments. There can be as many as 15 argument matrices. The SUM

SUM Function *continued*

function checks for missing values and does not include them in the accumulation. It returns 0 if all values are missing.

For example, the statements

```
a={2 1, 0 -1};
b=sum(a);
```

return the scalar

B	1 row	1 col	(numeric)

2

SUMMARY Statement

computes summary statistics for SAS data sets

Syntax

SUMMARY <CLASS *operand*> <VAR *operand*> <WEIGHT *operand*>
       <STAT *operand*> <OPT *operand*> <WHERE(*expression*)>;

where the *operands* used by most clauses take either a matrix name, a matrix literal, or an expression yielding a matrix name or value. The clauses and *operands* are discussed below.

Description

The SUMMARY statement computes statistics for numeric variables for an entire data set or a subset of observations in the data set. The statistics can be stratified by the use of class variables. The computed statistics are displayed in tabular form and optionally can be saved in matrices. Like most other IML data processing statements, the SUMMARY statement works on the current data set.

The following options are available with the SUMMARY statement:

CLASS *operand*
    specifies the variables in the current input SAS data set to be used to group the summaries. The *operand* is a character matrix containing the names of the variables, for example,

```
summary class {age sex};
```

Both numeric and character variables can be used as class variables.

VAR *operand*

    calculates statistics for a set of numeric variables from the current input data set. The *operand* is a character matrix containing the names of the variables. Also, the special keyword _NUM_ can be used as a VAR operand to specify all numeric variables. If the VAR clause is missing, the SUMMARY statement produces only the number of observations in each class group.

WEIGHT *operand*

    specifies a character value containing the name of a numeric variable in the current data set whose values are to be used to weight each observation. Only one variable can be specified.

STAT *operand*

    computes the statistics specified. The *operand* is a character matrix containing the names of statistics. For example, to get the mean and standard deviation, specify

```
summary stat{mean std};
```

Below is a list of the keywords that can be specified as the STAT *operand*:

CSS	computes the corrected sum of squares.
MAX	computes the maximum value.
MEAN	computes the mean.
MIN	computes the minimum value.
N	computes the number of observations in the subgroup used in the computation of the various statistics for the corresponding analysis variable.
NMISS	computes the number of observations in the subgroup having missing values for the analysis variable.
STD	computes the standard deviation.
SUM	computes the sum.
SUMWGT	computes the sum of the WEIGHT variable values if WEIGHT is specified; otherwise, IML computes the number of observations used in the computation of statistics.
USS	computes the uncorrected sum of squares.
VAR	computes the variance.

When the STAT clause is omitted, the SUMMARY statement computes these statistics for each variable in the VAR clause:

□ MAX

□ MEAN

□ MIN

□ STD.

SUMMARY Statement *(STAT operand continued)*

Note that NOBS, the number of observations in each CLASS group, is always given.

OPT *operand*

sets the PRINT or NOPRINT and SAVE or NOSAVE options. The NOPRINT option suppresses the printing of the results from the SUMMARY statement. The SAVE option requests that the SUMMARY statement save the resultant statistics in matrices. The *operand* is a character matrix containing one or more of the options.

When the SAVE option is set, the SUMMARY statement creates a class vector for each class variable, a statistic matrix for each analysis variable, and a column vector named _NOBS_. The class vectors are named by the corresponding class variable and have an equal number of rows. There are as many rows as there are subgroups defined by the interaction of all class variables. The statistic matrices are named by the corresponding analysis variable. Each column of the statistic matrix corresponds to a statistic requested, and each row corresponds to the statistics of the subgroup defined by the class variables. If no class variable has been specified, each statistic matrix has one row, containing the statistics of the entire population. The _NOBS_ vector contains the number of observations for each subgroup.

The default is PRINT NOSAVE.

WHERE *expression*

conditionally selects observations, within the *range* specification, according to conditions given in *expression*. The general form of the WHERE clause is

WHERE (*variable comparison-op operand*)

where

variable	is a variable in the SAS data set.
comparison-op	is one of the following comparison operators:

<	less than
<=	less than or equal to
=	equal to
>	greater than
>=	greater than or equal to
^=	not equal to
?	contains a given string
^?	does not contain a given string
=:	begins with a given string
=*	sounds like or is spelled similar to a given string.

operand   is a literal value, a matrix name, or an expression in parentheses.

WHERE comparison arguments can be matrices. For the following operators, the WHERE clause succeeds if *all* the elements in the matrix satisfy the condition:

    ^=   ^? < <= > >=.

For the following operators, the WHERE clause succeeds if *any* of the elements in the matrix satisfy the condition:

    = ? =: =*.

Logical expressions can be specified within the WHERE clause, using the AND (&) and OR (|) operators. The general form is

clause&*clause*   (for an AND clause)
clause | *clause*   (for an OR clause)

where *clause* can be a comparison, a parenthesized clause, or a logical expression clause that is evaluated using operator precedence.

    Note: The expression on the left-hand side refers to values of the data set variables, and the expression on the right-hand side refers to matrix values.

    See Chapter 6, "Working with SAS Data Sets," for an example using the SUMMARY statement.

SVD CALL

computes the singular value decomposition

returns the decomposition matrices

Syntax

CALL SVD(*u,q,v,a*);

where

u, q, and *v*   are the returned decomposition matrices.

a   is the input matrix that is decomposed as described below.

Description

The SVD subroutine decomposes a real $m \times n$ matrix $\mathbf{A}$ (where m is greater than or equal to n) into the form

$$\mathbf{A} = \mathbf{U}^* \text{diag}(\mathbf{Q})^* \mathbf{V}'$$

SVD CALL *continued*

where

$$U'U = V'V = VV' = I_n$$

and **Q** contains the singular values of **A**. **U** is $m \times n$, **Q** is $n \times 1$, and **V** is $n \times n$.

When m is greater than or equal to n, **U** consists of the orthonormal eigenvectors of **AA'**, and **V** consists of the orthonormal eigenvectors of **A'A**. **Q** contains the square roots of the eigenvalues of **A'A** and **AA'**, except for some zeros.

If m is less than n, a corresponding decomposition is done where **U** and **V** switch roles:

$$A = U^*diag(Q)^*V'$$

but

$$U'U = UU' = V'V = I_w \quad .$$

The singular values are sorted in descending order.

For information about the method used in the SVD subroutine, see Wilkinson and Reinsch (1971). Consider the following example (Wilkinson and Reinsch 1971, p. 149):

```
a={22  10   2    3    7,
   14   7  10    0    8,
   -1  13  -1  -11    3,
   -3  -2  13   -2    4,
    9   8   1   -2    4,
    9   1  -7    5   -1,
    2  -6   6    5    1,
    4   5   0   -2    2};
call svd(u,q,v,a);
```

The results are

```
         U              8 rows      5 cols     (numeric)

0.7071068 0.1581139 -0.176777  -0.06701   0.279804
0.5303301 0.1581139 0.3535534 -0.045208 -0.645372
0.1767767 -0.790569 0.1767767 0.5368704 -0.060458
        0 0.1581139 0.7071068 0.1086593  0.592536
0.3535534 -0.158114         0 -0.228736 0.2300372
0.1767767 0.1581139  -0.53033 0.5116134  0.212316
        0 0.4743416 0.1767767 0.5867386 -0.102189
0.1767767 -0.158114         0 -0.187346 0.2049688
```

```
Q                5 rows      1 col     (numeric)

                      35.327043
                             20
                      19.595918
                       1.281E-15
                       3.661E-16


V                5 rows      5 cols    (numeric)

0.8006408 0.3162278 -0.288675 0.4190955         0
0.4803845 -0.632456         0 -0.440509 -0.418548
0.1601282 0.3162278 0.8660254 0.0520045  -0.34879
        0 0.6324555 -0.288675 -0.676059 -0.244153
0.3202563         0 0.2886751 -0.412977 0.8022171
```

SWEEP Function

sweeps a matrix

returns a matrix

Syntax

SWEEP(*matrix,index-vector*)

where

matrix	is a numeric matrix or literal.
index-vector	is a numeric vector indicating the pivots.

Description

The SWEEP function sweeps *matrix* on the pivots indicated in *index-vector* to produce a new matrix. The values of the index vector must be less than or equal to the number of rows or the number of columns in *matrix*, whichever is smaller.

For example, suppose that **A** is partitioned into

$$
\begin{bmatrix} R & S \\ T & U \end{bmatrix}
$$

such that **R** is $q \times q$ and **U** is $(m-q) \times (n-q)$. Let

```
I = [1 2 3 . . . q]
```

SWEEP Function *continued*

Then, the statement

    s=sweep(A,I);

becomes

$$
\begin{bmatrix} R^{-1} & R^{-1}S \\ -TR^{-1} & U-T^{-1}S \end{bmatrix} \quad .
$$

The index vector can be omitted. In this case, the function sweeps the matrix on all pivots on the main diagonal 1:MIN(*nrow,ncol*).

The SWEEP function has sequential and reversibility properties when the submatrix swept is positive definite:

1. SWEEP(SWEEP(**A**,1),2)=SWEEP(**A**,{1 2})

2. SWEEP(SWEEP(**A**,**I**),**I**)=**A**

See Beaton (1964) for more information about these properties.

To use the SWEEP function for regression, suppose the matrix **A** contains

$$
\begin{bmatrix} X'X & X'Y \\ Y'X & Y'Y \end{bmatrix}
$$

where **X'X** is $k \times k$.

Then **B**=SWEEP(**A**,1. . .*k*) contains

$$
\begin{bmatrix} (X'X)^{-1} & (X'X)^{-1}X' \\ -Y'X(X'X)^{-1} & Y'(I-X(X'X)^{-1}X')Y \end{bmatrix}
$$

The partitions of **B** form the beta values, SSE, and a matrix proportional to the covariance of the beta values for the least-squares estimates of **B** in the linear model

$$
\mathbf{Y} = \mathbf{XB} + \varepsilon \quad .
$$

If any pivot becomes very close to zero (less than or equal to 1E−12), the row and column for that pivot are zeroed. See Goodnight (1979) for more information.

SYMSQR Function

converts a square matrix to a symmetric matrix

returns a column vector

Syntax

SYMSQR(*matrix*)

where *matrix* is a square numeric matrix.

Description

The SYMSQR function takes a square numeric matrix (size $n \times n$) and compacts the elements from the lower triangle into a column vector ($n^*(n+1)/2$ rows). The matrix is not checked for actual symmetry. Therefore, the statement

```
sym=symsqr({1 2, 3 4});
```

sets

```
        SYM         3 rows      1 col     (numeric)

                                1
                                3
                                4
```

Note that the 2 is lost since it is only present in the upper triangle.

T Function

transposes a matrix

returns a transposed matrix

Syntax

T(*matrix*)

where *matrix* is a numeric or character matrix or literal.

Description

The T (transpose) function returns the transpose of its argument. It is equivalent to the transpose operator as written with a transpose postfix operator (`), but since some keyboards do not support the backquote character, this function is provided as an alternate.

T Function *continued*

For example, the statements

```
x={1 2, 3 4};
y=t(x);
```

result in the matrix

```
         Y            2 rows      2 cols    (numeric)

                          1          3
                          2          4
```

TOEPLITZ Function

generates a Toeplitz or block-Toeplitz matrix

returns a matrix

Syntax

TOEPLITZ(*a*)

where *a* is either a vector or a numeric matrix.

Description

The TOEPLITZ function generates a Toeplitz matrix from a vector, or a block Toeplitz matrix from a matrix. A block Toeplitz matrix has the property that all matrices on the diagonals are the same. The argument *a* is an $(n^*p) \times p$ or $p \times (n^*p)$ matrix; the value returned is the $(n^*p) \times (n^*p)$ result.

The TOEPLITZ function uses the first $p \times p$ submatrix, A_1, of the argument matrix as the blocks of the main diagonal. The second $p \times p$ submatrix, A_2, of the argument matrix forms one secondary diagonal, with the transpose A_2' forming the other. The remaining diagonals are formed accordingly. If the first $p \times p$ submatrix of the argument matrix is symmetric, the result is also symmetric. If A is $(n^*p) \times p$, the first p columns of the returned matrix, **R**, will be the same as **A**. If **A** is $p \times (n^*p)$, the first p rows of **R** will be the same as **A**. The TOEPLITZ function is especially useful in time-series applications, where the covariance matrix of a set of variables with its lagged set of variables is often assumed to be a block Toeplitz matrix.

If

$$\mathbf{A} = \mathbf{A}_1 \mid \mathbf{A}_2 \mid \mathbf{A}_3 \mid \ldots \mid \mathbf{A}_n$$

and if **R** is the matrix formed by the TOEPLITZ function, then

$$
\begin{aligned}
\mathbf{R} = \ & \mathbf{A}_1 \mid \mathbf{A}_2 \mid \mathbf{A}_3 \mid \ldots \mid \mathbf{A}_n \\
& \mathbf{A}_2' \mid \mathbf{A}_1 \mid \mathbf{A}_2 \mid \ldots \mid \mathbf{A}_{-1} \\
& \mathbf{A}_3' \mid \mathbf{A}_2' \mid \mathbf{A}_1 \mid \ldots \mid \mathbf{A}_{n-2} \\
& \qquad \cdot \\
& \qquad \cdot \\
& \qquad \cdot \\
& \mathbf{A}_n' \mid \mathbf{A}_{n-1}' \mid \mathbf{A}_{n-2}' \mid \ldots \mid \mathbf{A}_1 \quad .
\end{aligned}
$$

If

$$
\begin{aligned}
\mathbf{A} = \ & \mathbf{A}_1 \\
& \mathbf{A}_2 \\
& \ \cdot \\
& \ \cdot \\
& \ \cdot \\
& \mathbf{A}_n
\end{aligned}
$$

and if **R** is the matrix formed by the TOEPLITZ function, then

$$
\begin{aligned}
\mathbf{R} = \ & \mathbf{A}_1 \mid \mathbf{A}_2' \mid \mathbf{A}_3' \mid \ldots \mid \mathbf{A}_n' \\
& \mathbf{A}_2 \mid \ldots \\
& \qquad \cdot \\
& \qquad \cdot \\
& \qquad \cdot \\
& \mathbf{A}_n \mid \mathbf{A}_{n-1} \mid \mathbf{A}_{n-2} \mid \ldots \mid \mathbf{A}_1 \quad .
\end{aligned}
$$

Three examples follow below.

```
r=toeplitz(1:5);
```

	R	5 rows	5 cols	(numeric)	
	1	2	3	4	5
	2	1	2	3	4
	3	2	1	2	3
	4	3	2	1	2
	5	4	3	2	1

TOEPLITZ Function *continued*

```
r=toeplitz({1 2 ,
            3 4 ,
            5 6 ,
            7 8});
```

```
        R          4 rows     4 cols    (numeric)

              1        2         5         7
              3        4         6         8
              5        6         1         2
              7        8         3         4
```

```
r=toeplitz({1 2 3 4 ,
            5 6 7 8});
```

```
        R          4 rows     4 cols    (numeric)

              1        2         3         4
              5        6         7         8
              3        7         1         2
              4        8         5         6
```

TRACE Function

sums diagonal elements

returns a scalar containing the trace

Syntax

TRACE(*matrix*)

where *matrix* is a numeric matrix or literal.

Description

The TRACE function produces a single numeric value that is the sum of the diagonal elements of *matrix*. For example, the statement

```
a=trace({5 2, 1 3});
```

produces the result

```
        A          1 row      1 col     (numeric)
```

TYPE Function

determines the type of a matrix

returns a single character value

Syntax

TYPE(*matrix*)

where *matrix* is a numeric or character matrix or literal.

Description

The TYPE function returns a single character value; it is **N** if the type of the matrix is numeric; it is **C** if the type of the matrix is character; it is **U** if the matrix does not have a value. Examples of valid statements follow.

The statements

```
a={tom};
r=type(a);
```

set R to **C**. The statements

```
free a;
r=type(a);
```

set R to **U**. The statements

```
a={1 2 3};
r=type(a);
```

set R to **N**.

UNIFORM Function

generates pseudo-random uniform deviates

returns a pseudo-random number

Syntax

UNIFORM(*seed*)

where *seed* is a numeric matrix or literal. The *seed* can be any integer value up to $2^{31}-1$.

Description

The UNIFORM function returns one or more pseudo-random numbers with a uniform distribution over the interval 0 to 1. The UNIFORM function returns a

UNIFORM Function *continued*

matrix with the same dimensions as the argument. The first argument on the first call is used for the seed, or if that argument is 0, the system clock is used for the seed. The function is equivalent to the DATA step function RANUNI. An example of a valid statement follows:

```
c=uniform(0);
```

UNION Function

performs unions of sets

returns a row vector containing the set of elements of the union of the arguments

Syntax

UNION(*matrix1*<,*matrix2*,...,*matrix15*>)

where *matrix* is a numeric or character matrix or quoted literal.

Description

The UNION function returns as a row vector the sorted set (without duplicates) which is the union of the element values present in its arguments. There can be up to 15 arguments, which can be either all character or all numeric. For character arguments, the element length of the result is the longest element length of the arguments. Shorter character elements are padded on the right with blanks. This function is identical to the UNIQUE function. For example, the statements

```
a=(1 2 4 5);
b=(3 4);
c=union(a,b);
```

set

C		1 row	5 cols	(numeric)
1	2	3	4	5

The UNION function can be used to sort elements of a matrix when there are no duplicates by calling UNION with a single argument.

UNIQUE Function

sorts and removes duplicates

returns a row vector containing the unique elements of the arguments

Syntax

UNIQUE(*matrix1*<,*matrix2*,...,*matrix15*>)

where *matrix* is a numeric or character matrix or quoted literal.

Description

The UNIQUE function returns as a row vector the sorted set (without duplicates) of all the element values present in its arguments. The arguments can be either all numeric or all character, and there can be up to 15 arguments specified. This function is identical to the UNION function, the description of which includes an example.

USE Statement

opens a SAS data set for reading

Syntax

USE *SAS-data-set* <VAR *operand*> <WHERE(*expression*)> <NOBS *name*>;

where

SAS-data-set	can be specified with a one-word name (for example, A) or a two-word name (for example, SASUSER.A). For more information on specifying SAS data sets, see Chapter 6, "SAS Files," in *SAS Language: Reference*.
operand	selects a set of variables.
expression	is evaluated for being true or false.
name	is the name of a variable to contain the number of observations.

Description

If the data set has not already been opened, the USE statement opens the data set for read access. The USE statement also makes the data set the current input data set so that subsequent statements act on it. The USE statement optionally can define selection criteria that are used to control access.

USE Statement *continued*

The VAR clause specifies a set of variables to use, where *operand* can be any of the following:

□   a literal containing variable names

□   the name of a matrix containing variable names

□   an expression in parentheses yielding variable names

□   one of the keywords described below:

ALL   for all variables

CHAR   for all character variables

NUM   for all numeric variables.

Below are examples showing each possible way you can use the VAR clause:

```
var {time1 time5 time9};    /* a literal giving the variables  */
var time;                   /* a matrix containing the names   */
var('time1':'time9');       /* an expression                   */
var _all_;                  /* a keyword                       */
```

The WHERE clause conditionally selects observations, within the *range* specification, according to conditions given in the clause. The general form of the WHERE clause is

WHERE(*variable comparison-op operand*)

where

variable   is a variable in the SAS data set.

comparison-op   is one of the following comparison operators:

      <   less than

      <=   less than or equal to

      =   equal to

      >   greater than

      >=   greater than or equal to

      ^=   not equal to

      ?   contains a given string

      ^?   does not contain a given string

      =:   begins with a given string

      =*   sounds like or is spelled similar to a given string.

operand is a literal value, a matrix name, or an expression in parentheses.

WHERE comparison arguments can be matrices. For the following operators, the WHERE clause succeeds if *all* the elements in the matrix satisfy the condition:

^= ^? < <= > >=

For the following operators, the WHERE clause succeeds if *any* of the elements in the matrix satisfy the condition:

= ? =: =*

Logical expressions can be specified within the WHERE clause using the AND (&) and OR (|) operators. The general form is

clause&clause (for an AND clause)
clause | clause (for an OR clause)

where *clause* can be a comparison, a parenthesized clause, or a logical expression clause that is evaluated using operator precedence.

Note: The expression on the left-hand side refers to values of the data set variables, and the expression on the right-hand side refers to matrix values.

The VAR and WHERE clauses are optional, and you can specify them in any order. If a data set is already open, all the options that the data set was first opened with are still in effect. To override any old options, the new USE statement must explicitly specify the new options. Examples of valid statements follow:

```
use class;
use class var{name sex age};
use class var{name sex age} where(age>10);
```

VALSET Call

performs indirect assignment

Syntax

CALL VALSET(*char-scalar,argument*);

where

char-scalar is a character scalar containing the name of a matrix.

argument is a value to which the matrix is set.

Description

The VALSET subroutine expects a single character string argument containing the name of a matrix. It looks up the matrix and moves the value of the second

VALSET Call *continued*

argument to this matrix. For example, the following statements find that the value of the argument **B** is **A** and then assign the value 99 to **A**, the indirect result:

```
b="A";
call valset(b,99);
```

The previous value of the indirect result is freed. The following statement sets **B** to 99, but the value of **A** is unaffected by this statement:

```
b=99;
```

VALUE Function

assigns values by indirect reference

returns the value of a matrix

Syntax

VALUE(*char-scalar*)

where *char-scalar* is a character scalar containing the name of a matrix.

Description

The VALUE function expects a single character string argument containing the name of a matrix. It looks up the matrix and moves its value to the result. For example, the statements

```
a={1 2 3};
b="A";
c=value(b);
```

find that the value of the argument **B** is **A** and then look up **A** and copy the value {1 2 3} to **C**.

C	1 row	3 cols	(numeric)
1	2	3	

VECDIAG Function

creates a vector from a diagonal

returns a column vector containing the diagonal elements of the argument

Syntax

VECDIAG(*square-matrix*)

where *square-matrix* is a square numeric matrix.

Description

The VECDIAG function creates a column vector whose elements are the main diagonal elements of *square-matrix*. For example, the statements

```
a={2 1, 0 -1};
c=vecdiag(a);
```

produce the column vector

```
        C            2 rows       1 col     (numeric)

                                    2
                                   -1
```

WINDOW Statement

opens a display window

Syntax

WINDOW <CLOSE=>*window-name* <*window-options*>
        <GROUP=*group-name field-specs*>
        <...GROUP=*group-name field-specs*>;

where the arguments and options are described below.

Description

The WINDOW statement defines a window on the display and can include a number of fields. The DISPLAY statement actually writes values to the window. The following fields can be specified in the WINDOW statement:

window-name
    specifies a name 1 to 8 characters long for the window. This name is displayed in the upper-left border of the window.

WINDOW Statement *continued*

CLOSE=*window-name*
    closes the window.

window-options
    control the size, position, and other attributes of the window. The attributes
    can also be changed interactively with window commands such as WGROW,
    WDEF, WSHRINK, and COLOR. The window options are described in
    "Window Options" below.

GROUP=*group-name*
    starts a repeating sequence of groups of fields defined for the window. The
    group-name specification is a name 1 to 8 characters long used to identify a
    group of fields in a later DISPLAY statement.

field-specs
    are a sequence of field specifications made up of positionals, field operands,
    formats, and options. These are described in the next section.

Window options

The following window options can be specified in the WINDOW statement:

CMNDLINE=*name*
    specifies the name of a variable in which the command line entered by the
    user will be stored.

COLOR=*operand*
    specifies the background color for the window. The *operand* is either a
    quoted character literal, a name, or an operand. The valid values are
    "WHITE", "BLACK", "GREEN", "MAGENTA", "RED", "YELLOW", "CYAN",
    "GRAY", and "BLUE". The default is BLACK.

COLUMNS=*operand*
    specifies the starting number of columns for the window. The *operand* is
    either a literal number, a variable name, or an expression in parentheses.
    The default is 78 columns.

ICOLUMN=*operand*
    specifies the initial starting column position of the window on the display.
    The *operand* is either a literal number or a variable name. The default is
    column 1.

IROW=*operand*
    specifies the initial starting row position of the window on the display. The
    operand is either a literal number or a variable name. The default is row 1.

MSGLINE=*operand*
    specifies the message to be displayed on the standard message line when the
    window is made active. The *operand* is almost always the name of a variable,
    but a character literal can be used.

ROWS=*operand*

>   determines the starting number of rows of the window. The *operand* is either a literal number, the name of a variable containing the number, or an expression in parentheses yielding the number. The default is 23 rows.

Field specifications

Both the WINDOW and DISPLAY statements allow field specifications, which have the general form:

<*positionals*> *field-operand* <*format*> <*field-options*>

where

positionals

>   are directives determining the position on the screen to begin the field. There are four kinds of positionals; any number of positionals are allowed for each field operand.

> # *operand*

>>   specifies the row position; that is, it moves the current position to column 1 of the specified line. The *operand* is either a number, a name, or an expression in parentheses.

> /

>>   specifies that the current position move to column 1 of the next row.

> @ *operand*

>>   specifies the column position. The *operand* is either a number, a name, or an expression in parentheses. The @ directive should come after the # position if # is specified.

> + *operand*

>>   specifies a skip of columns. The *operand* is either a number, a name, or an expression in parentheses.

field-operand

>   is a character literal in quotes or the name of a variable that specifies what is to go in the field.

format

>   is the format used for display, the value, and the informat applied to entered values. If no format is specified, then the standard numeric or character format is used.

field-options

>   specify the attributes of the field as follows:

> PROTECT=YES
> P=YES

>>   specifies that the field is protected; that is, you cannot enter values in the field. If the field operand is a literal, it is already protected.

> COLOR=*operand*

>>   specifies the color of the field. The *operand* is a literal character value in quotes, a variable name, or an expression in parentheses. The colors available are "WHITE", "BLACK", "GREEN", "MAGENTA", "RED", "YELLOW", "CYAN", "GRAY", and "BLUE". Note that the color

---

WINDOW Statement *continued*

specification is different from that of the corresponding DATA step value because it is an operand rather than a name without quotes. The default is "BLUE".

XMULT Function

performs accurate matrix multiplication

Syntax

XMULT(*matrix1,matrix2*)

where *matrix1* and *matrix2* are numeric matrices.

Description

The XMULT function computes the matrix product like the matrix multiplication operator (*) except XMULT uses extended precision to accumulate sums of products. You should use the XMULT function only when you need great accuracy.

XSECT Function

intersects sets

returns a row vector

Syntax

XSECT(*matrix1<,matrix2,...,matrix15>*)

where *matrix* is a numeric or character matrix or quoted literal.

Description

The XSECT function returns as a row vector the sorted set (without duplicates) of the element values that are present in all of its arguments. This set is the intersection of the sets of values in its argument matrices. When the intersection is empty, the XSECT function returns a null matrix (zero rows and zero columns). There can be up to 15 arguments, which must all be either character or numeric. For characters, the element length of the result is the same as the shortest of the element lengths of the arguments. For comparison purposes, shorter elements are padded on the right with blanks.

For example, the statements

```
a={1 2 4 5};
b={3 4};
c=xsect(a,b);
```

return the result shown:

```
          C           1 row      1 col     (numeric)

                                 4
```

Appendix **1** SAS/IML® Quick Reference

Table A1.1
Operators

Operator	Symbol	Syntax Type	Data Type
sign reverse	−	prefix	num
addition	+	infix	both
subtraction	−	infix	num
index creation	:	infix	num
matrix multiplication	*	infix	num
elementwise multiplication	#	infix	num
direct product	@	infix	num
matrix power	**	infix	num
elementwise power	##	infix	num
division	/	infix	num
horizontal concatenation	||	infix	both
vertical concatenation	//	infix	both
element maximum	<>	infix	both

(continued)

Table A1.1
(continued)

Operator	Symbol	Syntax Type	Data Type
element minimum	><	infix	both
logical AND	&	infix	num
logical OR	|	infix	num
logical NOT	^	prefix	num
less than	<	infix	both
greater than	>	infix	both
equal to	=	infix	both
less than or equal to	<=	infix	both
greater than or equal to	>=	infix	both
not equal to	^=	infix	both
transpose	`	postfix	both
subscripts	[ ]	postfix	both

Table A1.2
Subscript
Reduction
Operators

Operator	Action
+	addition
#	multiplication
<>	maximum
><	minimum
<:>	index of maximum
>:<	index of minimum
:	mean (different from the MATRIX procedure)
##	sum of squares

Table A1.3
Operator
Precedence

Priority Group	Operators
I (highest)	^ ` subscripts −(prefix) ## **
II	* # <> >< / @
III	+ −
IV	‖ // :
V	< <= > >= = ^=
VI	&
VII (lowest)	|

Table A1.4
IML Functions and Calls

Name	Description	Data Type
ABS	takes the absolute value	num
ALL	checks for all elements nonzero	num
ANY	checks for any nonzero element	num
APPLY	applies an IML module to arguments	
ARMACOV	computes an autocovariance sequence for an ARMA model	num
ARMALIK	computes the log-likelihood and residuals for an ARMA model	num
ARMASIM	simulates a univariate ARMA series	num
BLOCK	forms block-diagonal matrices	num
BRANKS	computes bivariate ranks	num
BTRAN	computes the block transpose	num
BYTE	translates numbers to ordinal characters	char
CHANGE	finds and replaces text in an array	char
CHAR	produces a character representation of a numeric matrix	num
CHOOSE	conditionally chooses and changes elements	both
CONCAT	performs elementwise string concatenation	char
CONTENTS	obtains the variables in a data set	char
CONVMOD	converts modules to character matrices	char
COVLAG	computes autocovariance estimates for a vector time series	num
CSHAPE	reshapes and repeats character values	char
CUSUM	calculates cumulative sums	num
CVEXHULL	finds a convex hull of a set of planar points	num
DATASETS	obtains the names of SAS data sets in a SAS data library	char
DELETE	deletes a data set	char
DESIGN	creates a design matrix	num
DESIGNF	creates a full-rank design matrix	num
DET	computes the determinant of a square matrix	num
DIAG	creates a diagonal matrix	num
DO	produces an arithmetic series	num
ECHELON	reduces a matrix to row-echelon form	num

(continued)

Table A1.4
(*continued*)

Name	Description	Data Type
EIGEN	computes eigenvalues and eigenvectors	num
EIGVAL	computes eigenvalues	num
EIGVEC	creates eigenvectors	num
EXP	calculates the exponential	num
FFT	computes the finite Fourier transform	num
GENEIG	computes eigenvalues and eigenvectors of a generalized eigenproblem	num
GINV	computes the generalized inverse	num
GSORTH	computes the Gram-Schmidt orthonormalization	num
HALF	computes Cholesky decomposition	num
HANKEL	generates a Hankel matrix	num
HDIR	performs a horizontal direct product	num
HERMITE	reduces a matrix to Hermite normal form	num
HOMOGEN	solves homogeneous linear systems	num
I	creates an identity matrix	num
IFFT	computes the inverse finite Fourier transform	num
INSERT	inserts one matrix inside another	both
INT	truncates a value	num
INV	computes the matrix inverse	num
INVUPDT	updates a matrix inverse	num
IPF	performs an iterative proportional fit	num
J	creates a matrix of identical values	num
LCP	solves the linear complementarity problem	num
LENGTH	finds the length of character matrix elements	char
LOC	finds nonzero elements of a matrix	num
LOG	takes the natural logarithm	num
LP	solves the linear programming problem	num
MARG	evaluates margin totals in a multiway contingency table	num
MAX	finds the maximum value of a matrix	both
MIN	finds the smallest element of a matrix	both
MOD	computes the modulo (remainder)	both
NAME	lists the names of arguments	both

(*continued*)

Table A1.4
(continued)

Name	Description	Data Type
NCOL	finds the number of columns of a matrix	both
NLENG	finds the size of an element	both
NORMAL	generates a pseudo-random normal deviate	both
NROW	finds the number of rows of a matrix	both
NUM	produces a numeric representation of a character matrix	char
OPSCAL	rescales qualitative data to be a least-squares fit to qualitative data	num
ORPOL	generates orthogonal polynomials	num
PGRAF	produces scatterplots	num
POLYROOT	finds zeros of a real polynomial	num
PRODUCT	multiplies matrices of polynomials	num
RANK	ranks elements of a matrix	num
RANKTIE	ranks matrix elements with tie-averaging	num
RATIO	divides matrix polynomials	num
REMOVE	discards elements from a matrix	both
RENAME	renames a SAS data set	char
REPEAT	creates a new matrix of repeated values	num
ROOT	performs the Cholesky decomposition of a matrix	num
ROWCAT	concatenates rows without using blank compression	char
ROWCATC	concatenates rows using blank compression	char
SETDIF	compares elements of two matrices	both
SHAPE	reshapes and repeats values	both
SOLVE	solves a system of linear equations	num
SOUND	produces a tone	num
SPLINE	evaluates points on the spline	num
SPLINEC	evaluates points on the spline	num
SPLINEV	evaluates points on a spline	num
SQRSYM	converts a symmetric matrix to a square matrix	num
SQRT	calculates the square root	char
SSQ	calculates the sum of squares of all elements	num

(continued)

Table A1.4
(*continued*)

Name	Description	Data Type
STORAGE	lists names of matrices and modules in storage	char
SUBSTR	takes substrings of matrix elements	char
SUM	sums all elements	num
SVD	computes the singular value decomposition	num
SWEEP	sweeps a matrix	num
SYMSQR	converts a square matrix to a symmetric matrix	num
T	transposes a matrix	both
TOEPLITZ	generates a Toeplitz or block-Toeplitz matrix	num
TRACE	sums diagonal elements	num
TYPE	determines the type of a matrix	char
UNIFORM	generates pseudo-random uniform deviates	num
UNION	performs unions of sets	both
UNIQUE	sorts and removes duplicates	both
VALSET	performs indirect assignment	char
VALUE	assigns values by indirect reference	char
VECDIAG	creates a vector from a diagonal	num
XMULT	performs accurate multiplication	num
XSECT	intersects sets	both

Table A1.5
Scalar Functions

Function	Description
ABS	takes the absolute value
EXP	calculates the exponential
INT	truncates a value
LOG	takes the natural logarithm
MOD	computes the modulo (remainder)
NORMAL	generates a pseudo-random normal deviate
SQRT	calculates the square root
UNIFORM	generates pseudo-random uniform deviates

Table A1.6
Reduction Functions

Function	Description
MAX	finds the maximum value of a matrix
MIN	returns the smallest element of a matrix
SSQ	calculates the sum of squares of all elements
SUM	sums all elements

Table A1.7
Matrix Inquiry Commands and Functions

Function	Description
ALL	checks for all elements nonzero
ANY	checks for any nonzero element
LOC	finds nonzero elements of a matrix
NCOL	finds the number of columns of a matrix
NLENG	finds the size of an element
NROW	finds the number of rows of a matrix
SHOW ALLNAMES	shows all attributes of all names with or without values
SHOW NAMES	shows attributes of all names having values
SHOW name	shows attributes of the matrix specified in name
TYPE	determines the type of a matrix

Table A1.8
Matrix Arithmetic Operators and Functions

Name	Description
*	performs matrix multiplication
@	takes the direct product of two matrices
**	raises a matrix to a power
CUSUM	calculates cumulative sums
HDIR	performs a horizontal direct product
TRACE	sums diagonal elements

Table A1.9
Matrix Manipulation and Reshaping Operators and Functions

Name	Description
`	transposes a matrix
[ ]	select submatrices
||	concatenates matrices horizontally
//	concatenates matrices vertically

(*continued*)

Table A1.9
(continued)

Name	Description
BLOCK	forms block-diagonal matrices
BTRAN	computes the block transpose
DIAG	creates a diagonal matrix
I	creates an identity matrix
INSERT	inserts one matrix inside another
J	creates a matrix of identical values
REMOVE	discards elements from a matrix
REPEAT	creates a new matrix of repeated values
SHAPE	reshapes and repeats values
SQRSYM	converts a symmetric matrix to a square matrix
SYMSQR	converts a square matrix to a symmetric matrix
T	transposes a matrix
VECDIAG	creates a vector from a diagonal

Table A1.10
Character
Manipulation
Functions

Function	Description
BYTE	translates numbers to ordinal characters
CHANGE	finds and replaces text in an array
CHAR	produces a character representation of a numeric matrix
CHOOSE	conditionally chooses and changes elements
CONCAT	performs elementwise string concatenation
CSHAPE	reshapes and repeats character values
LENGTH	finds the length of character matrix elements
NAME	lists the names of arguments
NUM	produces a numeric representation of a character matrix
PRODUCT	multiplies matrices of polynomials
RATIO	divides matrix polynomials
ROWCAT	concatenates rows without blank compression
ROWCATC	concatenates rows with blank compression
SUBSTR	takes substrings of matrix elements

Table A1.11
Set Functions

Function	Description
SETDIF	compares elements of two matrices
UNION	performs unions of sets
UNIQUE	sorts and removes duplicates
XSECT	intersects sets

Table A1.12
Linear Algebraic and Statistical Functions and Calls

Name	Description
CVEXHULL	finds a convex hull of a set of planar points
DESIGN	creates a design matrix
DESIGNF	creates a full-rank design matrix
DET	computes the determinant of a square matrix
ECHELON	reduces a matrix to row-echelon form
EIGEN	computes eigenvalues and eigenvectors
EIGVAL	computes eigenvalues
EIGVEC	creates eigenvectors
GENEIG	computes eigenvalues and eigenvectors of a generalized eigenproblem
GINV	computes the generalized inverse
GSORTH	computes the Gram-Schmidt orthonormalization
HALF	computes Cholesky decomposition
HERMITE	reduces a matrix to Hermite normal form
HOMOGEN	solves homogeneous linear systems
INV	computes the matrix inverse
INVUPDT	updates a matrix inverse
IPF	performs an iterative proportional fit
LCP	solves the linear complementarity problem
LP	solves the linear programming problem
MARG	evaluates marginal totals in a multiway contingency table
ORPOL	generates orthogonal polynomials
POLYROOT	finds zeros of a real polynomial
PRODUCT	multiplies matrices of polynomials
RANK	ranks elements of a matrix
RANKTIE	ranks matrix elements with tie-averaging
RATIO	divides matrix polynomials

(continued)

Table A1.12
(*continued*)

Name	Description
ROOT	performs the Cholesky decomposition of a matrix
SOLVE	solves a system of linear equations
SVD	computes the singular value decomposition
SWEEP	sweeps a matrix
T	transposes a matrix

Table A1.13
Time Series
Functions

Function	Description
ARMACOV	computes an autocovariance sequence for an ARMA model
ARMALIK	computes the log-likelihood and residuals for an ARMA model
ARMASIM	simulates a univariate ARMA series
COVLAG	computes autocovariance estimates for a vector time series
FFT	computes the finite Fourier transform
HANKEL	generates a Hankel matrix
IFFT	computes the inverse finite Fourier transform
TOEPLITZ	generates a Toeplitz or block-Toeplitz matrix

Table A1.14
Base SAS
Functions

Functions by Type	Description
Arithmetic functions	
SIGN	returns the sign of the argument or 0
Truncation functions	
CEIL	returns the smallest integer $>=$ argument
FLOOR	returns the largest integer $<=$ argument
FUZZ	returns the integer if the argument is within $1E-12$
ROUND	rounds a value to the nearest roundoff unit
TRUNC	returns a truncated numeric value of a specified length
Mathematical functions	
ARCOS	calculates the arccosine
ARSIN	calculates the arcsine
ATAN	calculates the arctangent
COSINE	calculates the cosine
DIGAMMA	computes the derivative of the log of the GAMMA function
ERF	is the error function
ERFC	returns the complement of the ERF function

(*continued*)

	Functions by Type	Description
Table A1.14 *(continued)*		
	GAMMA	produces the complete GAMMA function
	LGAMMA	calculates the natural logarithm of the GAMMA function of a value
	LOG2	calculates the logarithm to the base 2
	LOG10	produces the common logarithm
	SIN	calculates the sine
	TAN	calculates the tangent
Trigonometric and hyperbolic functions		
	COSH	calculates the hyperbolic cosine
	SINH	calculates the hyperbolic sine
	TANH	calculates the hyperbolic tangent
Probability functions		
	PROBCHI	calculates the chi-square distribution function
	PROBF	calculates the F distribution function
	PROBIT	calculates the inverse normal distribution function
	PROBNORM	calculates the normal distribution function
	PROBT	calculates the *t* distribution function
	POISSON	calculates the Poisson probability distribution function
	PROBBETA	calculates the beta probability distribution function
	PROBBNML	calculates the binomial probability distribution function
	PROBGAM	calculates the gamma probability distribution function
	PROBHYPR	calculates the hypergeometric probability distribution function
	PROBNEGB	calculates the negative binomial probability distribution function
Quantile functions		
	BETAINV	calculates the inverse beta distribution function
	CINV	calculates the quantile for the chi-square distribution
	FINV	calculates the quantile for the *F* distribution
	GAMINV	calculates the inverse gamma distribution function
	TINV	calculates the quantile for the *t* distribution
Sample statistic functions		
	CSS	calculates the corrected sum of squares of all arguments
	CV	calculates the coefficient of variation

(continued)

Table A1.14
(continued)

Functions by Type	Description
KURTOSIS	gives the kurtosis of all arguments
MEAN	computes the arithmetic mean (average)
N	returns the number of nonmissing arguments
NMISS	returns the number of missing values
RANGE	returns the range
SKEWNESS	gives the skewness
STD	calculates the standard deviation
STDERR	calculates the standard error of the mean
USS	calculates the uncorrected sum of squares
VAR	calculates the variance

Random number functions

Functions by Type	Description
RANBIN	generates an observation from a binomial distribution
RANCAU	generates a Cauchy deviate
RANEXP	generates an exponential deviate
RANGAM	generates an observation from a gamma distribution
RANNOR	generates a normal deviate
RANPOI	generates an observation from a Poisson distribution
RANTBL	generates deviates from a tabled probability mass function
RANTRI	generates an observation from a triangular distribution
RANUNI	generates a uniform deviate

Financial functions

Functions by Type	Description
COMPOUND	calculates compounded value parameters
DACCDB	calculates accumulated declining balance depreciation
DACCDBSL	calculates accumulated declining balance converting to straight-line depreciation
DACCSL	calculates accumulated straight-line depreciation
DACCSYD	calculates accumulated sum-of-years'-digits depreciation
DACCTAB	calculates accumulated depreciation from specified tables
DEPDB	calculates declining balance depreciation
DEPDBSL	calculates declining balance converting to straight-line depreciation
DEPSL	calculates straight-line depreciation
DEPSYD	calculates sum-of-years'-digits depreciation

(continued)

	Functions by Type	Description
Table A1.14 *(continued)*	DEPTAB	calculates depreciation from specified tables
	INTRR	calculates internal rate of return as a fraction
	IRR	calculates internal rate of return as a percentage
Financial functions		
	MORT	calculates mortgage loans
	NETPV	calculates net present value as a fraction
	NPV	calculates net present value with rate expressed as a percentage
	SAVING	calculates future value of periodic saving
Character functions		
	COLLATE	generates a string of characters in collating sequence
	COMPRESS	removes characters from a character variable argument
	INDEX	searches for a pattern of characters
	INDEXC	finds the first occurrence of any one of a set of characters
	LEFT	left-aligns a character string
	RANK	returns the position of a character in the ASCII collating sequence
	REVERSE	reverses characters
	RIGHT	right-aligns a character string
	SCAN	scans for words
	TRANSLATE	changes characters
	TRIM	removes trailing blanks
	UPCASE	converts to uppercase
	VERIFY	validates a character value
Date and time functions		
	DATE	returns today's date as a SAS date value
	DATEJUL	converts a Julian date to a SAS date value
	DATEPART	extracts the date part of a SAS datetime value or literal
	DATETIME	returns the current date and time of day
	DAY	returns the day of the month from a SAS date value
	DHMS	returns a SAS datetime value from date, hour, minute, and second
	HMS	returns a SAS time value from hour, minute, and second
	HOUR	returns the hour from a SAS datetime or time value or literal

(continued)

Table A1.14 (*continued*)	

Functions by Type	Description
INTCK	returns the number of time intervals
Date and time functions	
INTNX	advances a date, time, or datetime value by a given interval
JULDATE	returns the Julian date from a SAS date value or literal
MDY	returns a SAS date value from month, day, and year
MINUTE	returns the minute from a SAS time or datetime value or literal
MONTH	returns the month from a SAS date value or literal
QTR	returns the quarter from a SAS date value or literal
SECOND	returns the second from a SAS time or datetime value or literal
TIME	returns the current time of day
TIMEPART	extracts the time part of a SAS datetime value or literal
TODAY	returns the current date as a SAS date value
WEEKDAY	returns the day of the week from a SAS date value or literal
YEAR	returns the year from a SAS date value
YYQ	returns a SAS date value from the year and quarter
State and ZIP code functions	
FIPNAME	converts a FIPS code to the state name (all uppercase)
FIPNAMEL	converts a FIPS code to the state name in upper- and lowercase
FIPSTATE	converts a FIPS code to the two-character postal code
STFIPS	converts a two-character postal code to the FIPS code
STNAME	converts a state postal code to the state name (all uppercase)
STNAMEL	converts a state postal code to the state name (upper- and lowercase)
ZIPFIPS	converts a ZIP code to the FIPS code
ZIPNAME	converts a ZIP code to the state name (all uppercase)
ZIPNAMEL	converts a ZIP code to the state name (upper- and lowercase)
ZIPSTATE	converts a ZIP code to the two-letter state code

Table A1.15
General Purpose Commands

Command	Description
FINISH	denotes the end of a module
FREE	frees matrix storage space
LOAD	loads modules and matrices from library storage
REMOVE	removes matrices from storage
RESET	sets processing options
PRINT	prints matrix values
QUIT	exits from IML
RUN	executes statements in a module
SAVE	saves data
SHOW	prints system information
START	denotes the start of a module
STORE	stores matrices and modules in library storage
MATTRIB	associates printing attributes with matrices

Table A1.16
Control Statements

Statement	Description
IF-THEN/ELSE	conditionally executes statements
DO-END	groups statements as a unit
iterative DO-END	iteratively executes a DO group
GOTO	jumps to a new statement
LINK	jumps to another statement
RETURN	returns to caller
STOP	stops execution of statements
ABORT	stops execution and exits IML
PAUSE	interrupts module execution
RESUME	resumes execution
PUSH	pushes SAS statements into the command input stream
QUEUE	queue lines into command input stream
EXECUTE	execute lines immediately

Table A1.17
RESET Statement Options

Option	Description
PRINT	prints all results automatically
NAME	controls printing of the matrix name and the default row and column names
FLOW	traces the flow of execution with messages
DETAILS	causes more details to be shown
FW=	specifies the field width for printing matrices
FUZZ	specifies to print very small numbers as zero
STORAGE=	specifies the matrix library storage
DEFLIB=	is the default libname for SAS data sets
LOG	routes output to the log rather than the output scroll
ALL	prints intermediate results automatically
CENTER	centers output
LINESIZE=	specifies the line size
PAGESIZE=	specifies the page size
SPACES=	specifies the number of spaces between output
SPILL	specifies that a spill file be used

Table A1.18
Data Processing Commands

Command	Description
Opening and closing	
EDIT	opens a SAS data set for editing
CREATE	create a new SAS data set
CLOSE	closes a SAS data set
SETIN	makes a data set current for input
SETOUT	makes a data set current for output
USE	opens a SAS data set for reading
Showing and resetting	
SHOW DATASETS	shows data sets currently active
SHOW CONTENTS	shows contents of the current data set
RESET DEFLIB=	sets up the default libname
Input and output	
LIST	displays observations of a data set
READ	reads observations from a data set
REPLACE	replaces values in observations and updates observations

(continued)

Table A1.18
(continued)

Command	Description
APPEND	adds records at the end of the data set
FIND	finds observations
DELETE	marks records for deletion
PURGE	removes observations marked for deletion and renumbers records

External files

Command	Description
INFILE	opens a file for input
INPUT	inputs data
FILE	opens or points to an external file
PUT	writes data to an external file
CLOSEFILE	closes an input or output file

Applications

Command	Description
INDEX	indexes a variable in a data set
SORT	sorts a SAS data set
SUMMARY	computes summary statistics for SAS data sets

Display commands

Command	Description
WINDOW	opens a display window
DISPLAY	displays fields in a window

Call routines and functions

Command	Description
DATASETS	obtains members in a data library
CONTENTS	obtains variables in a member
RENAME	renames a SAS data set
DELETE	deletes (erases) a SAS data set

Table A1.19
*Graphics
Commands (Call
Routines)*

Call	Description
GBLKVP	defines a blanking viewport
GBLKVPD	deletes the blanking viewport
GCLOSE	closes the graphics segment
GDELETE	deletes a graphics segment
GDRAW	draws a polyline
GDRAWL	draws individual lines
GGRID	draws a grid
GINCLUDE	includes a graphics segment
GOPEN	opens a graphics segment
GPIE	draws pie slices
GPIEXY	converts from polar to world coordinates
GPOINT	plots points
GPOLY	draws and fills a polygon
GPORT	defines a viewport
GPORTPOP	pops the viewport
GPORTSTK	stacks the viewport
GSCALE	calculates round numbers for labeling axes
GSCRIPT	writes multiple text strings with special fonts
GSET	sets attributes for a graphics segment
GSHOW	shows a graph
GSTART	initializes the graphics system
GSTOP	deactivates the graphics system
GSTRLEN	finds the string length
GTEXT	places text horizontally on a graph
GVTEXT	places text vertically on a graph
GWINDOW	defines the data window
GXAXIS	draws a horizontal axis
GYAXIS	draws a vertical axis

Appendix **2** SAS/IML® Software Compared with the MATRIX Procedure

Introduction

This appendix introduces the IML procedure to those who already know the MATRIX procedure. PROC MATRIX is an older matrix programming language available with Version 5 SAS/STAT software. It is not available with Release 6.06 SAS/STAT or SAS/IML software.

Features of SAS/IML Software

Enhanced features of SAS/IML software over the features of the MATRIX procedure are summarized below:

- The IML procedure can be run interactively. Although IML code can be run in batch mode, its real strength is its interactivity.

- The IML procedure automatically allocates space and memory as needed.

- The IML procedure allows larger applications to be run.

- The IML procedure offers many new language constructs, commands, and options, providing additional power and flexibility in programming.

- The IML procedure has distinct character processing ability.

- The IML procedure has a graphics subsystem.

- The IML procedure offers programmable windows.

- □ The IML procedure offers external file input and output.

- □ The IML procedure has a more extensive set of operators and built-in functions.

- □ The IML procedure has modular programming features.

- □ The IML procedure interacts with SAS data sets more easily.

- □ The IML procedure offers library storage of matrices and modules.

- □ The IML procedure has elaborate error and interrupt handling mechanisms.

PROC MATRIX still works in Version 5 SAS software on IBM® OS environments, and you can still obtain documentation for PROC MATRIX (SAS Technical Report P-135), which is an extract of the documentation for PROC MATRIX in *SAS User's Guide: Statistics*.

PROC IML is similar to PROC MATRIX in many ways. The rest of this appendix highlights the differences between PROC IML and PROC MATRIX.

Literals

With PROC IML, literals with more than one element must be enclosed in braces. For example, you define the 2×3 matrix **A** with the statement

```
a={1 2 3,
   4 5 6};
```

Rows are separated by commas rather than slashes. These changes correct several problems. For example, the sign change and subtraction operators can no longer be confused with signs in numeric literals. Consider the following PROC MATRIX statement:

```
b=5 -7;
```

If the intent were to create a two-element matrix **B** with elements 5 and -7, PROC MATRIX would instead perform the subtraction $(5-7)$, and the result would be a matrix **B** with a single element, 2. You can define the matrix as you want in IML with the statement

```
b={5 -7};
```

The term -7 generates a sign change operator on 7 rather than a negative 7. In addition, the slash (/) can now be used for indicating division within a matrix.

Character Matrices and Literals

A matrix can now have character values. Each element is a string that can be up to 32767 bytes long. Each element in a character matrix is the same size. Literals are given in single or double quotes. For example, the following statement

```
a="coffee";
```

generates a character matrix **A**. Literals with several elements are enclosed in braces ({ }) the same as numeric matrix literals. Strings inside the braces need to be enclosed in quotes only when you want to preserve uppercase and lowercase distinctions or embedded blanks.

Character-valued matrices in PROC MATRIX are fixed in size and are not recognized by most operations as character-valued. For example, the PRINT command does not know a matrix is character-valued unless a character format is used.

Subscripts

Subscripts are written using square brackets ([ ]) as a postfix operator with one or two arguments enclosed in the brackets. Below are several examples:

```
a=b[row,column];        /*right-hand-side selection*/
a[row,column]=b;        /*left-hand-side insertion */
a=b[elements];          /*right-hand-side selection*/
a[elements]=b;          /*left-hand-side insertion */
```

Either argument can be left empty to signify all rows or columns. Right-hand-side subscripts can be reduction operators.

This change is needed to eliminate the ambiguity in PROC MATRIX between subscripted matrices and functions with arguments. The rule of first use does not work now that resolution is delayed until after parsing.

The subscript operator for finding the mean of the selected elements is a colon (:) in IML, not a period (.). This is necessary because periods are parsed as missing values.

Single subscripts are allowed. They refer to the matrix element in row-major order. For vectors with only one row or column, this accesses the expected element. For example, consider the following statement:

```
a={ 3  1  7,
    2 -1  0};
```

The statement

```
b=a[5];
```

yields the value $\begin{bmatrix} -1 \end{bmatrix}$

The statement

```
c=a[{2 5 1}];
```

yields the vector $\begin{bmatrix} 1 \\ -1 \\ 3 \end{bmatrix}$

The following statement is allowed and replaces the fifth element in **A**, -1, with the value 10:

```
a[5]=10;
```

Special Character Operators

With PROC IML, the division operator is a slash (/) rather than #/ used in PROC MATRIX. The transpose operator is a backquote (`) rather than a regular single quote (').

Commands Converted to Subroutines

The EIGEN, SVD, and GS commands have been converted into subroutines in IML. For example, to generate eigenvectors and eigenvalues, call the EIGEN subroutine:

```
call eigen(m,e,a);
```

NOTE Statement

IML replaces the NOTE command with the PRINT statement. The PRINT statement accepts quoted strings and commas in addition to matrix names. Separate each PRINT argument from the next with a comma unless you mean for them to print side-by-side. Extra commas cause extra spacing lines, and slashes (/) force a new page. The PRINT statement also affords more control over labeling and formatting printed matrices.

Error Diagnostics

With PROC IML, syntax errors are reported with a list of items that can possibly correct the error.

Input and Output to SAS Data Sets

With PROC IML, many new commands are provided for SAS data set input and output. The FETCH statement has been implemented as the READ INTO command, and the OUTPUT statement has been implemented as the APPEND FROM command in order to be consistent with the new data management commands.

SHAPE Function

With PROC IML, the second argument of SHAPE is the number of rows, not columns. Columns is an optional third argument.

Division by Zero

With IML, if you divide by zero, the result is set to missing and a warning is issued.

Appendix **3** The MATIML Procedure

Introduction

The MATRIX procedure implements a programming language that uses matrix algebra operators. PROC MATRIX works under IBM mainframe releases of SAS software up through Version 5, but it is not included as part of Version 6.

In order to run existing PROC MATRIX programs, they must be converted to PROC IML code. The MATIML procedure provides an aid to translating PROC MATRIX code into PROC IML code.

Although PROC MATRIX and PROC IML are similar, there are a number of incompatibilities that necessitate translating PROC MATRIX code into code compatible with IML. In particular, the punctuation for subscripts and multi-element literal values is different. You should not regard the translation process as fully automatic. Some limitations may apply, in which case verification of the results is necessary.

Appendix 2, "SAS/IML Software Compared with the MATRIX Procedure," summarizes the differences between PROC MATRIX and PROC IML.

PROC MATIML Statement

The MATIML procedure is invoked by the PROC MATIML statement. The only change you need to make to translate PROC MATRIX code is to replace the PROC MATRIX statement with the PROC MATIML statement when you invoke the procedure:

PROC MATIML *options*;
matrix statement;

   .
   .
   .

matrix statement;

where *options* refers to any options you used in the PROC MATRIX statement and *matrix statement* is PROC MATRIX code.

How to Use PROC MATIML for Translation

First, you should specify an output file by adding a FILENAME statement at the top of the file containing the PROC MATRIX code:

```
filename matiml 'iml_code';
```

The translated IML code will be placed in the file IML_CODE. If you do not specify an output file, the translated IML code goes to the SAS print (listing) file.

Next, change the PROC MATRIX statement to the PROC MATIML statement and submit the program for execution. PROC MATIML can be run interactively or in batch mode.

Finally, examine the SAS log for any errors, warnings, or both that occurred during the translation. If no errors occur, the translated code is ready for execution under Version 6 SAS/IML software.

Limitations of PROC MATIML and Suggested Solutions

Macros

Applications that use macros cannot be completely and reliably translated by the MATIML procedure. However, PROC MATIML can aid in manual translation of such applications.

For example, macro references can be commented out before translation using PROC MATIML and can be restored after the translation. Alternatively, every macro body can be run separately through PROC MATIML, and the translated pieces can be reassembled to produce the IML application. Macros can also be replaced with IML modules (defined using the START and FINISH statements).

LINK and GOTO Statements

When the LINK and RETURN statements or the GOTO statement is used, the *rule of first use* might not work. The *rule of first use* states that if a matrix **X** is already defined, **X**(i,j) is considered a submatrix and translated as **X**$[i,j]$. If **X** is not defined, **X**(i,j) is assumed to be a function call and translated as **X**(i,j) unless it occurs on the left-hand side of an assignment statement. Also $(\mathbf{X})(i,j)$ is always translated as **X**$[i,j]$ and **X**.(i,j) is always translated as **X**(i,j), regardless of the definition of **X**.

A subscripted matrix may be incorrectly translated as a function call. This is not serious because, when the translated code is run through IML, IML immediately detects these errors and issues the message "Function XXX not found". You can then correct XXX(i,j) to XXX$[i,j]$.

For example, in the following code, the MATIML procedure will translate X(1,1) as a function call. The correct translation is the submatrix X[1,1].

```
proc matrix;
link label1;
c=x(1,1);
print c;
stop;
label1:
x= 1 2 3/4 5 6;
return;
```

PROC MATIML does print a CHECK warning to the SAS log and the output file, whenever the LINK or GOTO statements are encountered. The quickest way to resolve this problem is to run the translated IML code as is and make corrections if an "ERROR: Function *XXX* not found" message is encountered.

Functions

The functions LINPROG, QP, and SOLVIT are not available in IML. The LINPROG function can be replaced by the LP subroutine, and the QP function can be replaced by the LCP subroutine. The SOLVIT function can be replaced by the SOLVE function, which is the same except that it uses regular precision, not extended precision. In any case, PROC MATIML prints a CHECK warning to the SAS log and the output file about the unavailability of these functions.

PROC MATRIX Statement Options

The LIST, DUMP, and ERRMAX options of the PROC MATRIX statement are not available in IML. PROC MATIML prints a CHECK warning to the SAS log and the output file when these options are requested. These options are primarily debugging aids. While they are not available in IML, IML has other, more flexible and powerful debugging features.

The DUMP option prints the values of all defined matrices when an error occurs. In IML, the module MAIN pauses each time an error is encountered. You can submit the SHOW ALLNAMES statement to list all defined matrices and then print any matrices you want with a PRINT statement.

The RESET statement with the PRINTALL and FLOW options in IML achieves the same result as the LIST option in the PROC MATRIX statement.

The ERRMAX option in the PROC MATRIX statement sets the maximum number of errors that can occur before execution stops. In IML, the module MAIN pauses each time an error occurs, and you can resume execution as often as you want by submitting a RESUME statement. There is no limit to the number of RESUME statements you can submit. While the module is in a paused state, you can even correct errors before resuming.

Details of Translation

The MATIML procedure translates PROC MATRIX code into the IML module MAIN. The translation always begins with a START MAIN statement and ends with FINISH MAIN and RUN MAIN statements. The RUN statement in IML causes module MAIN to execute. You can execute the translated code repeatedly by simply reissuing the RUN command.

Translation Items

The items listed below are converted by the MATIML procedure. The nature of the conversion is described for each item.

comments
> are copied into the output code. Both comments using /* and */ and statement comments starting with an asterisk (*) and ending with a semicolon are copied.

division operator (#/)
> is translated to /.

FETCH and OUTPUT statements
> The FETCH statement is translated into a pair of IML statements: a USE statement followed by a READ statement. The OUTPUT statement is translated into either the pair of statements CREATE followed by APPEND or the pair EDIT followed by APPEND. If several OUTPUT statements refer to the same data set, that is, if the OUT= value is textually identical on each of the statements, the first OUTPUT statement generates a CREATE and APPEND pair of IML statements and all other OUTPUT statements referring to the data set produce EDIT and APPEND pairs of statements.
>
> The translator incorporates any data set options in FETCH and OUTPUT statements appropriately into the translated code. If TYPE=CHAR or NUM appears, it is translated into a scope clause of VAR _CHAR_ or VAR _NUM_ in the READ command. Before the CREATE statement, PROC MATIML includes a CLOSE DATASET command. This helps avoid the error of creating a data set which is already open. If the data set being created is not open, the CLOSE command prints a note to that effect. The note can be ignored safely.

function calls
> have the dot between the name and the open-parenthesis (connoting a function call in PROC MATRIX) removed.

functions
> The functions MRATIO, RATIO, PRODUCT, ROWSSQ, and ROWSUM are translated into their IML equivalents.

global statements
> are copied directly into the output code.

GO TO statement
> is translated to a single word, GOTO.

labels
> are translated and properly resolved in the output code.

LIST statement
> is translated to

```
show names space;
```

lowercase code
> is translated to uppercase code, except for characters in quotes.

macros
> are not treated by the translator. They are processed by the SAS macro facility and then processed by the translator. You may have to hand edit the macro code before translating the PROC MATRIX code with the MATIML procedure.

math commands
> The SVD, EIGEN, and GS commands are translated into the corresponding subroutines.

NOTE statement
> is translated to a PRINT statement.

numeric literals
> are translated to the IML convention that surrounds literals with braces and separates rows with commas. Where possible, minus signs are treated as signs on numbers rather than as subtraction operators. PROC MATRIX literals are a stream of numbers without braces or other enveloping punctuation. Slashes separate one row from the next.

PRINT statement
> has commas inserted before each operand and brackets placed around the ROWNAME=, COLNAME=, and FORMAT= options.

SHAPE function
> receives an extra 0 as a second argument. This is necessary because the second argument for PROC MATRIX is the number of columns, and for IML, it is the number of rows; the 0 instructs IML to figure out the number of rows from the number of columns. This translation works even if you nest calls.

subscripts
> have the parentheses translated into square brackets. The PROC MATRIX *rule of first use* says that a name followed by a parenthesis connotes a subscript if the first use of the name has been as a variable not followed by parentheses. Otherwise, the construct is regarded as a function call.

transpose operator (')
> is translated into a backquote (`) unless the single quote is in a NOTE statement or at the start of an expression for a character literal assignment.

Example

The example below illustrates the translation of PROC MATRIX statements to IML statements.

The PROC MATRIX code, with the PROC MATIML statement in place of the PROC MATRIX statement, is as follows:

```
filename matiml 'matiml.out';
proc matiml print fw=5 flow list;

    /*-literals-*/
a=1;
b= 1 2 3 / 5 6 7;
c=-1 2;

    /*-char literals and transpose-*/
a='abc' 'def' 'geh';
b=a'; /* forward quote */
c=b` ; /* backward quote */

    /*-test functions vs. subscripts-*/
b=a(1,2); /* subscript*/
c=sqrt(2); /* function */
d=a(1,2); /* subscript*/
e=a.(1,2); /* function */
r=sqrt.(2); /* function */
r=sqrt(a(1,2));
a(1,2)=a(sqrt(2),1);
c=a(+,#);

    /*shape func. adds zero argument*/
r=shape(a,2);
    /*-if and do-*/
if true=1 then do i=1 to 5 while(b);
   hi=i;
   end;
else print a;

do while(true);
   do i=1 to 4 by -1;
      j=i;
   end;
end;

    /*-misc statements-*/
note "here";
list;

goto 1;
go to 1;

print a rowname=r colname=c format=f.;
```

```
svd a b c d;
gs a b c d;
eigen a b c;

   /*-i/o statements-*/
fetch a;
fetch a data=mem;
fetch a rowname=r data=lib.mem colname=c;

output a;
output a out=mem;
output a rowname=r out=lib.mem colname=c;

label: note label;
```

The module created by the translation is shown below:

```
PROC IML; /*---Start of MATRIX/IML Translation---*/
RESET AUTONAME  PRINT FW=5 FLOW
/* CHECK: Option LIST not available in IML */;

START MAIN;
   /*-literals-*/
  A={1};
  B={ 1 2 3, 5 6 7};
  C={-1 2};
   /*-character literals and transpose-*/
  A={'abc' 'def' 'geh'};
  B=A`;
     /* FORWARD quote */
  C=B`;
     /* BACKWARD quote */
   /*-test functions vs. subscripts-*/
  B= A[{1},{2}];
     /* SUBSCRIPT*/
  C= SQRT({2});
     /* FUNCTION */
  D= A[{1},{2}];
     /* SUBSCRIPT*/
  E= A({1},{2});
     /* FUNCTION */
  R= SQRT({2});
     /* FUNCTION */
  R= SQRT( A[{1},{2}]);
  A[{1},{2}]= A[ SQRT({2}),{1}];
  C= A[+,#];
   /*-shape function adds zero argument-*/
  R= SHAPE(A,0,{2});
   /*-if and do-*/
  IF ( TRUE={1}) THEN  DO I={1} TO{ 5} WHILE(B);
     H=I;
     END;
  ELSE  PRINT A;
```

```
DO WHILE(TRUE);
  DO I={1} TO{ 4} BY{-1};
    J=I;
    END;
  END;
 /*-misc statements-*/
PRINT  "here";
SHOW SPACE NAMES;

/* CHECK: Rule of First may not work with GOTO and LINK/RETURN */
/*        statements. Subscripted matrix may get translated to */
/*        a function call.                                     */
GOTO L;

/* CHECK: Rule of First may not work with GOTO and LINK/RETURN */
/*        statements. Subscripted matrix may get translated to */
/*        a function call.                                     */
GOTO L;
PRINT A[ ROWNAME=R][ COLNAME=C][ FORMAT=F.];
CALL SVD( A, B, C, D);
CALL GSORTH(  A, B, C, D);
CALL EIGEN( A, B, C);


 /*-i/o statements-*/
USE _LAST_ ;
READ ALL INTO A ;
USE MEM ;
READ ALL INTO A ;
USE LIB.MEM ;
READ ALL INTO A [ROWNAME=R COLNAME=C];
_TMP_ROW = 'ROW1    ' : compress('ROW'+char(nrow(A)));
CLOSE _DATA_;
CREATE _DATA_ ( RENAME=(_TMP_ROW=ROW  )) FROM A [ROWNAME=_TMP_ROW ];
APPEND FROM A [ROWNAME=_TMP_ROW];
_TMP_ROW = 'ROW1    ' : compress('ROW'+char(nrow(A)));
CLOSE MEM;
CREATE MEM ( RENAME=(_TMP_ROW=ROW  )) FROM A [ROWNAME=_TMP_ROW ];
APPEND FROM A [ROWNAME=_TMP_ROW];
CLOSE LIB.MEM;
CREATE LIB.MEM ( RENAME=(R  =ROW  )) FROM A [ROWNAME=R COLNAME=C ];
APPEND FROM A [ROWNAME=R];
LABEL: PRINT " LABEL";

FINISH MAIN;

RUN MAIN;  /*---End of MATRIX/IML Translation---*/
```

Appendix **4** Alternative Characters

Table A4.1 presents alternative characters you can use in SAS/IML software.

Table A4.1
Alternative Characters in SAS/IML Software

Condition	Character	Alternative
OPTIONS CHARCODE specified	[	(|
	]	|)
	|	?/
	{	?(
	}	?)
	^	?=
OPTIONS CHARCODE not specified	{	¢
	}	!

If you do not have a backquote character (ˋ), which is the transpose operator in IML, you can use the T function. For example, the following two statements both assign **B** the value of the transpose of **A**:

```
b=aˋ;
b=t(a);
```

References

Ansley, C. (1979), "An Algorithm for the Exact Likelihood of a Mixed Autoregressive-moving Average Process," *Biometrika*, 66, 59–65.

Armitage, P. (1975), *Sequential Medical Trials*, Oxford: Blackwell.

Bassett, G. and Koenker, R. (1982), "An Empirical Quantile Function for Linear Models with IID Errors," *Journal of the American Statistical Association*, 77, 407–415.

Beaton, A. (1964), "The Use of Special Matrix Operators in Statistical Calculus," *Research Bulletin*, Princeton: Educational Testing Service.

Bishop, Y.M., Fienberg, S.E., and Holland, P.W. (1975), *Discrete Multivariate Analysis: Theory and Practice*, Cambridge, MA: MIT Press.

Box, G.E.P. and Jenkins, G.M. (1976), *Time Series Analysis: Forecasting and Control*, Oakland, CA: Holden-Day.

Brinkman, N.D. (1981), "Ethanol Fuel—A Single-Cylinder Engine Study of Efficiency and Exhaust Emissions," *Society of Automotive Engineers Transactions*, 90, 1410–1424.

Charnes, A., Frome, E.L., and Yu, P.L. (1976), "The Equivalence of Generalized Least Squares and Maximum Likelihood Estimation in the Exponential Family," *Journal of the American Statistical Association*, 71, 169–172.

Cox, D.R. (1970), *The Analysis of Binary Data*, New York: Halsted Press.

Emerson, P.L. (1968), "Numerical Construction of Orthogonal Polynomials from a General Recurrence Formula," *Biometrics*, 24, 695.

Forsythe, G.E., Malcolm, M.A., and Moler, C.B. (1967), *Computer Solution of Linear Algebraic Systems*, Chapter 17, Englewood Cliffs, NJ: Prentice-Hall, Inc.

Gentleman, W.M. and Sande, G. (1966), "Fast Fourier Transforms—for Fun and Profit," *AFIPS Proceedings of the Fall Joint Computer Conference*, 19, 563–578.

Golub, G.H. (1969), "Matrix Decompositions and Statistical Calculations," *Statistical Computation*, eds. R.C. Milton and J.A. Nelder, New York: Academic Press, Inc.

Goodnight, J.H. (1978), SAS Technical Report R-106, *The Sweep Operator: Its Importance in Statistical Computing*, Cary, NC: SAS Institute Inc.

Goodnight, J.H. (1979) "A Tutorial on the SWEEP Operator," *The American Statistician*, 33, 149–158.

Graybill, F.A. (1969), *Introduction to Matrices with Applications in Statistics*, Belmont, CA: Wadsworth, Inc.

Grizzle, J.E., Starmer, C.F., and Koch, G.G. (1969), "Analysis of Categorical Data by Linear Models," *Biometrics*, 25, 489–504.

Hadley, G. (1963), *Linear Programming*, Reading, MA: Addison-Wesley Publishing Company, Inc.

Jenkins, M.A. and Traub, J.F. (1970), "A Three-stage Algorithm for Real Polynomials using Quadratic Iteration," *SIAM Journal of Numerical Analysis*, 7, 545–566.

Jennrich, R.I. and Moore, R.H. (1975), "Maximum Likelihood Estimation by Means of Nonlinear Least Squares," *American Statistical Association, 1975 Proceedings of the Statistical Computing Section*, 57–65.

Kaiser, H.F. and Caffrey, J. (1965), "Alpha Factor Analysis," *Psychometrika*, 30, 1–14.

Kastenbaum, M.A. and Lamphiear, D.E. (1959), "Calculation of Chi-Square to Test the No Three-Factor Interaction Hypothesis," *Biometrics*, 15, 107–122.

Koenker, R. and Bassett, G., (1978), "Regression Quantiles," *Econometrica,* 46, 33–50.

Kruskal, J.B. (1964), "Nonmetric Multidimensional Scaling," *Psychometrika,* 29, 1–27, 115–129.

Landis, J.R. and Lepkowski, J.M. (1984), "Tutorial on the Analysis of Categorical Data from Complex Sample Surveys," unpublished manuscript.

McDowell, A., Engel, A., Massey, J.T., and Maurer, K. (1981), *Plan and Operation of the Second National Health and Nutrition Examination Survey,* 1976–80, Vital and Health Statistics, Series 1, No. 15, DHHS Publication No. (PHS) 81–1317, Public Health Service, Washington: U.S. Government Printing Office.

McLeod, I. (1975), "Derivation of the Theoretical Autocovariance Function of Autoregressive-Moving Average Time Series," *Applied Statistics,* 24, 255–256.

Monro, D.M. and Branch, J.L. (1976), "Algorithm AS 117. The Chirp Discrete Fourier Transform of General Length," *Applied Statistics,* 26, 351–361.

Nelder, J. A. and Wedderburn, R.W.M. (1972), "Generalized Linear Models," *Journal of The Royal Statistical Society,* A.3, 370.

Nobel, B. (1969), *Applied Linear Algebra,* Englewood Cliffs, NJ: Prentice-Hall.

Nussbaumer, H.J. (1982), *Fast Fourier Transform and Convolution Algorithms,* Second Edition, New York: Springer-Verlag.

Pizer, S.M. (1975), *Numerical Computing and Mathematical Analysis,* Chicago: Science Research Associates, Inc.

Rao, C.R. and Mitra, S.K. (1971), *Generalized Inverse of Matrices and Its Applications,* New York: John Wiley & Sons, Inc.

Reinsch, Christian H. (1967), "Smoothing by Spline Functions," *Numerische Mathematik,* 10, 177–183.

Rodriguez, R.N. (1985), "A Comparison of the ACE and MORALS Algorithms in an Application to Engine Exhaust Emissions Modeling," *Computer Science and Statistics: Proceedings of the Sixteenth Symposium on the Interface,* ed. L. Billard, Amsterdam: North-Holland Publishing Company.

Sall, J.P. (1977), "Matrix Algebra Notation as a Computer Language," 1977 Statistical Computing Section of the American Statistical Association, Washington, DC, 342–344.

SAS Institute Inc., SAS Technical Report P-135, *The Matrix Procedure: Language and Applications,* Cary, NC, 1985.

SAS Institute Inc., *SAS User's Guide: Statistics, 1982 Edition,* Cary, NC, 1982.

Singleton, R.C. (1969), "An Algorithm for Computing the Mixed Radix Fast Fourier Transform," *IEEE Transactions on Audio and Electroacoustics,* AU-17, 93–103.

Stanish, W. (1985), "Categorical Data Analysis Strategies Using SAS Software," *Computer Science and Statistics: Proceedings of the Seventeenth Symposium on the Interface,* ed. David M. Allen, Amsterdam: North-Holland Publishing Company.

Stoer, J. and Bulirsch, R. (1980), *Introduction to Numerical Analysis,* New York: Springer-Verlag.

Wilkinson, J.H. and Reinsch, C. (eds.), (1971), *Linear Algebra, Handbook for Automatic Computation,* Volume 2, New York: Springer-Verlag.

Young, F.W. (1981), "Quantitative Analysis of Qualitative Data," *Psychometrika,* 46, 357–388.

Index

Your Turn

If you have comments or suggestions about *SAS/IML Software: Usage and Reference, Version 6, First Edition* or SAS/IML software, please send them to us on a photocopy of this page.

Please return the photocopy to the Publications Division (for comments about this book) or the Technical Support Division (for suggestions about the software) at SAS Institute Inc., SAS Campus Drive, Cary, NC 27513.